Designing Audio Effect Plug-Ins in C++

Designing Audio Effect Plug-Ins in C++

With Digital Audio Signal Processing Theory

Will Pirkle

Focal Press
Taylor & Francis Group
NEW YORK AND LONDON

First published 2013
by Focal Press
70 Blanchard Road, Suite 402, Burlington, MA 01803

Simultaneously published in the UK
by Focal Press
2 Park Square, Milton Park, Abingdon, Oxon OX14 4RN

Focal Press is an imprint of the Taylor and Francis Group, an Informa business

© 2013 Taylor and Francis

The right of Will Pirkle to be identified as author of this work has been asserted by him/her in accordance with sections 77 and 78 of the Copyright, Designs and Patents Act 1988.

All rights reserved. No part of this book may be reprinted or reproduced or utilised in any form or by any electronic, mechanical, or other means, now known or hereafter invented, including photocopying and recording, or in any information storage or retrieval system, without permission in writing from the publishers.

Notices
Knowledge and best practice in this field are constantly changing. As new research and
experience broaden our understanding, changes in research methods, professional practices, or medical treatment may become necessary.

Practitioners and researchers must always rely on their own experience and knowledge in evaluating and using any information, methods, compounds, or experiments described herein. In using such information or methods they should be mindful of their own safety and the safety of others, including parties for whom they have a professional responsibility.

Product or corporate names may be trademarks or registered trademarks, and are used only for identification and explanation without intent to infringe.

Library of Congress Cataloging-in-Publication Data
Pirkle, William C., author.
 Designing audio effect plug-ins in C++ with digital audio signal processing theory/Will Pirkle.
  pages cm
 Includes bibliographical references and index.
  ISBN 978-0-240-82515-1 (paperback)
1. Computer sound processing. 2. Plug-ins (Computer programs)
3. C++ (Computer program language) I. Title.
 MT723.P57 2013
  006.4'5—dc23
                    2012024241

ISBN: 978-0-240-82515-1 (pbk)
ISBN: 978-0-123-97882-0 (ebk)

Typeset in Times
Project Managed and Typeset by: diacriTech

Printed and bound in the United States of America by Sheridan Books, Inc. (a Sheridan Group Company).

Dedicated to

my father and mother
C.H. Pirkle
and
J.V. Pirkle

Contents

Introduction xvii

Chapter 1: Digital Audio Signal Processing Principles 1
   1.1 Acquisition of Samples 1
   1.2 Reconstruction of the Signal 3
   1.3 Signal Processing Systems 4
   1.4 Synchronization and Interrupts 5
   1.5 Signal Processing Flow 6
   1.6 Numerical Representation of Audio Data 7
   1.7 Using Floating-Point Data 9
   1.8 Basic DSP Test Signals 10
        1.8.1 DC and Step 10
        1.8.2 Nyquist 11
        1.8.3 ½ Nyquist 11
        1.8.4 ¼ Nyquist 12
        1.8.5 Impulse 12
   1.9 Signal Processing Algorithms 13
   1.10 Bookkeeping 13
   1.11 The One-Sample Delay 15
   1.12 Multiplication 16
   1.13 Addition and Subtraction 17
   1.14 Algorithm Examples and the Difference Equation 18
   1.15 Gain, Attenuation, and Phase Inversion 18
   1.16 Practical Mixing Algorithm 19
   Bibliography 20

Chapter 2: Anatomy of a Plug-In 21
   2.1 Static and Dynamic Linking 21
   2.2 Virtual Address Space and DLL Access 22
   2.3 C and C++ Style DLLs 24
   2.4 Maintaining the User Interface 25
   2.5 The Applications Programming Interface 27
   2.6 Typical Required API Functions 29
   2.7 The RackAFX Philosophy and API 31
        2.7.1 __stdcall 31
   Bibliography 34

Chapter 3: Writing Plug-Ins with RackAFX 35
3.1 Building the DLL 35
3.2 Creation 36
3.3 The GUI 36
3.4 Processing Audio 37
3.5 Destruction 38
3.6 Your First Plug-Ins 38
    3.6.1 Project: Yourplugin 39
    3.6.2 Yourplugin GUI 39
    3.6.3 Yourplugin.h File 39
    3.6.4 Yourplugin.cpp File 40
    3.6.5 Building and Testing 40
    3.6.6 Creating and Saving Presets 40
    3.6.7 GUI Designer 40
3.7 Design a Volume Control Plug-In 40
3.8 Set Up RackAFX for Use 41
3.9 Setup Preferences 43
    3.9.1 Project: Volume 44
    3.9.2 Volume GUI 45
    3.9.3 Configure a Slider Control 46
    3.9.4 Volume.h File 48
    3.9.5 Volume.cpp File 50
3.10 Design a Volume-in-dB Plug-In 54
    3.10.1 Project: VolumedB 56
    3.10.2 VolumedB GUI 56
    3.10.3 VolumedB.h File 56
    3.10.4 VolumedB.cpp File 57
3.11 Design a High-Frequency Tone Control Plug-In 58
    3.11.1 Project: SimpleHPF 60
    3.11.2 SimpleHPF GUI 60
    3.11.3 SimpleHPF.h File 60
    3.11.4 SimpleHPF.cpp File 62
3.12 Design a High-Frequency Tone Control with Volume Plug-In 66
    3.12.1 Project: SimpleHPF 66
    3.12.2 SimpleHPF GUI 66
    3.12.3 SimpleHPF.h File 66
    3.12.4 SimpleHPF.cpp File 67
3.13 The User Plug-In Menu in RackAFX 69

Chapter 4: How DSP Filters Work 71
4.1 First-Order Feed-Forward Filter 74
4.2 Design a General First-Order Feed-Forward Filter 84
4.3 First-Order Feed-Back Filter 88
4.4 Design a General First-Order Feed-Back Filter 89

        4.4.1 Project FeedBackFilter 89
        4.4.2 FeedBackFilter GUI 89
        4.4.3 FeedBackFilter.h File 89
        4.4.4 FeedBackFilter.cpp File 90
    4.5 Observations 94
        4.5.1 General 94
        4.5.2 Feed-Forward Filters 95
        4.5.3 Feed-Back Filters 95
    Bibliography 95

Chapter 5: Basic DSP Theory 97
    5.1 The Complex Sinusoid 97
    5.2 Complex Math Review 100
    5.3 Time Delay as a Math Operator 102
    5.4 First-Order Feed-Forward Filter Revisited 103
        5.4.1 Negative Frequencies 104
        5.4.2 Frequencies Above and Below ±Nyquist 106
    5.5 Evaluating the Transfer Function H(ω) 106
        5.5.1 DC (0 Hz) 107
        5.5.2 Nyquist (π) 108
        5.5.3 ½ Nyquist ($\pi/2$) 109
        5.5.4 ¼ Nyquist ($\pi/4$) 109
    5.6 Evaluating $e^{j\omega}$ 112
    5.7 The z Substitution 114
    5.8 The z Transform 114
    5.9 The z Transform of Signals 116
    5.10 The z Transform of Difference Equations 117
    5.11 The z Transform of an Impulse Response 118
    5.12 The *Zeros* of the Transfer Function 119
    5.13 Estimating the Frequency Response: *Zeros* 121
    5.14 Filter Gain Control 122
    5.15 First-Order Feed-Back Filter Revisited 123
    5.16 The *Poles* of the Transfer Function 124
        5.16.1 DC (0 Hz) 128
        5.16.2 Nyquist (π) 128
        5.16.3 ½ Nyquist ($\pi/2$) 129
        5.16.4 ¼ Nyquist ($\pi/4$) 130
    5.17 Second-Order Feed-Forward Filter 132
        5.17.1 DC (0 Hz) 139
        5.17.2 Nyquist (π) 139
        5.17.3 ½ Nyquist ($\pi/2$) 140
        5.17.4 ¼ Nyquist ($\pi/4$) 140
    5.18 Second-Order Feed-Back Filter 142
        5.18.1 DC (0 Hz) 148
        5.18.2 Challenge 149

- 5.19 First-Order Pole-Zero Filter: The Shelving Filter 149
  - 5.19.1 DC (0 Hz) 155
  - 5.19.2 Challenge 155
- 5.20 The Bi-Quadratic Filter 157
- Bibliography 162

Chapter 6: Audio Filter Designs: IIR Filters 163

- 6.1 Direct z-Plane Design 163
- 6.2 Single Pole Filters 164
  - 6.2.1 First-Order LPF and HPF 164
- 6.3 Resonators 165
  - 6.3.1 Simple Resonator 165
  - 6.3.2 Smith-Angell Improved Resonator 168
- 6.4 Analog Filter to Digital Filter Conversion 170
  - 6.4.1 Challenge 178
- 6.5 Effect of Poles or Zeros at Infinity 178
- 6.6 Generic Bi-Quad Designs 181
  - 6.6.1 First-Order LPF and HPF 182
  - 6.6.2 Second-Order LPF and HPF 183
  - 6.6.3 Second-Order BPF and BSF 184
  - 6.6.4 Second-Order Butterworth LPF and HPF 184
  - 6.6.5 Second-Order Butterworth BPF and BSF 185
  - 6.6.6 Second-Order Linkwitz-Riley LPF and HPF 186
  - 6.6.7 First- and Second-Order APF 188
- 6.7 Audio Specific Filters 188
  - 6.7.1 Modified Bi-Quad 189
  - 6.7.2 First-Order Shelving Filters 189
  - 6.7.3 Second-Order Parametric/Peaking Filter: Non-Constant-Q 191
  - 6.7.4 Second-Order Parametric/Peaking Filter: Constant-Q 192
  - 6.7.5 Cascaded Graphic EQ: Non-Constant-Q 194
  - 6.7.6 Cascaded Graphic EQ: Constant-Q 195
- 6.8 Design a Resonant LPF Plug-In 196
  - 6.8.1 Project: ResonantLPF 197
  - 6.8.2 ResonantLPF GUI 197
  - 6.8.3 ResonantLPF.h File 198
  - 6.8.4 ResonantLPF.cpp File 199
- 6.9 The Massberg Analog-Matched Low-Pass Filter 201
  - 6.9.1 First-Order Massberg LPF 201
  - 6.9.2 Second-Order Massberg LPF 203
- Bibliography 204
- References 205

Chapter 7: Delay Effects and Circular Buffers 207

- 7.1 The Basic Digital Delay 209
- 7.2 Digital Delay with Wet/Dry Mix 214

  7.2.1 Frequency and Impulse Responses...214
  7.2.2 The Effect of Feedback 218
 7.3 Design a DDL Module Plug-In...224
  7.3.1 Project: DDLModule 225
  7.3.2 DDLModule GUI 225
  7.3.3 DDLModule.h File 226
  7.3.4 DDLModule.cpp File 226
  7.3.5 Declare and Initialize the Delay Line Components 228
  7.3.6 DDLModule.h File 230
  7.3.7 DDLModule.cpp File 230
 7.4 Modifying the Module to Be Used by a Parent Plug-In 233
  7.4.1 DDLModule.h File 233
  7.4.2 DDLModule.cpp File 234
 7.5 Modifying the Module to Implement Fractional Delay......................................235
  7.5.1 DDLModule.cpp File 238
 7.6 Design a Stereo Digital Delay Plug-In 239
  7.6.1 Project: StereoDelay..239
  7.6.2 StereoDelay GUI 241
  7.6.3 StereoDelay.h File 241
  7.6.4 StereoDelay.cpp File 242
 7.7 Design a Stereo Crossed-Feedback Delay Plug-In..244
 7.8 Enumerated Slider Variables...245
  7.8.1 Constructor 246
  7.8.2 PrepareForPlay() 246
  7.8.3 UserInterfaceChange() 246
  7.8.4 ProcessAudioFrame()..247
 7.9 More Delay Algorithms 248
  7.9.1 Advanced DDL Module 248
  7.9.2 Delay with LPF in Feedback Loop 248
  7.9.3 Multi-Tap Delay 249
  7.9.4 Ping-Pong Delay 250
  7.9.5 LCR Delay...250
 Bibliography 251

Chapter 8: Audio Filter Designs: FIR Filters *253*

 8.1 The IR Revisited: Convolution 253
 8.2 Using RackAFX's Impulse Convolver 258
  8.2.1 Loading IR Files...258
  8.2.2 Creating IR Files 259
  8.2.3 The IR File Format 261
 8.3 Using RackAFX's FIR Designer 262
 8.4 The Frequency Sampling Method...263
  8.4.1 Linear-Phase FIR Using the Frequency Sampling Method.........................263
 8.5 Complementary Filter Design for Linear Phase FIR Filters...............................266
 8.6 Using RackAFX's Frequency Sampling Method Tool 267

8.7 Designing a Complementary Filter 269
8.8 The Optimal (Parks-McClellan) Method 270
8.9 Using RackAFX's Optimal Method Tool 271
8.10 Design a Convolution Plug-In 273
    8.10.1 Project: Convolver 275
    8.10.2 Convolver.h File 275
    8.10.3 Convolver.cpp File 276
8.11 Numerical Method FIR Filters 281
    8.11.1 Moving Average Interpolator 282
    8.11.2 Lagrange Interpolator 284
    8.11.3 Median Filter 284
Bibliography 287

Chapter 9: Oscillators 289

9.1 Direct Form Oscillator 289
    9.1.1 Initial Conditions 290
9.2 Design a Direct Form Oscillator Plug-In 292
    9.2.1 Project: DirectOscillator 292
    9.2.2 DirectOscillator GUI 292
    9.2.3 DirectOscillator.h File 294
    9.2.4 DirectOscillator.cpp File 295
    9.2.5 Improving the Oscillator Design 297
9.3 The Gordon-Smith Oscillator 299
9.4 Wave Table Oscillators 301
9.5 Design a Wave Table Oscillator Plug-In 303
    9.5.1 Project: WTOscillator 303
    9.5.2 WTOscillator GUI 303
    9.5.3 WTOscillator.h File 304
    9.5.4 WTOscillator.cpp File 305
9.6 Adding More Wave Tables 308
    9.6.1 WTOscillator.h File 308
    9.6.2 WTOscillator.cpp File 309
    9.6.3 WTOscillator GUI 310
    9.6.4 WTOscillator.h File 310
    9.6.5 WTOscillator.cpp File 311
9.7 Band-Limited Additive Wave Tables 312
    9.7.1 WTOscillator GUI 313
    9.7.2 WTOscillator.h File 313
    9.7.3 WTOscillator.cpp File 314
    9.7.4 Saw-Tooth 317
    9.7.5 Square Wave 317
9.8 Additional Oscillator Features (LFO) 320
    9.8.1 WTOscillator.h File 320
    9.8.2 WTOscillator.cpp File 321
    9.8.3 WTOscillator.h File 321
    9.8.4 WTOscillator.cpp File 322

    9.9 Bipolar/Unipolar Functionality 324
        9.9.1 WTOscillator GUI 324
        9.9.2 WTOscillator.cpp File 325
    Bibliography 326

Chapter 10: Modulated Delay Effects 327
    10.1 The Flanger/Vibrato Effect 328
    10.2 The Chorus Effect 331
    10.3 Design a Flanger/Vibrato/Chorus Plug-In 334
        10.3.1 Project: ModDelayModule 335
        10.3.2 ModDelayModule GUI 336
        10.3.3 ModDelayModule.h File 336
        10.3.4 ModDelayModule.cpp File 337
        10.3.5 PrepareForPlay() 340
        10.3.6 Challenge 342
    10.4 Design a Stereo Quadrature Flanger Plug-In 342
        10.4.1 Project: StereoQuadFlanger 342
        10.4.2 StereoQuadFlanger GUI 342
        10.4.3 StereoQuadFlanger.h File 342
        10.4.4 StereoQuadFlanger.cpp File 343
        10.4.5 Challenges 345
    10.5 Design a Multi-Unit LCR Chorus Plug-In 345
        10.5.1 Project: StereoLCRChorus 346
        10.5.2 StereoLCRChorus GUI 346
        10.5.3 StereoLCRChorus.h File 346
        10.5.4 StereoLCRChorus.cpp File 347
    10.6 More Modulated Delay Algorithms 350
        10.6.1 Stereo Cross-Flanger/Chorus (Korg Triton®) 350
        10.6.2 Multi-Flanger (Sony DPS-M7®) 350
        10.6.3 Bass Chorus 350
        10.6.4 Dimension-Style (Roland Dimension D®) 351
        10.6.5 Deca-Chorus (Sony DPS-M7®) 354
    Bibliography 355

Chapter 11: Reverb Algorithms 357
    11.1 Anatomy of a Room Impulse Response 358
        11.1.1 RT_{60}: The Reverb Time 359
    11.2 Echoes and Modes 360
    11.3 The Comb Filter Reverberator 364
    11.4 The Delaying All-Pass Filter Reverberator 368
    11.5 More Delaying All-Pass Filter Reverberators 370
    11.6 Schroeder's Reverberator 372
    11.7 The Low-Pass Filter–Comb Reverberator 373
    11.8 Moorer's Reverberator 375
    11.9 Stereo Reverberation 376
    11.10 Gardner's Nested APF Reverberators 377

11.11 Modulated APF and Comb/APF Reverb 381
11.12 Dattorro's Plate Reverb 382
11.13 Generalized Feedback Delay Network Reverbs 385
11.14 Other FDN Reverbs 389
11.15 An Example Room Reverb 391
11.16 RackAFX Stock Objects 394
    11.16.1 COnePoleLPF 394
    11.16.2 CDelay 395
    11.16.3 CCombFilter 396
    11.16.4 CLPFCombFilter 396
    11.16.5 CDelayAPF 398
11.17 Design the Room Reverb 398
    11.17.1 Project: Reverb 398
    11.17.2 Reverb GUI 398
    11.17.3 Reverb.h 402
    11.17.4 Reverb.cpp 403
11.18 Challenge 408
Bibliography 409
References 409

Chapter 12: Modulated Filter Effects *411*

12.1 Design a Mod Filter Plug-In: Part I Modulated f_c 412
    12.1.1 Project: ModFilter 413
    12.1.2 ModFilter GUI 413
    12.1.3 ModFilter.h File 413
    12.1.4 ModFilter.cpp File 416
12.2 Design a Mod Filter Plug-In: Part II, Modulated f_c, Q 419
    12.2.1 ModFilter GUI 419
    12.2.2 ModFilter.h File 419
    12.2.3 ModFilter.cpp File 420
12.3 Design a Mod Filter Plug-In: Part III, Quad-Phase LFOs 423
    12.3.1 ModFilter GUI 423
    12.3.2 ModFilter.cpp File 424
12.4 Design an Envelope Follower Plug-In 425
12.5 Envelope Detection 428
    12.5.1 Project EnvelopeFollower 430
    12.5.2 EnvelopeFollower GUI 430
    12.5.3 EnvelopeFollower.h File 431
    12.5.4 EnvelopeFollower.cpp File 432
12.6 Design a Phaser Plug-In 436
    12.6.1 Project Phaser 440
    12.6.2 Phaser GUI 440
    12.6.3 Phaser.h File 440
    12.6.4 Phaser.cpp File 441

    12.7 Design a Stereo Phaser with Quad-Phase LFOs 446
        12.7.1 Phaser GUI 446
        12.7.2 Phaser.h File 446
        12.7.3 Phaser.cpp File 447
    Bibliography 451
    References 451

Chapter 13: Dynamics Processing 453

    13.1 Design a Compressor/Limiter Plug-In 457
        13.1.1 Project: DynamicsProcessor 458
        13.1.2 DynamicsProcessor: GUI 458
        13.1.3 DynamicsProcessor.h File 459
        13.1.4 DynamicsProcessor.cpp File 460
        13.1.5 DynamicsProcessor.cpp File 465
    13.2 Design a Downward Expander/Gate Plug-In 466
        13.2.1 DynamicsProcessor.h File 466
        13.2.2 DynamicsProcessor.cpp File 466
    13.3 Design a Look-Ahead Compressor Plug-In 468
        13.3.1 DynamicsProcessor: GUI 469
        13.3.2 DynamicsProcessor.h File 470
        13.3.3 DynamicsProcessor.cpp File 470
    13.4 Stereo-Linking the Dynamics Processor 472
        13.4.1 DynamicsProcessor: GUI 472
        13.4.2 DynamicsProcessor.cpp File 473
    13.5 Design a Spectral Compressor/Expander Plug-In 475
        13.5.1 Project: SpectralDynamics 476
        13.5.2 SpectralDynamics: GUI 476
        13.5.3 Additional Slider Controls 477
        13.5.4 Spectral Dynamics Buttons 477
        13.5.5 Spectral Dynamics Metering 477
        13.5.6 SpectralDynamics.h File 478
        13.5.7 SpectralDynamics.cpp File 479
    13.6 Alternate Side-Chain Configurations 486
    Bibliography 487
    References 487

Chapter 14: Miscellaneous Plug-Ins 489

    14.1 Design a Tremolo/Panning Plug-In 489
        14.1.1 Project: TremoloPanner 490
        14.1.2 TremoloPanner: GUI 490
    14.2 Design a Ring Modulator Plug-In 494
        14.2.1 Project: RingModulator 494
        14.2.2 RingModulator: GUI 494
        14.2.3 RingModulator.h File 495
        14.2.4 RingModulator.cpp File 495

 14.3 Design a Wave Shaper Plug-In 497
  14.3.1 Project: WaveShaper 498
  14.3.2 WaveShaper: GUI 498
 Bibliography 500

Appendix A: The VST® and AU® Plug-In APIs 501
 A.1 Compiling as a VST Plug-In in Windows 501
 A.2 Wrapping Your RackAFX Plug-In 503
 A.3 Comparison of Objects/Methods/GUIs 505
 A.4 VST Plug-In without Wrapping 506
  A.4.1 Default GUI 507
  A.4.2 Signal Processing 509
 A.5 VST Plug-In with RackAFX Wrapping 512
  A.5.1 Default GUI 512
 A.6 AU Overview 514
  A.6.1 Default GUI 515
  A.6.2 Signal Processing 516

Appendix B: More RackAFX Controls and GUI Designer 519
 B.1 The Alpha Wheel and LCD Control 519
 B.2 The Vector Joystick Control 521
 B.3 Using the sendUpdateGUI() Method 525
 B.4 Using GUI Designer 525

Index 531

Introduction

When I started teaching in the Music Engineering Technology Program at the University of Miami in 1996, we were writing signal processing algorithms in digital signal processing (DSP) assembly language and loading them on to DSP evaluation boards for testing. We had also just begun teaching a software plug-in programming class, since computers were finally at the point where native processing was feasible. I began teaching Microsoft's DirectX® in 1997 and immediately began a book/manual on converting DSP algorithms into C++ code for the DirectX platform. A year later I had my first manuscript of what would be a thick DirectX programming book. However, I had two problems on my hands: first, DirectX is bloated with Common Object Model (COM) programming, and it seemed like the lion's share of the book was really devoted to teaching basic COM skills rather than converting audio signal processing algorithms into C++, creating a graphical user interface (GUI), and handling user input. More importantly, developers had dropped DirectX in favor of a new, lean, cross-platform compatible plug-in format called Steinberg VST®, written in "straight" C++ without the need for operating system (OS) specific components. After taking one look at the Virtual Studio Technology (VST) applications programming interface (API), I immediately dropped all DirectX development, wrote VST plug-ins all summer, and then switched to teaching it the following year. And, I put my now-obsolete manuscript on the shelf.

After half a dozen semesters teaching VST programming, a few things had become clear. For any given project, the students were spending more time dealing with setting up and maintaining a GUI than they were with the audio signal processing code. Instead of being on the Internet to look for cool processing algorithms, they were searching for knob and switch bitmaps. While I can certainly appreciate a nice-looking GUI, I was trying to teach audio signal processing and not graphic design. I next spent some time trying to come up with some kind of way to speed up the GUI programming, and I wrote a helper program that let students define lists of variable names, data types, minimum, maximum, and initial values. Then it would synthesize the code for handling part of the GUI interaction. The actual GUI itself was another problem, and I spent years trying to come up with a way to free the students (and myself) from dealing with the hassles of designing and maintaining a GUI. Around 2004, as a result of a consulting job, I needed to come up with a simple C++ based audio signal processing API—my own plug-in format. I would also need to write my own plug-in client, the software that would load and process data through my plug-ins. I was determined to write an API that was far simpler and leaner than even VST. And, I wanted my client software to handle 100% of the GUI details so I could focus on the signal processing work for my consulting client. This would also make great software to teach audio signal processing and plug-in programming simultaneously.

Since 2009 I've been using my RackAFX™ software in the classroom at both the graduate and undergraduate levels. My students never cease to amaze me with what they design. Currently, RackAFX runs on the Windows® OS and generates plug-ins that are compatible with it and optionally VST as well. You can develop your plug-in in RackAFX, then use it in any Windows VST client. RackAFX runs in tandem with Microsoft Visual Studio compilers, even the free ones. It sets up your project files, writes GUI code for you, and allows you to remotely launch some compiler functions from its control surface. Once you understand how the RackAFX API works, learning other commercial plug-in APIs will be much easier because most plug-in APIs share similar methods and set-up sequences. And, since RackAFX plug-ins are written in C++, the plug-in objects can be compiled on different platforms and embedded (or wrapped) to operate in just about any signal processing environment. We are currently running RackAFX plug-ins in Apple's iOS® in one of my other classes.

In the fall of 2010 I began writing and handing out chapters, one at a time, to students in a graduate-level class. The chapters combined DSP theory, audio algorithms, and programming in the proportions I had always wanted, including difference equations and derivations that were fully worked out. The RackAFX API requires zero GUI programming. It isn't tied to a licensed product that I have to worry will become obsolete or will change in complexity or underlying design. Most importantly, if you can learn to write RackAFX plug-ins, you can learn to write for just about any other API on the platform you choose. See Appendix A for more information.

Those initial grad students helped shape the direction and flow of the book (perhaps without knowing it). I wanted the book to be aimed at people with programming skills who wanted to get into audio signal processing or the other way around. Academic types that are tired of using mathematical software to try to do real-time signal processing should also benefit. The API does not require a steep learning curve.

Chapter 1 presents the number systems, sample indexing, and basic block diagram algorithm components. Chapter 2 explains how plug-ins are packaged in software and Chapter 3 immediately starts you writing plug-ins. Chapters 4 and 5 are the basic DSP theory chapters. Feel free to skip these chapters if you already have a strong DSP background or if you don't care about it. Chapter 6 is a cookbook of infinite impulse response (IIR) and audio specific filter designs. Chapter 7 introduces delay lines and delay effects. The circular buffering they require is necessary for Chapter 8's finite impulse response (FIR) tools and convolution plug-in.

Chapter 9 introduces oscillator design, which is needed in Chapter 10's modulated delay effects: flanger, chorus and vibrato effects. Chapter 11 includes the analysis of a collection of reverberator modules and an example reverb design. Chapter 12 returns to the modulation theme with modulated filter effects, including my own phaser, while Chapter 13 introduces dynamics processor design. Chapter 14 is a clean-up chapter of miscellaneous effects algorithms. The two appendices can be read any time after Chapter 3, where you design your first plug-ins. Appendix A deals with the VST and Audio Units (AU)® formats, their similarities and differences, and how to use RackAFX objects inside them. Appendix B shows how to use the GUI designer to drag and drop GUI elements on your control surface—0% GUI coding, guaranteed. And, if you have the tools and skills to make and maintain your own GUI, there are options for that as well.

My colleagues Ken Pohlmann, Joe Abbati, Colby Leider, Chris Bennett, and Rey Sanchez had nothing but encouragement and enthusiasm for the book/software effort. Like many of the students, they watched RackAFX evolve—even change names—from a fairly bland, gray academic interface into a highly customizable personal plug-in development lab with a drag-and-drop GUI designer. Of course those features and enhancements came from suggestions, comments, and critiques from the numerous students who used the software in class. The API would not be as robust today without their invaluable input and feedback. Stefan Sullivan tirelessly beta-tested the software, even in the middle of his thesis work; his comments led to many of the features in the current product. The RackAFX power users (students who went above and beyond the parameters of their projects and pushed the envelope of the API) include Mike Chemistruck, Greg Dion, Felipe Espic, Chen Zhang, Tim Schweichler, Phil Minnick, Sherif Ahmed, Scott Dickey, Matan Ben-Asher, Jay Coggin, Michael Everman, and Jordan Whitney.

I hope you enjoy learning the audio effects theory and plug-in design from the book and that each chapter opens up a new treasure box of cool audio gems you can use to make your own truly unique audio effects in your own audio effect laboratory. The software FAQs, sample code, and tutorials are all available at the website www.willpirkle.com and you are encouraged to upload your own plug-ins and videos as well. I can't wait to hear what you cook up in your lab!

Will Pirkle
June 1, 2012

CHAPTER 1
Digital Audio Signal Processing Principles

The first affordable digital audio devices began appearing in the mid-1980s. Digital signal processing (DSP) mathematics had been around since the 1960s and commercial digital recordings first appeared in the early 1970s, but the technology did not become available for widespread distribution until about 15 years later when the advent of the compact disc (CD) ushered in the age of digital audio. Digital sampling refers to the acquisition of data points from a continuous analog signal. The data points are sampled on a regular interval known as the sample period or sample interval. The inverse of the sample period is the sampling frequency. A compact disc uses a sampling frequency of 44,100 Hz, producing 44,100 discrete samples per channel each second, with a sample interval of about 22.7 microseconds (μS). While digital sampling applies to many different systems, this book is focused on only one of those applications: audio.

During the course of this book, you will learn both DSP theory and applications. This is accomplished by experimenting with your own DSP algorithms at the same time as you learn the theory using the included software, RackAFX™. The goal is to understand how the DSP algorithms translate into C++ code. The resulting plug-ins you write can be used to process audio in real time. Because plug-ins are software variations on hardware designs, it's worth examining how the hardware systems operate, the typical audio coding formats, and the algorithm building blocks you will be using to implement your own software versions.

1.1 Acquisition of Samples

The incoming analog audio signal is sampled with an analog-to-digital converter (ADC or A/D). A/Ds must accurately sample and convert an incoming signal, producing a valid digitally coded number that represents the analog voltage within the sampling interval. This means that for CD audio, a converter must produce an output every 22.7 μS. Figure 1.1 shows the block diagram of the input conversion system with LFP, A/D, and encoder.

Violating the Nyquist criteria will create audible errors in the signal in the form of an erroneously encoded signal. Frequencies higher than Nyquist will fold back into the

> The *sampling theorem* states that a continuous analog signal can be sampled into discrete data points and then reconstructed into the original analog signal without any loss of information—including inter-sample fluctuations—if and only if the input signal has been band-limited so that it contains no frequencies higher than one-half the sample rate, also known as the *Nyquist frequency* or *Nyquist rate*. Band-limited means low-pass filtered (LPF). Band-limiting the input signal prior to sampling is also known as adhering to the *Nyquist criteria*.

spectrum. This effect is called aliasing because the higher-than-Nyquist frequencies are encoded "in disguise" or as an "alias" of the actual frequency. This is easiest explained with a picture of an aliased signal, shown in Figure 1.2.

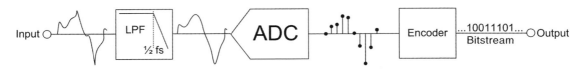

Figure 1.1: The input conversion system ultimately results in numerical coding of the band-limited input signal.

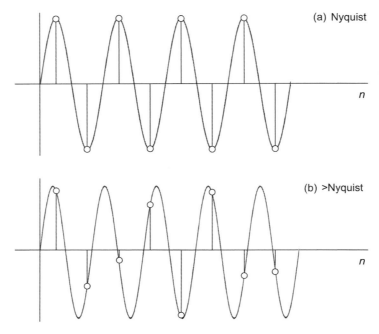

Figure 1.2: (a) The Nyquist frequency is the highest frequency that can be encoded with two samples per period. (b) Increasing the frequency above Nyquist but keeping the sampling interval the same results in an obvious coding error—the aliased signal is the result.

Once the aliased signal is created, it can never be removed and will remain as a permanent error in the signal. The LPF that band-limits the signal at the input is called the *anti-aliasing filter*. Another form of aliasing occurs in the movies. An analog movie camera takes 30 pictures (frames) per second. However, it must often film objects that are rotating at much higher rates than 30 per second, like helicopter blades or car wheels. The result is visually confusing—the helicopter blades or car wheels appear to slow down and stop, then reverse directions and speed up, then slow down and stop, reverse, and so on. This is the visual result of the high-frequency rotation aliasing back into the movie as an erroneous encoding of the actual event.

1.2 Reconstruction of the Signal

The digital-to-analog converter (DAC or D/A) first decodes the bit-stream, then takes the sampled data points or impulses and converts them into analog versions of those impulses. The D/A output is then low-pass filtered to produce the final analog output signal, complete with all the inter-sample fluctuations. As with the A/D diagram, the decoder and D/A are both inside the same device (a chip). Figure 1.3 shows the conceptual block diagram of the decoding end of the system.

The output filter is called the *reconstruction filter* and is responsible for re-creating the seemingly missing information that the original sampling operation did not acquire—all the inter-sample fluctuations. The reason it works is that low-pass filtering converts impulses into smeared out, wiggly versions. The smeared out versions have the shape $f(x) = \sin(x)/x$ which is also called the sinc() function and somewhat resembles the profile of a sombrero, as shown in Figure 1.4.

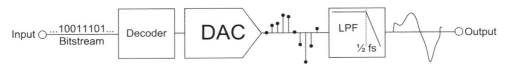

Figure 1.3: The digital bit-stream is decoded and converted into an analog output using a low-pass filter to reconstruct the analog signal.

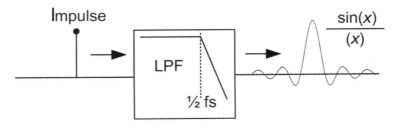

Figure 1.4: The ideal reconstruction filter creates a smeared output with a damped oscillatory shape. The amplitude of the $\sin(x)/x$ shape is proportional to the amplitude of the input pulse.

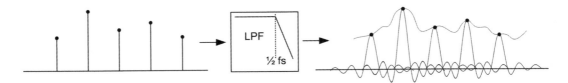

Figure 1.5: The sin(x)/x outputs of the LPF are summed together to reconstruct the original band-limited input waveform; the inter-sample information has been reconstructed.

When a *series* of impulses is filtered, the resulting set of sin(x)/x pulses overlap with each other and their responses all add up linearly. The addition of all the smaller curves and damped oscillations reconstructs the inter-sample curves and fluctuations (Figure 1.5).

1.3 Signal Processing Systems

Signal processing systems combine data acquisition devices with microprocessors to run mathematical algorithms on the audio data. Those algorithms are the focus of this book. Today's plug-ins are descendents of stand-alone hardware effects processors which are based on DSP chips. A DSP chip is a highly specialized processor designed mainly to run DSP algorithms. A DSP would function poorly as a central processing unit (CPU) for a personal computer because it only has a small but highly specialized instruction set. DSP devices (or just "DSPs") feature a core processor designed to multiply and accumulate data because this operation is fundamental to DSP algorithms. Because this process is repeated over and over, modern DSPs use pipelining to fetch the next data while simultaneously processing the current data. This technique greatly improves the efficiency of the system. A typical signal processing system consists of the following components (Figure 1.6):

- Data acquisition devices (A/D and D/A)
- A DSP chip
- Memory to store the algorithm program (program memory)
- Memory to store the algorithm data (data memory)
- A user interface (UI) with buttons, knobs, switches, and graphic displays

Music synthesis devices are an embedded version of a CPU+DSP arrangement. The microcontroller or microprocessor manages the UI and keyboard, while the DSP performs the audio synthesis and effects algorithms. Practical implementations now involve multiple DSPs, some of which are custom designed specifically for sound synthesis. An example block diagram is shown in Figure 1.7.

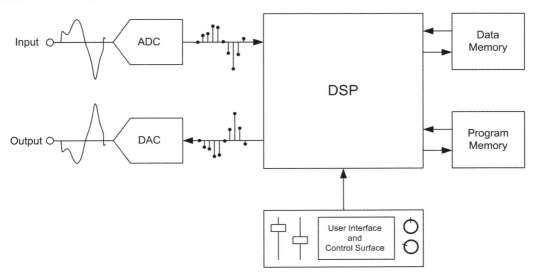

Figure 1.6: A simple signal processing system. The algorithm in this case is inverting the phase of the signal; the output is upside-down.

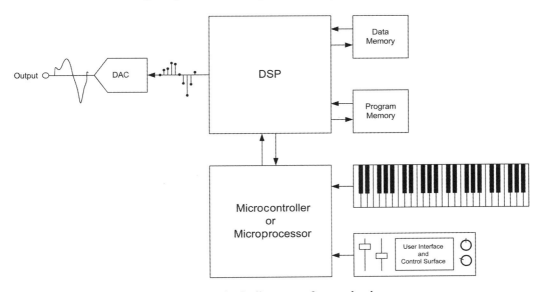

Figure 1.7: Block diagram of a synthesizer.

1.4 Synchronization and Interrupts

There are two fundamental modes of operation when dealing with incoming and outgoing audio data: synchronous and asynchronous. In synchronous operation, all audio input and output data words are synchronized to the same clock as the DSP. These are typically simple systems with a minimum of input/output (I/O) and peripherals. More-complex systems usually involve asynchronous operation, where the audio data is not synchronized to the DSP. Moreover, the audio itself might not be synchronous, that is, the input and output bit-streams might not operate on the same clock. A purely synchronous system will be more foolproof, but less flexible.

An asynchronous system will almost always be interrupt-based. In an interrupt-based design, the processor enters a wait-loop until a processor interrupt is toggled. The processor interrupt is just like a doorbell. When another device such as the A/D has data ready to deliver, it places the data in a predesignated buffer, and then it rings the doorbell by toggling an interrupt pin. The processor services the interrupt with a function that picks up the data, and then goes on with its processing code. The function is known as an *interrupt-service routine* or an *interrupt handler*. The interrupt-based system is the most efficient use of processor time because the processor can't predict the exact clock cycle when the data will be ready at the input.

Another source of interrupts is the UI. Each time the user changes a control, clicks a button, or turns a knob, the updated UI control information needs to be sent to the DSP so it can alter its processing to accommodate the new settings. This is also accomplished via interrupts and interrupt handlers. The interrupt is sometimes labeled INT, INTR, or IRQ (interrupt request line) in block diagrams or schematics.

1.5 Signal Processing Flow

Whether the processing is taking place on a DSP chip or in a plug-in, the overall processing flow, also known as the signal processing loop, remains basically the same. For a DSP chip, the processes in the loop will be encoded in the program memory's set of instructions, often in a nearly pure-inline fashion for optimization of the processor's cycles. For a plug-in, each processing block more or less takes on the form of a function in code, allowing for maximum flexibility.

The processing loop in Figure 1.8 consists of

- A one-time initialization function to set up the initial state of the processor and prepare for the arrival of data interrupts
- An infinite wait-loop, which does nothing but wait for an interrupt to occur
- An interrupt handler which decodes the interrupt and decides how to process—or ignore—the doorbell
- Data reading and writing functions for both control and data information
- A processing function to manipulate and create the audio output
- A function to set up the variables for the next time around through the loop, altering variables if the UI control changes warrant it

Digital Audio Signal Processing Principles 7

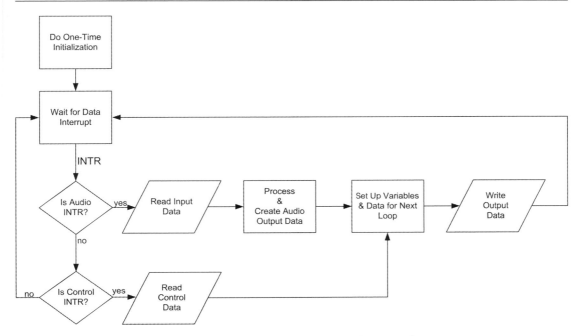

Figure 1.8: The flowchart for an audio signal processing system.

For the sampling theorem to hold true, the audio data must be arriving and leaving on a strictly timed interval, although it may be asynchronous with the DSP's internal clock. This means that when the DSP does receive an audio INTR it must do all of the audio processing and handle any UI interrupts before the next audio INTR arrives in one sample interval. The interrupt handling scheme is prioritized such that the audio data interrupt is the most important. Thus, while servicing the audio data interrupt, the DSP might be set up to ignore all other interrupts (except the reset interrupt) until the audio data is finished processing.

1.6 Numerical Representation of Audio Data

The audio data can be numerically coded in several different ways. Basic digital audio theory reveals that the number of quantization levels available for coding is found by Equation 1.1.

$$q = 2^N \qquad (1.1)$$

where N is the bit depth of the system.

Thus, an 8-bit system can encode 2^8 values or 256 quantization levels. A 16-bit system can encode 65,536 different values. Figure 1.9 shows the hierarchy of encoded audio data. As a system designer, you must first decide if you are going to deal with unipolar (unsigned) or bipolar (signed) data. After that, you need to decide on the data types.

8 Chapter 1

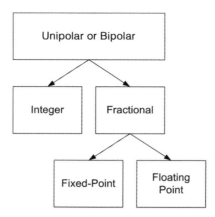

Figure 1.9: The hierarchy of numerical coding.

- Unipolar or *unsigned* data is in the range of 0 to +Max, or −Min to 0, and only has one polarity (+ or −) of data, plus the number zero (0).
- Bipolar or *signed* data varies from −Min to +Max and is the most common form today. It also includes the number zero (0).
- Integer data is represented with integers and no decimal place. Unsigned integer audio varies from 0 to +65,535 for 16-bit systems. Signed integer audio varies from −32,768 to +32,767 for the same 16-bit audio. In both cases, there are 65,536 quantization levels.
- Fractional data is encoded with an integer and fractional portion, combined as int.frac with the decimal place separating the parts (e.g. 12.09).

Within the fractional data subset are fixed- and floating-point types. Fixed-point data fixes the number of significant digits on each side of the decimal point and is combined as int-sig-digits.frac-sig-digits. "8.16" data would have 8 significant digits before the decimal place and 16 afterwards. Floating-point data has a moving mantissa and exponent which code the values across a range predetermined by the Institute of Electrical and Electronics Engineers (IEEE). The positive and negative portions are encoded in 2's complement so that the addition of exactly opposite values (e.g., −0.5 and +0.5) always results in zero. Figure 1.10 reveals how the quantization levels are encoded. The dashed lines represent the minimum and maximum values.

A fundamental problem is that the number of quantization levels will always be an even number since 2^N is always even. In bipolar systems, you would very much like to have the number zero (0) encoded as the number 0. If you do that, then you use up one of your quantization levels for it. This leaves an odd number of levels which cannot be split exactly in half. That creates the anomaly you see in Figure 1.10—there are always more negative (−) quantization levels than positive ones for bipolar coding. For the unipolar case, there is no value which exactly codes the zero or midpoint level; in Figure 1.10 it is midway between 127 and 128.

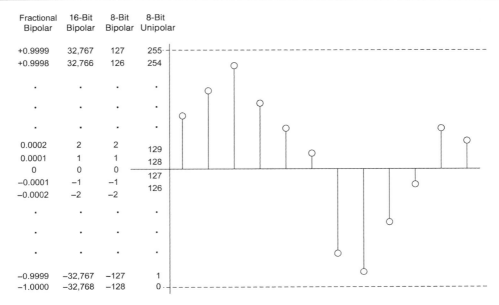

Figure 1.10: A comparison of several different types of data representations. The floating-point version is fixed for a range of −1.0 to +0.9999, though any range can be used.

This slight skewing of the data range is unavoidable if you intend on using the number zero in your algorithms, and that is almost guaranteed. In some systems the algorithm limits the negative audio data to the second most negative value. This is because phase inversion is common in processing algorithms, either on purpose or in the form of a negative-valued coefficient or multiplier. If a sample came in with a value of −32,768 and it was inverted, there would be no positive version of that value to encode. To protect against that, the −32,768 is treated as −32,767. The audio data that travels from the audio hardware adapter (DSP and sound card) as well as that stored in WAV files is signed-integer based. However, for audio signal processing, we prefer to use floating-point representation.

1.7 Using Floating-Point Data

In many audio systems, the DSP and plug-in data is formatted to lie on the range of −1.0 to +1.0 (which is simplifying the real range of −1.0 to +0.9999). In fact the plug-ins you code in this book will all use data that is on that same range. The reason has to do with overflow. In audio algorithms, addition and multiplication are both commonplace. With integer-based numbers, you can get into trouble quickly if you mathematically combine two numbers that result in a value that is outside the range of known numbers.

Consider the 16-bit integer bipolar format on the range of −32,768 to +32,767. Most of the values on this range, when multiplied together, will result in a product that is outside these limits. Addition and subtraction can cause this as well, but only for half the possible values. However, numbers between -1.0 and $+1.0$ have the interesting property that their product is always a number in that range. Converting an integer value to a fractional value along the normalized range of -1.0 to $+1.0$ in an N-bit digital audio system is easy, as is the reverse conversion shown in Equation 1.2:

$$\text{Fraction} = \frac{\text{Integer}}{2^N} \qquad (1.2)$$
$$\text{Integer} = \text{Fraction} * 2^N$$

where N is the bit depth of the system.

1.8 Basic DSP Test Signals

You need to know the data sequences for several fundamental digital signals in order to begin understanding how the DSP theory works. The basic signal set consists of

- Direct Current (DC) and step: DC is a 0Hz signal
- Nyquist
- ½ Nyquist
- ¼ Nyquist
- Impulse

The first four of these signals are all you need to get a ballpark idea of the frequency response of some basic DSP filters. The good news is that all the sequences are simple to remember.

1.8.1 DC and Step

The DC or 0 Hz and step responses can both be found with the DC/step input sequence: {...0, 0, 1, 1, 1, 1, 1, 1, 1, 1...}.

This signal in Figure 1.11 contains two parts: the step portion where the input changes from 0 to 1 and the DC portion where the signal remains at the constant level of 1.0 forever. When

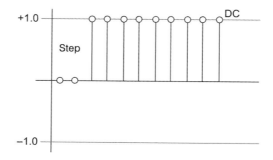

Figure 1.11 Representation of the DC/Step sequence.

you apply this signal to your DSP filter and examine the output, you will get two pieces of information. The step portion will tell you the transient attack time and the DC portion will give you the response at DC or 0 Hz.

1.8.2 Nyquist

The Nyquist input sequence represents the Nyquist frequency of the system and is independent of the actual sample rate. The Nyquist sequence is $\{\ldots -1, +1, -1, +1, -1, +1, -1, +1 \ldots\}$.

The Nyquist frequency signal in Figure 1.12 is the highest frequency that can be encoded. It contains the minimum of two samples per cycle with each sample representing the maximum and minimum values. The two-sample minimum is another way of stating the Nyquist frequency as it relates to the sampling theorem.

1.8.3 ½ Nyquist

The ½ Nyquist input sequence in Figure 1.13 represents the ½ Nyquist frequency of the system and is independent of the actual sample rate. The signal is encoded with four samples

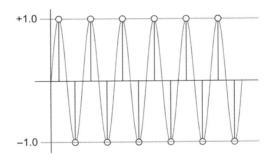

Figure 1.12 The Nyquist sequence.

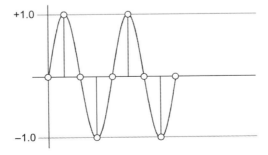

Figure 1.13 The ½ Nyquist sequence has four samples per cycle.

per cycle, twice as many as Nyquist. The ½ Nyquist sequence is {...−1, 0, +1, 0, −1, 0, +1, 0, −1, 0, +1, 0, −1, 0, +1, 0...}.

1.8.4 ¼ Nyquist

The ¼ Nyquist input sequence in Figure 1.14 represents the ¼ Nyquist frequency of the system and is independent of the actual sample rate. It is encoded with eight samples per cycle. The ¼ Nyquist sequence is {...0.0, 0.707, +1.0, 0.707, 0.0, −0.707, −1.0, −0.707, 0.0...}.

1.8.5 Impulse

The impulse shown in Figure 1.15 is a single sample with the value 1.0 in an infinitely long stream of zeros. The *impulse response* of a DSP algorithm is the output of the algorithm after applying the impulse input. The impulse sequence is {...0, 0, 0, 0, 1, 0, 0, 0, 0,...}.

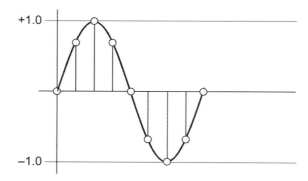

Figure 1.14 ¼ Nyquist sequence has eight samples per cycle.

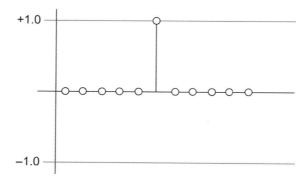

Figure 1.15 The impulse is a single nonzero sample value in a sea of zeros.

1.9 Signal Processing Algorithms

In the broadest sense, an algorithm is a set of instructions that completes a predefined task. The signal processing loop in Figure 1.8 is a picture of an algorithm for processing audio and control (UI) data in real time. In the specialized case of audio signal processing, an algorithm is a set of instructions that operates on data to produce an audio output bit-stream. Most of the exercises in this book involve processing incoming audio data and transforming it into a processed output. However, synthesizing a waveform to output also qualifies and in this special case, there is no real-time audio input to process. Most of the plug-ins in this book use the effects model, where an input sequence of samples is processed to create an output sequence, as shown in Figure 1.16.

Conventions and rules:

- $x(n)$ is always the input sequence; the variable n represents the location of the nth sample of the x-sequence.
- $y(n)$ is always the output sequence; the variable n represents the location of the nth sample of the y-sequence.
- $h(n)$ is the impulse response of the algorithm; a special sequence that represents the algorithm output for a single sample input or impulse.
- For real-time processing, the algorithm must accept a new input sample (or set of samples), do the processing, then have the output sample(s) available before the next input arrives; if the processing takes too long, clicks, pops, glitches, and noise will be the real-time result.

1.10 Bookkeeping

You can see that there are already three sequences to deal with: the input, output, and impulse response, all of which are coded with the same variable n to keep track of the location of samples within the sequence. The first step is to decide how to use n to do this bookkeeping task. Using it to represent the absolute position in the sequence would quickly become tiresome—how do you deal with indexing numbers like $x(12,354,233)$?

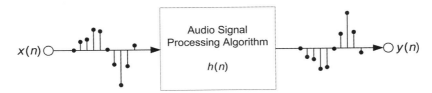

Figure 1.16: An audio signal processing algorithm that converts an input sequence $x(n)$ into an output sequence $y(n)$.

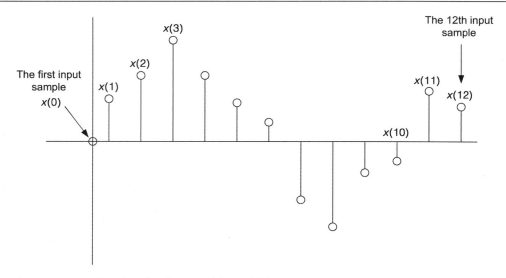

Figure 1.17: Using the absolute position within the sequence is one way to keep track, but the index values are going to get very large very quickly.

Figure 1.17 shows an input signal, $x(n)$, starting from $t = 0$ or $x(0)$. The $x(0)$ sample is the first sample that enters the signal processing algorithm. In the grand scheme of things, $x(0)$ will be the oldest input sample ever. Indexing the numbers with absolute position is going to be a chore as the index values are going to become large, especially at very high sample rates.

Another problem with dealing with the absolute position of samples is that algorithms do not use the sample's absolute position in their coding. Instead, algorithms use the position of the current sample and make everything relevant to that sample. On the next sample period, everything gets reorganized in relation to the current sample again. It might sound confusing at first, but it is a much better method of keeping track of the samples and more importantly, defining the I/O characteristics of the algorithm, called the *transfer function*. Figure 1.18 shows the input signal frozen at the current time, $x(n)$, and the other samples are indexed based on its location. One sample period later (Figure 1.19) you can see the frame of reference has shifted to the right and that $x(n)$ has now become $x(n - 1)$.

Bookkeeping rules:

- The current sample is labeled "n."
- Previous samples are negative, so $x(n - 1)$ would be the previous input sample.
- Future samples are positive, so $x(n + 1)$ would be the next input sample relative to the current one.
- On the *next* sample interval, everything is shuffled and referenced to the new current sample, $x(n)$.

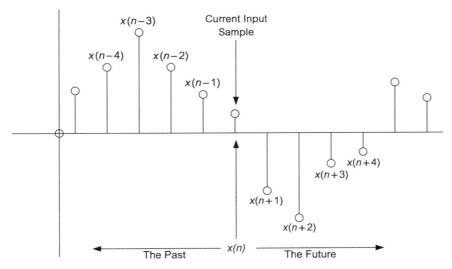

Figure 1.18: DSP algorithms use the current sample location as the reference location and all other samples are indexed based on that sample. Here you can see the current state of the algorithm frozen in time at the current input sample $x(n)$.

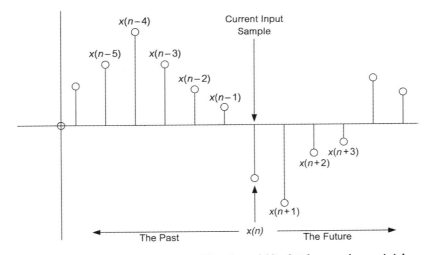

Figure 1.19: One sample period later, everything has shifted. The previous $x(n)$ is now indexed as $x(n-1)$ and what was the next sample, $x(n+1)$ now becomes the current sample.

1.11 The One-Sample Delay

Whereas analog processing circuits like tone-controls use capacitors and inductors to alter the phase and delay of the analog signal, digital algorithms use time-delay instead. You will uncover the math and science behind this fact later on in Chapters 4 and 5 when you start to use it. In algorithm diagrams, a delay is represented by a box with the letter z inside. The z-term will have an exponent such as z^{-5} or z^{+2} or z^0—the exponent codes the delay in samples following the same bookkeeping rules, with negative ($-$) exponents representing a delay in time backward (past samples) and positive ($+$) representing delay in forward time (future samples). You call z the *delay operator* and as it turns out, time-delay will be treated as a mathematical operation.

You are probably asking yourself how you can have a positive delay toward the future, and the answer is that you can't for real-time signal processing. In real-time processing you never know what sample is going to come in next. However, in non-real-time processing (for example, an audio file that you are processing offline) you do know the future samples because they are in the file. Figure 1.20 shows two common ways to denote a one-sample delay in an algorithm block diagram.

Delay rules:

- Each time a sample goes into the delay register (memory location), the previously stored sample is ejected.
- The ejected sample can be used for processing or deleted.
- The delay elements can be cascaded together with the output of one feeding the input of the next to create more delay time.

If a sample $x(n)$ goes into the one-sample delay element, then what do you call the sample that is ejected? It's the previous sample that came in, one sample interval in the past. So, the output sample is $x(n-1)$. In Figure 1.21 you can see how delay elements can be cascaded with outputs taken at multiple locations generating multiple samples to use in the algorithm.

1.12 Multiplication

The next algorithm building block is the *scalar multiplication* operation. It is a sample-by-sample operator that simply multiplies the input samples by a coefficient. The multiplication operator is used in just about every DSP algorithm around. Figure 1.22 shows the multiplication operator in action. The inputs are simply scaled by the coefficients.

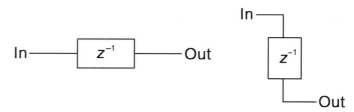

Figure 1.20: Two common ways to draw a delay; the one-sample delay is represented with the z^{-1}. Both versions are equivalent.

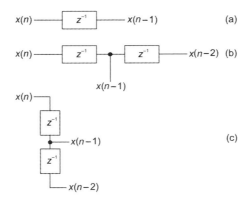

Figure 1.21: Three delay algorithms: (a) one-sample delay, (b) two one-sample delays cascaded, producing two different outputs, $x(n-1)$ and $x(n-2)$, notice that (c) is functionally identical to (b).

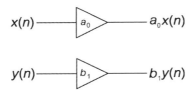

Figure 1.22: The multiplication operator is displayed as a triangle and a coefficient letter or number inside or just above it.

1.13 Addition and Subtraction

Addition and subtraction are really the same kind of operation—subtracting is the addition of a negative number. There are several different algorithm symbols to denote addition and subtraction. The operation of *mixing* signals is really the mathematical operation of addition. Figure 1.23 shows several ways of displaying the addition and subtraction operation in block diagrams.

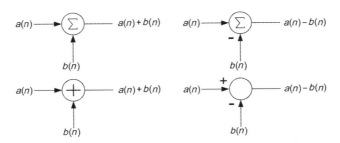

Figure 1.23: Addition and subtraction diagrams for two input sequences $a(n)$ and $b(n)$; these are all commonly used forms of the same functions.

1.14 Algorithm Examples and the Difference Equation

By convention, the output sequence of the DSP algorithm is named $y(n)$ and the mathematical equation that relates it to the input is called the *difference equation*. Combining the operations will give you a better idea of what the difference equation looks like. Figure 1.24 shows the difference equations for several combinations of algorithm building blocks. The output $y(n)$ is a mathematical combination of inputs.

1.15 Gain, Attenuation, and Phase Inversion

As shown in Figure 1.25, a simple coefficient multiplier will handle the three basic audio processing functions of gain, attenuation, and inversion. If the coefficient is a negative number, phase inversion will be the result. If the coefficient has a magnitude less than 1.0, attenuation will occur, while amplification occurs if the magnitude is greater than 1.0.

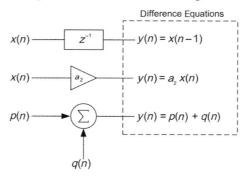

Figure 1.24: The difference equations relate the input(s) to the output via mathematical operations

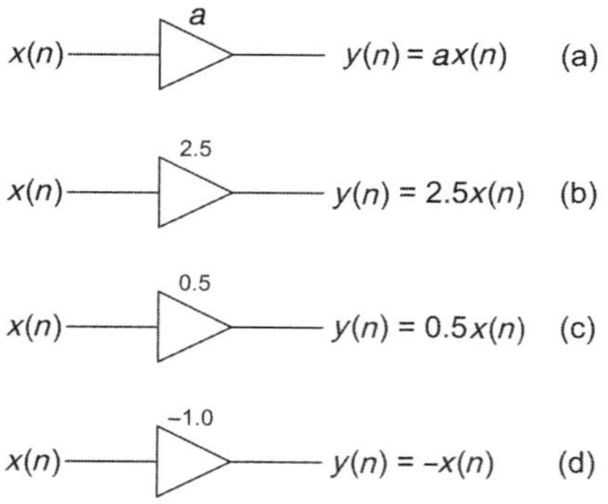

Figure 1.25: Examples of simple multiplier algorithms. Notice the different notation with the coefficient placed outside the triangle; this is another common way to designate it. (a) Simple scalar multiplication by an arbitrary value "a". (b) Gain is accomplished with a coefficient magnitude greater than one. (c) Attenuation reduces the size of the input value with a coefficient that is less than one. (d) Phase inversion turns the signal upside down by using a negative coefficient; a value of -1.0 perfectly inverts the signal.

1.16 Practical Mixing Algorithm

A problem with mixing multiple channels of digital audio is the possibility of overflow or creating a sample value that is outside the range of the system. You saw that by limiting the bipolar fractional system to the bounds of -1.0 to $+1.0$, multiplication of any of these numbers always results in a number that is smaller than either, and always within the same range of -1.0 to $+1.0$. However, addition of signals can easily generate values outside the ± 1 limits. In order to get around this problem, N-channel mixing circuits incorporate attenuators to reduce the size of the inputs, where the attenuation value is 1/N. When mixing two channels, the attenuators each have a value of ½ while a three-channel mixer would have attenuators with a value of 1/3 on each mixing branch. If all channels happen to have a maximum or minimum value at the same sample time, their sum or difference will still be limited to ± 1. Figures 1.26 and 1.27 show the generalized mixing and weighted-sum algorithms.

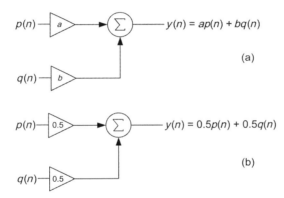

Figure 1.26: (a) Generalized mixer/summer with a separate coefficient on each line and (b) a normalized mixer which will not overflow or clip.

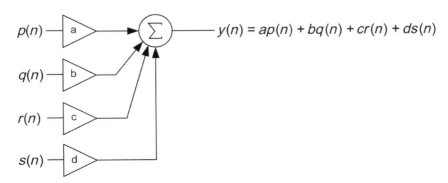

Figure 1.27: An example of a weighted-sum algorithm; each sample from each channel has its own weighting coefficient, a–d.

In the next chapter, you will be introduced to the anatomy of a plug-in from a software point of view. In Chapters 6 through 14, you will learn how DSP theory allows you to combine these building blocks into filters, effects, and oscillators for use in your own plug-ins.

Bibliography

Ballou, G. 1987. *Handbook for Sound Engineers*, pp. 898–906. Indiana: Howard W. Sams & Co.
Jurgens, R. K., ed. 1997. *Digital Consumer Electronics Handbook*, Chapter 2. New York: McGraw-Hill.
Kirk, R. and Hunt, A. 1999. *Digital Sound Processing for Music and Multimedia*, Chapter 1. Massachusetts: Focal Press.
KORG, Inc. 1991. *KORG Wavestation SR Service Manual.* Tokyo, Japan: KORG Inc.
Limberis, A. and Bryan, J. 1993. An architecture for a multiple digital-signal processor based music synthesizer with dynamic voice allocation. *Journal of the Audio Engineering Society*, Preprint No. 3699.
Pohlmann, K. C. 2011. *Principles of Digital Audio*, pp. 16–30. New York: McGraw-Hill.
Stearns, S. D. and Hush, D. R. 1990. *Digital Signal Analysis*, pp. 44–52. Englewood Cliffs, NJ: Prentice-Hall.

CHAPTER 2
Anatomy of a Plug-In

A plug-in is a software component that interfaces with another piece of software called the client in order to extend the client's capabilities in some way. For example, Internet browsers use plug-ins that implement added functions like searching and text messaging. In computer audio systems, a plug-in is usually an audio effect of some kind. However, a plug-in could also implement an oscilloscope or frequency analyzer function. Synthesizer plug-ins extend the client's functionality by working as musical instruments.

In order to start writing plug-ins, you need to know how the plug-in connects to and communicates with the client. Windows® plug-ins are typically packaged as dynamic link library, or DLL, files. Apple® plug-ins are packaged in a *bundle* which is configured as a *component*. The main difference between the two is in the lingo their designers use to describe them. Rather then try to accommodate both DLL and component labels during this discussion, we will just use DLL to describe both in conceptual form. Specific programmatic differences in the formats will be addressed and can be found on the website www.willpirkle.com.

2.1 Static and Dynamic Linking

C++ compilers include sets of precompiled libraries of functions for you to use in your projects. Perhaps the most common of these is the math library. If you try to use the sin() method you will typically get an error when you compile stating that "sin() is not defined." In order to use this function you need to link to the library that contains it. The way you do this is by placing #include <math.h> at the top of your file. Depending on your compiler, you might also need to tell it to link to math.lib. When you do this, you are statically linking to the math.h library, a precompiled set of math functions in a .lib file. Static linking is also called implicit linking. When the compiler comes across a math function, it replaces the function call with the precompiled code from the library. In this way, the extra code is compiled *into* your executable. You cannot un-compile the math functions. Why would you do this? Suppose a bug is found in the sin() function and the math.h library has to be re-compiled and redistributed. You would then have to re-compile your software with the new math.h library to get the bug fix.

The solution is to link to the functions at run time. This means that these precompiled functions will exist in a separate file, which your executable will know about and communicate with, but only after it starts running. This kind of linking is called dynamic linking or explicit

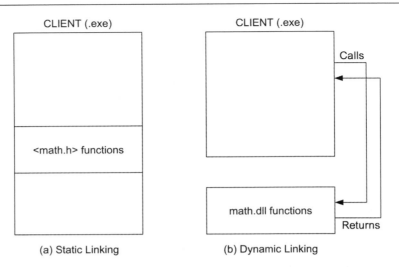

Figure 2.1: (a) In static linking the functions are compiled inside the client. (b) In dynamic linking the functions are located in a different file. This requires a communications mechanism between the client and DLL to call functions and return information.

linking and is shown in Figure 2.1. The file that contains the precompiled functions is the DLL. It is more complicated because extra steps must be taken during run-time operation rather than relying on code compiled directly into the executable. The advantage is that if a bug is found in the library, you only need to redistribute the newly compiled file rather than re-compiling your executable. The other advantage is that the way this system is set up—a client that connects to a component at run time—works perfectly as a way to extend the functionality of a client without the client knowing anything about the component when it is compiled. This is simplified when the DLL is loaded into the same virtual address space as the client. If you already understand this, you can skip the next section; otherwise, read on.

2.2 Virtual Address Space and DLL Access

When you start a program (client) in a Windows 32-bit operating system (OS), the first thing that happens is the OS creates a new virtual machine and virtual address space. The maximum amount of memory that can be accessed by a 32-bit microprocessor is 2^{32} or about 4 gigabytes of memory. The client executable believes it has a full 4 gigabytes of memory and that its executable is loaded into part of this memory block. This block of memory is also called the process address space. The OS is responsible for maintaining this illusion for the client as well as all the other executables that are running on the system. This is done through virtualization of the microprocessor, memory, and peripherals.

While that topic could be the subject of another book, the main thing you need to know is that typically when the client loads the DLL and begins the communication process, the DLL is loaded into the same virtual address space as the client. This means that the client code might as well have the DLL code compiled into it since the addressing requires no translation. It should be noted that a DLL *can* be loaded into another process space or even on another machine across a network. In this case it is called an "out of process DLL"; however, inter-process communication is complicated and requires OS and/or network function calls. We will not be considering out of process DLLs here. With the DLL in the same process address space as the client, there is no extra overhead and the communication uses in-process addressing. Both types are shown in Figure 2.2.

In order to use the code in a DLL the client must perform two activities:

1. Load the DLL into the process address space.
2. Establish the communication mechanism for functions to be called.

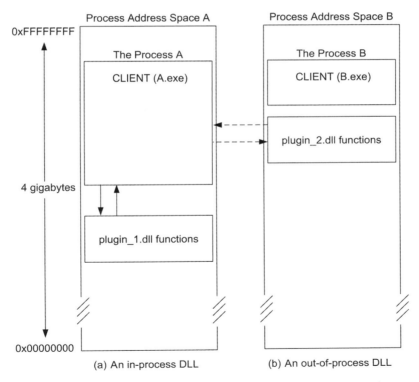

Figure 2.2: The process address space is a 4-gigabyte memory map starting at address 0x00000000 and going to 0xFFFFFFFF. When the client is launched it is placed inside the memory space. (a) An in-process DLL communicates directly with the client while (b) an out-of-process DLL communication requires OS intervention.

In the Windows OS, the three functions that are used to complete these tasks are

1. LoadLibrary(): Loads the DLL into the process address space.
2. GetProcAddress(): Retrieves pointers to functions in the DLL.
3. FreeLibrary(): Unloads the DLL from the process address space.

2.3 C and C++ Style DLLs

A DLL written in the C programming language consists of a set of stand-alone functions. There is no *main*() function. The functions can be defined and implemented in one .c file or can be broken into an interface file (.h) and implementation file (.c)—either way, the DLL performs a set of isolated functions. A problem with using the C programming language to write a DLL is the persistence of data. In C (and C++) the curly brackets { } define the *scope* of a function.

A fundamental problem is that the data declared inside a function cannot persist from one call to the next. One solution involves using global variables, which is generally frowned upon. A better solution to this problem is for the DLL to dynamically declare a data structure (e.g., using malloc()) that will hold all the persistent variables and then pass a pointer to this data structure back to the client to maintain. During subsequent calls to the DLL, the client passes the pointer back to the DLL as a function argument so that it may operate on the persistent data. This is shown in Figure 2.3. When the DLL is no longer needed, it clears the memory by deleting the structure.

In the C++ programming language, the *class* data type defines an object that is a collection of member variables and member functions which can operate on those variables. By packaging a plug-in as a C++ class, you get several advantages; first, all of the benefits of C++ (encapsulation, polymorphism, etc.) are available during the coding process. Second, rather than allocating a data structure and returning a pointer to it, the DLL can create a new instance of the plug-in object and pass a pointer to the *object* to the client. Now, the client can simply call functions on the object—it does not have to communicate with the DLL again until it is time to either unload the DLL or, better yet, create another instance of the plug-in object. This leads to a third advantage over the C-based DLL: the client can create multiple plug-ins easily. The DLL can serve up multiple instances of the object. Sometimes, the plug-in is referred to as a *server* and this becomes another kind of client-server system. This is illustrated in Figure 2.4.

> Any variable declared inside a function, after the first open curly bracket { is only defined for the duration of the function. After the closing curly bracket } is encountered, the variable ceases to exist.

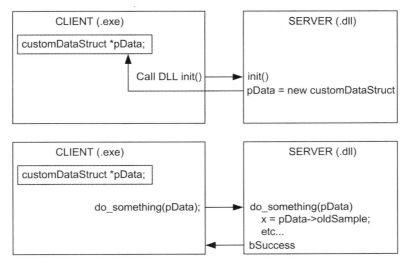

Figure 2.3: In a C-style DLL, the client first calls an initialization function and the DLL returns a pointer to a dynamically allocated data structure (pData), which the client stores. Later, the client calls a function do_something() and passes the pointer back to the DLL as a function argument. The DLL accesses the stored data, uses it, and then typically responds with a success flag.

2.4 Maintaining the User Interface

Most plug-ins have an associated graphical user interface (GUI or UI) with controls for manipulating the device. There are several schemes, but in general, when a new instance of the plug-in is created, a new child window is created with the GUI embedded in it. Whenever a GUI control changes state, a function on the plug-in is called. The client or GUI passes the plug-in information about which control changed and the new value or state of the control. The plug-in then handles the message by updating its internal variables to affect the change in signal processing. Generally, the GUI appearance (the position of the sliders or knobs or the states of switches) is controlled automatically by the client or the GUI code itself. There are three different ways the GUI can be handled:

1. The client creates, maintains, and destroys the GUI and forwards control-change messages to the plug-in.
2. The client creates, maintains, and destroys the GUI but the GUI communicates directly with the plug-in.
3. The plug-in creates, maintains, destroys, and communicates directly with the GUI, independent of the client.

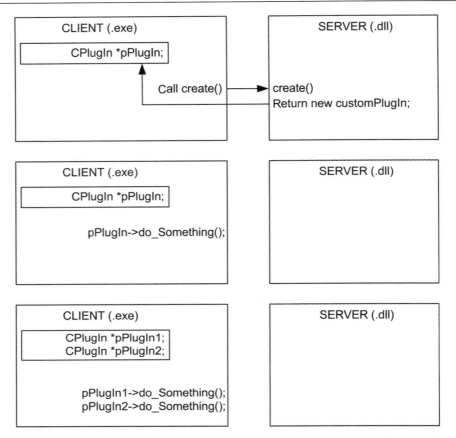

Figure 2.4: In a C++ plug-in, the client calls a creation function and the DLL (server) creates the object and passes a pointer to it back to the client. The client uses this pointer for future calls to the plug-in without having to bother communicating with the DLL. The client might also create multiple instances of the plug-in and then use the resulting pointers to implement processing functions.

Figures 2.5 through 2.7 show the three basic GUI scenarios. The first difference is in who creates, maintains, and destroys the GUI. When the client creates and maintains the GUI, it creates it as a child window which has the benefit that if the client is minimized or closed, the child windows will be hidden or destroyed accordingly. Therefore, the first two scenarios are more common. Some plug-ins can accommodate stand-alone GUI operation in addition to the client-child configuration. The second difference is in how the communication flows: indirectly routed through the client or directly from the GUI to the plug-in. The RackAFX software uses the second paradigm; the client creates the GUI but the GUI communicates directly with the plug-in when controls are modified.

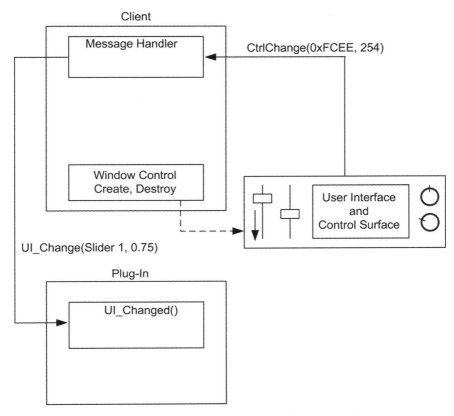

Figure 2.5: In this scenario, the client maintains the GUI and receives control change messages from it; the client then optionally translates that information before forwarding it to the plug-in.

2.5 The Applications Programming Interface

In order for the client-server scheme to work, both the client and DLL/plug-in must agree on the naming of the functions. This includes the creation function and all of the functions that the client will be able to call on the plug-in object. The plug-in might have other functions that the client doesn't know about, but they must agree on a basic set of them. Additionally, rules must be set up to define the sequence of function calls; the plug-in's author (that's you) will need to understand how the client intends to use the object.

The client must make sure that once it establishes these rules it adheres to them in future versions to avoid breaking the plug-in. On the other hand, the plug-in must also promise to implement the basic required functions to make the plug-in work. So, you can see that there is an implied *contract* between the client and DLL server. This contract is the *applications*

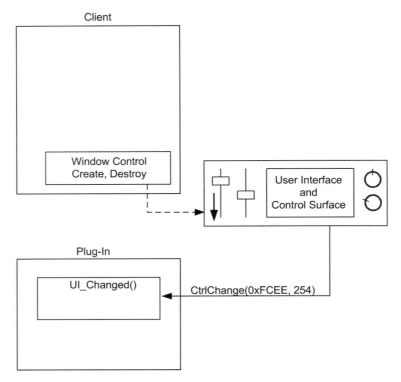

Figure 2.6: In this scenario, the client maintains the GUI, which communicates directly with the plug-in.

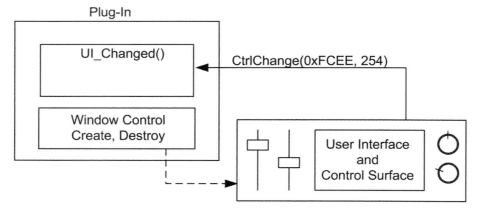

Figure 2.7: In another scenario, the plug-in maintains and communicates directly with the GUI without interference from the client.

programming interface or API. It is a definition of the functions an object must implement to be considered a proper plug-in as well as any additional functions that may be called or overridden. It defines the function prototypes and describes how the functions will be called and used. The API is written by the client manufacturer and is made available to programmers who want to create plug-ins for that target client. Some examples are Direct-X®, VST®, AU®, and AAX®. Each of these manufacturers publishes an API that describes the contract with the plug-in developers. However, the basic operation and core functionality are generally the same.

C++ is especially useful here. Since the plug-in is an instance of a C++ object, the client manufacturer can specify that the plug-in is a derived class of a special base class that it defines. The base class is made to be abstract, containing virtual functions that the derived class overrides. These virtual functions provide the common functionality of the plug-in. There are two options here:

1. The manufacturer defines the base class as abstract and then provides default implementations of the virtual functions. Typically, the default implementations do nothing but return a success code. The plug-in authors then override whichever methods they need to. For example, the plug-in might not care about responding to Musical Instrument Digital Interface (MIDI) messages, so the default implementation of the MIDI function will suffice.
2. The manufacturer defines the base class as a *pure abstract* base class by making one or more of the virtual functions *pure* virtual functions. A pure abstract base class cannot be instantiated; only derived classes that implement all the pure virtual functions can. This forms a binding contract between the plug-in developer and the client manufacturer since the derived class won't work properly unless it implements the pure abstract functions that the client specifies.

The RackAFX software uses the first method, supplying default implementations for all virtual functions. As the plug-in author, you only override the functions you need to. But what are the typical required functions and were do they come from?

2.6 Typical Required API Functions

Plug-ins are designed after the hardware devices that they replace. The audio processing loop is the same as the hardware version you saw in Chapter 1. Figure 2.8 shows a software variation on that flowchart.

Although the various plug-in APIs are different in their implementations, they share a common set of basic operations. Table 2.1 details the responsibility of each function.

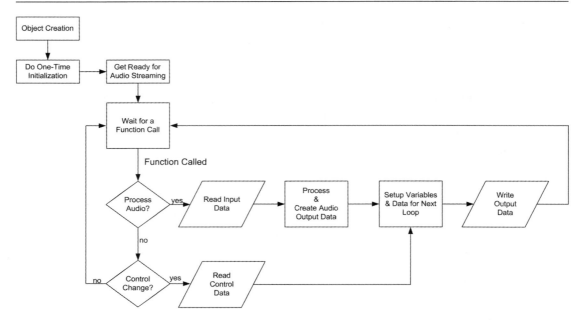

Figure 2.8: The software version of the DSP audio-processing loop. Most APIs also include functions to get or set information about the controls.

Table 2.1: The typical core operations that plug-in APIs share.

Function	Responsibility
One-time initialization	Called once when the plug-in is instantiated, this function implements any one-time-only initialization, usually consisting of initializing the plug-in variables, GUI, and allocating memory buffers dynamically.
Destruction	Called when the plug-in is to be destroyed, this function de-allocates any memory declared in the one-time initialization and/or in other functions that allocate memory. If there are any owned child-windows, the plug-in destroys them here.
Prepare for streaming	Called after the user has hit the play button or started audio streaming but before the data actually flows. This function is usually used to flush buffers containing old data or initialize any variables such as counters that operate on a per-play basis (not found in some APIs).
Process audio	The main function that does the actual signal processing. This function receives audio data, processes it, and writes out the result. This is the heart of the plug-in.
Get UI control info	Called to get information about a UI control—its name or label.
Set UI control info	Called when the plug-in needs to change the control information like its name or label.
Get UI control value	Called to get the value for the UI control that will set its appearance on the control surface.
Set UI control value	Called when the user makes a change to the plug-in's controls, this function is the message handler to deal with the user input. It usually causes a change or update in the plug-in's internal variables.

2.7 The RackAFX Philosophy and API

The fundamental idea behind the RackAFX software is to provide a platform for rapidly developing real-time audio signal processing plug-ins with a minimum of coding, especially with regard to the UI. In fact, most of the details of the connection between the RackAFX plug-in and the RackAFX UI screen are hidden from the developer so that he or she may concentrate more on the audio signal processing part and less on the UI details.

The RackAFX API specifies that the plug-in must be written in the C++ language and therefore takes advantage of the base class/derived class paradigm. The RackAFX API specifies a base class called CPlugIn from which all plug-ins are derived.

- RackAFX will automatically write C++ code for you to create a blank plug-in by creating a derived class of CPlugIn.
- As you add and remove controls from the control surface, the RackAFX client will automatically update your C++ code accordingly.
- This lets you focus on the signal processing and not the UI, making it a great tool for both rapid plug-in development and for teaching how to write plug-ins.
- After learning RackAFX, you will be able to understand other companies' APIs and learn to write plug-ins in their formats quickly and easily.
- Because the plug-in objects you create are written in C++, you can easily move them around between other APIs or computer platforms. You can wrap them to work easily in other systems too.

You only need to implement five functions in RackAFX to create a plug-in:

1. Constructor
2. Destructor
3. prepareForPlay()
4. processAudioFrame()
5. userInterfaceChange()

Figure 2.9 shows where these functions fit into the real-time audio processing loop.

2.7.1 __stdcall

> In the RackAFX code you will see the qualifier __stdcall preceding each function prototype as well as implementation. The __stdcall calling convention is there for future compatibility with other compilers as well as other third-party software. The __stdcall is a directive that lets the compiler know how the stack will be cleaned up after function calls; it has no effect on the math, logic, or audio processing, so you can essentially ignore it.

Chapter 2

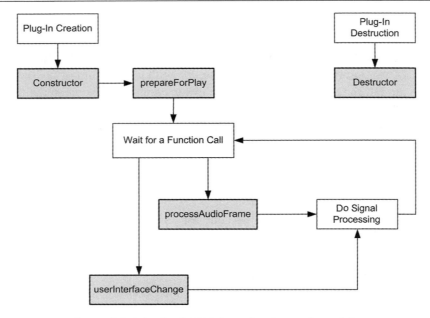

Figure 2.9: The RackAFX C++ plug-in version of the real-time audio processing loop in Figure 2.8.

Here is part of the interface file for the CPlugIn object plugin.h, which defines the contract or base class object interface for the primal six methods:

```
/*
        RackAFX(TM) Rapid Plug-In Development (RaPID) Client
        Applications Programming Interface
        Base Class Object Definition
        Copyright(C) Will Pirkle 2002-2012

        In order to write a RackAFX Plug-In, you need to create a C++ object that is
        derived from the CPlugIn base class. Your Plug-In must implement the
        constructor, destructor and virtual Plug-In API Functions below.
*/

// RackAFX abstract base class for DSP filters
class CPlugIn
{
public:
        // Plug-In API Member Methods:
        // The followung 5 methods must be impelemented for a meaningful Plug-In
        //
        // 1. One Time Initialization
        CPlugIn();
```

```
// 2. One Time Destruction
virtual ~CPlugIn(void);

// 3. The Prepare For Play Function is called just before audio streams
virtual bool __stdcall prepareForPlay();

// 4. processAudioFrame() processes an audio input to create an audio
output
virtual bool __stdcall processAudioFrame(float* pInputBuffer,
                                         float* pOutputBuffer,
                                         UINT uNumInputChannels,
                                         UINT uNumOutputChannels);
// 5. userInterfaceChange() occurs when the user moves a control.
virtual bool __stdcall userInterfaceChange(int nControlIndex);
```

The five functions in Table 2.2 are the core RackAFX API—implement them and you have a legitimate RackAFX plug-in. Best of all, the RackAFX plug-in designer will write and provide default implementations of all these functions for you. You need only to go in and alter them to change your plug-in's behavior. See Appendix A for a comparison of the RackAFX API and other commercially available formats as well as notes on using RackAFX plug-in objects inside API wrappers for other formats.

Table 2.2: The RackAFX API core functions.

RackAFX Function	Remarks
CPlugIn () Parameters: • none	The constructor for the plug-in object, this function is the one-time initialization function.
~CPlugIn () Parameters: • none	The destructor for the plug-in.
prepareForPlay() Parameters: • none	Function is called just before audio begins streaming. The audio file's sample rate, bit depth, and channel counts are extracted and then set by the client just before calling this method.
processAudioFrame() Parameters: • pInputBuffer • pOutputBuffer • uNumInputChannels • uNumOutputChannels	The most important function in the API; this is where the audio processing is handled. You might do all the processing in this function or use it to call sub-functions. You are responsible for writing the data to the output buffer via pOutputBuffer. A pointer to one frame of audio input data. A frame is a set of channels as defined by uNumInputChannels. A pointer to one frame of audio output data. A frame is a set of channels as defined by uNumOutputChannels. The number of input channels in this frame of data. Currently, this value will be either 1 (mono) or 2 (stereo). The number of output channels in this frame of data. Currently, this value will be either 1 (mono) or 2 (stereo).
userInterfaceChange() Parameters: • nControlIndex	Function is called after the user changes a control in the RackAFX UI. RackAFX will automatically update the variable linked to this control prior to calling this method. The index of the control that was moved and whose value RackAFX has changed.

Bibliography

Apple Computers, Inc. 2011. *The Audio Unit Programming Guide*. https://developer.apple.com/library/mac/#documentation/MusicAudio/Conceptual/AudioUnitProgrammingGuide/Introduction/Introduction.html. Accessed August 7, 2012.

Bargen, B. and Donnelly, P. 1998. *Inside DirectX*, Chapter 1. Redmond, WA: Microsoft Press.

Coulter, D. 2000. *Digital Audio Processing*, Chapters 7–8. Lawrence, KS: R&D Books.

Petzold, C. 1999. *Programming Windows*, Chapter 21. Redmond, WA: Microsoft Press.

Richter, J. 1995. *Advanced Windows*, Chapters 2 and 11. Redmond, WA: Microsoft Press.

Rogerson, D. 1997. *Inside COM*, Chapters 1–2. Redmond, WA: Microsoft Press.

Steinberg.net. *The Steinberg VST API*. http://www.steinberg.net/nc/en/company/developer/sdk_download_portal.html. (Note: you must create a free developer's account to download the API.) Accessed August 7, 2012.

CHAPTER 3
Writing Plug-Ins with RackAFX

The RackAFX plug-in designer will help you write your plug-in. When you create a new RackAFX project, it will set up a new Visual C++ project folder for you and populate your project with all the files you will need. It will automatically create a new derived class based on the name of your project. When you set up graphical user interface (GUI) controls like sliders and buttons, it will write and maintain the code for you. You will be switching back and forth between RackAFX and your C++ compiler. There are buttons on the RackAFX GUI that will let you jump to the compiler as well as launch compiler functions like rebuilding and debugging. You will use RackAFX to maintain your GUI and your compiler to write the signal processing code.

3.1 Building the DLL

RackAFX sets up your compiler to deliver your freshly built dynamic link libraries (DLL) to the /PlugIns folder in the RackAFX application directory. If you ever want to see, move, or delete your DLL you can find this folder by using the menu item PlugIn > Open PlugIns Folder. Each time you create a new project, RackAFX creates a pass-through plug-in by default; you are urged to build and test the DLL right after creating the new project to check your audio input/output (I/O) and any other connections. You then write over the pass-through code with your own processing algorithm.

After a successful build, you use RackAFX to test and debug the plug-in. You tell RackAFX to load the DLL and create your plug-in. The client needs to handle four basic operations during the lifecycle of your component:

1. Creation of the plug-in
2. Maintaining the UI
3. Playing and processing audio through the plug-in
4. Destruction of the plug-in

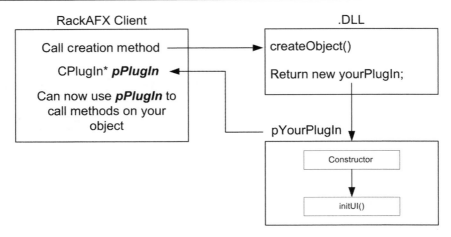

Figure 3.1: The *new* operator in createObject() dynamically creates your plug-in, which calls your constructor; the constructor in turn calls initUI() to create and initialize the user controls.

3.2 Creation

When you load a plug-in in RackAFX, you are actually passing the system a path to the DLL you've created. RackAFX uses an operating system (OS) function call to load the DLL into its process space. Once the DLL is loaded, RackAFX first runs a compatibility test, then requests a pointer to the creation method called createObject(). It uses this pointer to call the method and the DLL returns a newly created instance of your plug-in cast as the `PlugIn*` base class type. From that point on, the RackAFX client can call any of the base class methods on your object. Figure 3.1 shows the flow of operation during the creation phase.

Your constructor is where all your variables will be initialized. The very first line of code in the constructor has been written for you; it calls initUI(), which is a method that handles the creation and set up of your GUI controls. You never need to modify the initUI() method; RackAFX maintains this code for you.

3.3 The GUI

When you set up GUI elements such as sliders and buttons, RackAFX adds member variables to the .h file of your derived plug-in class. Each slider or button group controls one variable in your code. You set up each control with minimum, maximum, and initial values as well as supplying the variable name and data type. As the user moves the control, RackAFX calculates the variable's new value and delivers it to your plug-in automatically, updating it in real time. In some cases, this is all you will need and there is nothing left to write. In other cases, you will

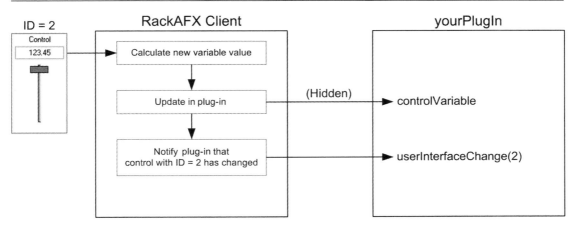

Figure 3.2: The sequence of events when the user moves the control with ID = 2 starts with a hidden change to the underlying linked variable, followed by a call to userInterfaceChange() passing the control's ID value as the parameter.

need to perform more calculations or logic processing in addition to just changing the control variable. So, in addition to changing and updating your internal GUI variable, RackAFX will also call the userInterfaceChange() method on your plug-in, shown in Figure 3.2.

3.4 Processing Audio

When the user loads an audio file and hits the Play button, a two-step sequence of events occurs. First, the client calls prepareForPlay() on the plug-in. The plug-in will do any last initializations it needs before audio begins streaming. prepareForPlay() is one of the most important functions to deal with. Your plug-in has variables declared in it (see PlugIn.h) that contain information about the currently selected audio file:

```
// information about the current playing-wave file
int m_nNumWAVEChannels;
int m_nSampleRate;
int m_nBitDepth;
```

Just prior to calling prepareForPlay(), the client sets these values on your plug-in object. The reason this is done at this point is that the user can load multiple audio file types of varying channels (mono or stereo), sample rates, and bit depths at any time; thus, this is a per-play method. Many algorithms require these values to be known before certain things can be created or initialized. Almost all filtering plug-ins require the sample rate in order to calculate their parameters correctly. After prepareForPlay() returns, audio begins streaming. When audio streams, the client repeatedly calls processAudioFrame(), passing it input and output buffer pointers as shown in Figure 3.3. This continues until the user hits Pause or Stop.

38 Chapter 3

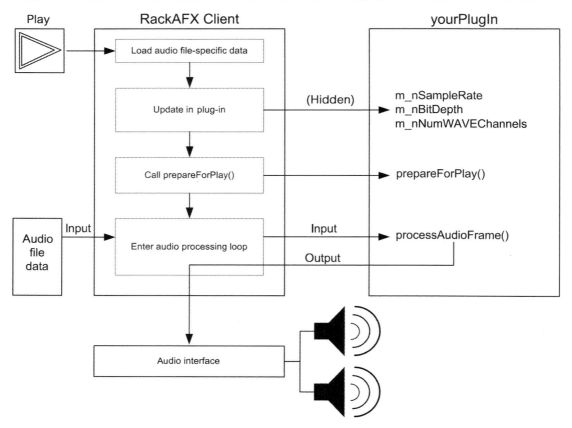

Figure 3.3: The sequence of events during the play/process operation; audio data from the file is processed in the plug-in and sent to the audio adapter for monitoring.

3.5 Destruction

When the user unloads the DLL either manually or by loading another plug-in, the client first deletes the plug-in object from memory, which calls the destructor. Any dynamically declared variables or buffers need to be deleted here. After destruction, the client unloads the DLL from the process space.

3.6 Your First Plug-Ins

You can break the audio processing down into two very fundamental types:

1. Processing that only works on the current audio sample; requires no memory elements
2. Processing that requires the current *and* past audio samples; requires memory elements

We'll start with the first type and make a simple volume control. After that, we'll design a simple Audio Equalizer (EQ) control that will require memory elements. You will need the following installed on your computer:

- RackAFX
- Microsoft Visual C++ Express® 2008 or 2010 (both are free from Microsoft)
- Microsoft Visual C++ Professional® 2008 or 2010

There is no advantage to having the full version of Visual C++ (aka VC++) in RackAFX programming unless you plan on using your own GUI resources. Make sure that Visual C++ is installed on the same machine as RackAFX. See the website www.willpirkle.com for updates on supported compiler platforms. Once you understand the flow of writing and testing your plug-ins, you will find that you can move easily and swiftly through the rest of the book's projects because they all follow the same design pattern and the design chapters will use the same conventions for each project.

3.6.1 Project: Yourplugin

The first step will always be the creation of a new project. In this phase, RackAFX creates the C++ project directory and files along with a derived class based on the project name.

3.6.2 Yourplugin GUI

Next, you lay out your GUI controls based on the algorithm you are following and decide on the variable data types and names that will connect the GUI elements to your plug-in. This generally starts with writing the difference equation(s) for the algorithm. Variables in the difference equation will map to member variables and GUI controls in your plug-in. Abstracting the GUI from the algorithm requires that you decide which parameters you want the user to be able to adjust, as well as the variable names, min, max, and initial values, and data types. You can change your mind later and remove GUI elements or alter their parameters. A full-featured GUI designer allows you to finalize your plug-in's controls and package them in a neat and clean GUI. Often during the prototyping phase, you set up many sliders and controls and work on the voicing of the plug-in. Then, you hide some of them for the final product, only allowing the user access to a specific set of controls over a specific set range of values. This last step is part of the voicing of the final product.

3.6.3 Yourplugin.h File

The next phase involves adding your own member variables and member methods to the derived plug-in class. The variables and methods will depend on the algorithm you are implementing. In the first plug-in there are no extra variables or methods to supply.

3.6.4 Yourplugin.cpp File

In this step, you will add the new member method implementations (if you have any). Then, you will step through the .cpp file, altering and filling in the base class methods, typically in this order:

1. Constructor
2. prepareForPlay()
3. userInterfaceChange()
4. processAudioFrame()

Once these methods have been implemented, you will technically have a working plug-in.

3.6.5 Building and Testing

Finally, you will build the DLL, finding and fixing any issues. After the build succeeds, you can load it into the RackAFX client. You can use audio files, the oscillator, or your sound adapter input as the audio stimulus for your plug-in. You can run your plug-in in three modes: Insert, Mono > Stereo AUX, or Stereo > Stereo AUX to mimic the various ways a plug-in is used in a digital audio workstation (DAW).

3.6.6 Creating and Saving Presets

The presets are created and maintained on the main RackAFX UI. After you load your plug-in you can move the controls as you like and then save them as a preset. You use the Save Preset button on the toolbar. The presets will be saved inside a file until the next time you compile your plug-in; after that, the presets will be built into the DLL. You can add, modify, or delete the presets any time the plug-in is loaded.

3.6.7 GUI Designer

Once you have debugged and finished your plug-in, you can optionally use the GUI designer to create a compact, visually appealing GUI. See Appendix B and the website www.willpirkle.com for the latest details on using the GUI designer tools. In general, the book skips the GUI designer phase because it is so open ended; you are free to layout your final GUI however you wish. Check the website often for the latest news and example GUIs as well as video tutorials.

3.7 Design a Volume Control Plug-In

The easiest meaningful plug-in to write is a volume control, which uses a single scaling variable, a_2, depicted in Figure 3.4.

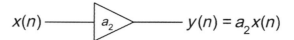

Figure 3.4: The volume control block diagram.

Coefficients in a block diagram (or transfer function or algorithm) become *float member variables* in your plug-in code.

- $a_2 = 0$: Mute
- $a_2 = 1.0$: Max volume

The output samples $y(n)$ are a scaled version of the input $x(n)$ and the scaling factor is named a_2. a_2 is called a *coefficient* in the algorithm. The algorithm states that a_2 will vary between 0 (mute) and 1.0 (full volume).

3.8 Set up RackAFX for Use

Start the RackAFX software. You will start in prototype view, where you will see a blank control surface as shown in Figure 3.5. Your GUI may look slightly different or have different background images.

The control surface is what you use to create your UI. It is full of assignable controls you can connect to your plug-in's variables. The surface consists of:

- 40 assignable sliders (continuous controls).
- Universal LCD control with 1024 more continuous controls.
- Project controls (open, load, edit, rebuild, debug, jump to C++).
- Vector joystick (advanced, see website for more details).
- Assignable buttons.
- 10 assignable LED meters.
- Plug-in routing controls to test insert and aux effect modes.
- Prototype tab, the main GUI.
- GUI designer tab, which opens the designer for editing; you must have GUI controls declared first. See the website for more details.

The menu and toolbar consist of two parts: the left and right side. The left side (Figure 3.6) implements the majority of the software functionality, while the right side (Figure 3.7) maintains lists.

Chapter 3

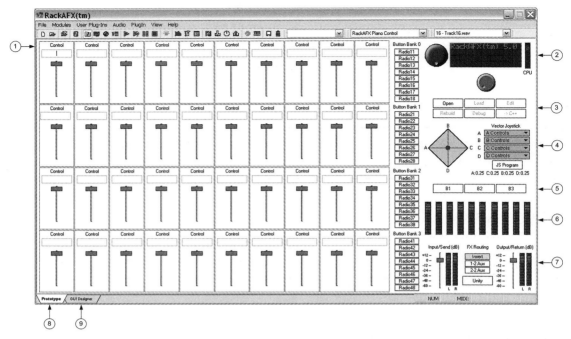

Figure 3.5: When you start RackAFX, it opens in prototype view. It features the control surface and plug-in routing controls.

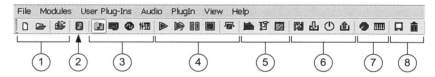

Figure 3.6: The menu and toolbar on the left handle most of your plug-in development.

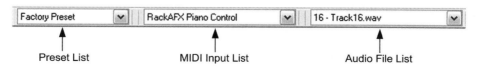

Figure 3.7: The dropdown boxes on the right let you store and recall presets, choose a MIDI input controller (advanced), and keep track of the audio files you have been using.

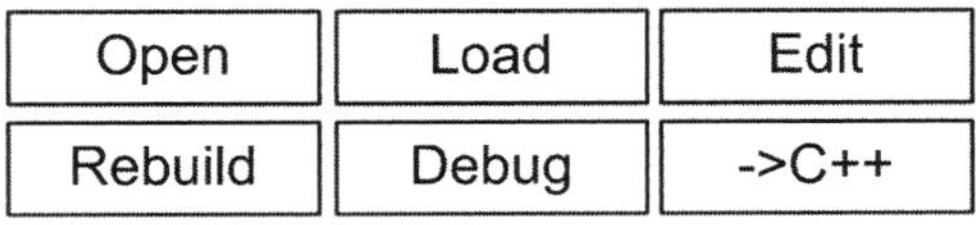

Open	Load	Edit
Rebuild	Debug	->C++

Figure 3.8: The Project/Compiler buttons make it easy to work with projects and control your compiler.

The menu items include:

- File: Manage projects by creating, editing or clearing the project.
- Modules: Built-in plug-ins that you can use for analysis and testing.
- User plug-ins: Each new plug-in you design gets loaded into this menu; you can audition or show off your plug-in in a standalone fashion.
- Audio: Manage all audio commands.
- Plug-in: Tools for loading/unloading and presets.
- View: Access the different windows.
- Help: Help information.

The toolbar items include:

1. New project, open project folder, open audio file
2. Set up low-level audio
3. Audio input mode: File, analog audio input, oscillator, user oscillator/synth
4. Transport controls: Play, loop, pause, stop, bypass
5. Additional windows: Analyzer, block diagram, status window
6. Plug-in tools: Synchronize code, load, reset, unload
7. GUI windows: Custom GUI, RackAFX MIDI Piano
8. Presets: Save, delete

Finally, there is a bank of buttons that allow you to manipulate your projects as well as control the C++ compiler shown in Figure 3.8. The buttons are set up as follows:

- Open: Open an existing project.
- Load: Load/unload the DLL from process space.
- Edit: Change an existing project's settings.
- Rebuild: Rebuild the project.
- Debug: Launch the debugger.
- ->C++: Jump to the C++ compiler and restore if minimized.

3.9 Setup Preferences

Before you start working on projects, take some time to configure your preferences. This is where you will choose your C++ compiler and set your default directories. Choose View > Preferences to get the interface shown in Figure 3.9.

44 Chapter 3

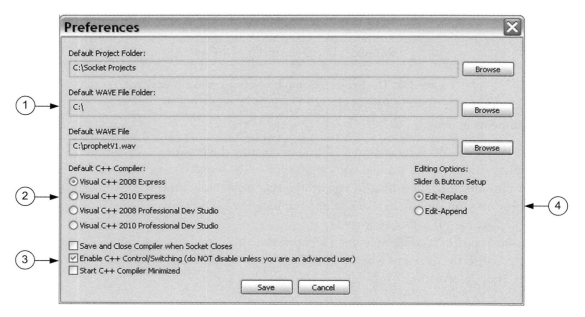

Figure 3.9: The preferences interface.

In the preferences you need to:

1. Choose your default folders for projects, WAVE files, and default WAVE files. You can use whatever directory you want for your project folder and you can also open projects from any other folder at any time; the default is simply for conveniently grouping all your projects together.
2. Choose a Visual C++ compiler.
3. Set the C++ options. Enable C++ Control/Switching should be left on for all but the most advanced users. C++ control/switching allows RackAFX to control Visual Studio, save files, launch the debugger, and so on.
4. Set up the edit options when entering information in the GUI slider/button dialogs.

3.9.1 Project: Volume

Create a new project with the menu, toolbar, or Ctrl+N and name it "Volume." The New/Edit Project window will appear and you can enter your project name. As you enter the name, the .h and .cpp files will be automatically named.

Notice the two C++ files listed in Figure 3.10—these are the interface (.h) and implementation (.cpp) files that RackAFX will create for you. They will contain a derived object named CVolume which will contain the bulk of your plug-in code. When you hit OK, several things happen. If you have C++ Control enabled in your preferences (it's the default) then you will

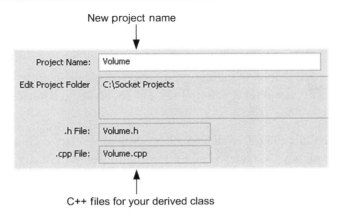

Figure 3.10: The top section of the New/Edit Project window. Notice that your project name becomes the name of a C++ object, so you will receive errors if you name the project in a way that produces illegal C++ syntax. Below this are more advanced settings that we will cover later.

see your C++ compiler start up. In Visual C++ you will see a new project and solution named "Volume." If you expand the Volume project then you can see the files that RackAFX wrote for you. Your derived class is contained in Volume.h and Volume.cpp. Before continuing, it's worth taking a peek into the RackAFXDLL.cpp file and locating the creation mechanism createObject():

```
//RackAFX Creation Function
DllExport CPlugIn* createObject()
{
        CPlugIn* pOb = new CVolume; // ***
        return pOb;
}
```

This is the method that the client calls on your DLL—you create the plug-in object with the *new* operator and return the pointer to the client. The DllExport specifier is OS-specific for calling a method in a DLL.

3.9.2 Volume GUI

You need to make sure you know the difference equation for each algorithm you want to implement. The difference equation relates the input and output samples and is what you are going to convert to C++ in order to make your plug-in. For this example, the equation is

$$y(n) = a_2 x(n) \tag{3.1}$$

46 Chapter 3

> Each slider or button control on the UI will map to and control a *member variable* in your plug-in.

In RackAFX, you can see that all the sliders and buttons are disabled; the sliders don't move and the edit boxes won't accept text. You first need to set up some controls to create your UI or control surface.

Now, decide how to handle the difference equation. Notice the use of Hungarian notation for handling the variable names. See the website for more information if you are not familiar with this kind of naming convention—you are certainly free to name your variables whatever you like.

- Let's have only one volume coefficient and share it between the channels so that each channel has the same volume control.
- Let's have one slider control the volume coefficient.
- The coefficient a_2 will become a float member variable in the code; let's decide to name it "m_fVolume."
- We'll name the slider "Volume" on the GUI and link it to the variable m_fVolume.
- Slider minimum = 0.
- Slider maximum = 1.0.
- Initial setting = 0.75.

3.9.3 Configure a Slider Control

Choose the first slider in the upper left and right-click just inside the outer bounding box; a slider properties box appears (Figure 3.11). Note: Your dialog may look slightly different.

You need to fill out the slider properties with the proper values. You will notice that the uControlID value is 0 for this slider. This is the ID number that will link the slider to a

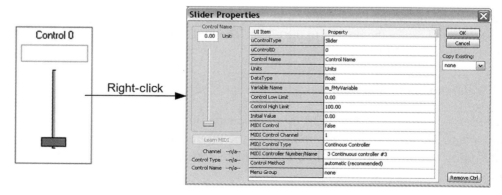

Figure 3.11: Right-click inside the bounding box of a slider and the slider properties window appears. This is how you configure the slider and link it to a variable.

variable in the object. You cannot edit this cell. Start with the control name and enter "Volume." Hit Enter to advance to the next cell. For this version of the plug-in there are no units, so use backspace to delete it. The next cell is one of the most important—it is the data type for the variable that the slider will be linked with; the choices are available from a drop-down list. As shown in Figure 3.12, you can select the data type with the mouse, or you can just type the first letter (e.g., "d" for double) while the box is highlighted.

You can create variables of the following types:

- float
- double
- integer
- enum: An enumerated Unsigned Integer (UINT) using a list of strings for the enum, for example {LPF, HPF, BPF, Notch}. We will work with enumerated UINTs later in the book.

We decided to use the *float* data type and to name the variable m_fVolume—it is really important to decide on a name and stick with it. Changing the variable name later can be tricky, so do yourself a favor and plan ahead. The completed slider properties are shown in Figure 3.13. The low and high limits are set and the initial value is set to 0.75. Do not worry about the MIDI control or other properties for this slider; see the website for details on these enhancements. After you are finished, choose the OK button to save the code.

Our control surface is pretty simple—it just consists of a single volume slider. If you realize you've made a mistake or left something out, just right-click in the slider box and fix the problem. You can also remove the slider by clicking the Remove Ctrl button on the properties window.

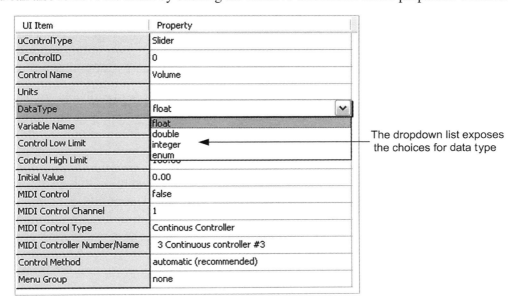

Figure 3.12: The data type is selected from a dropdown list control.

UI Item	Property
uControlType	Slider
uControlID	0
Control Name	Volume
Units	
DataType	float
Variable Name	m_fVolume
Control Low Limit	0.00
Control High Limit	1.00
Initial Value	0.75
MIDI Control	false
MIDI Control Channel	1
MIDI Control Type	Continous Controller
MIDI Controller Number/Name	3 Continuous controller #3
Control Method	automatic (recommended)
Menu Group	none

Figure 3.13: The completed slider properties.

As you add, edit, or remove controls from the main UI you will notice that RackAFX will flash to the compiler and back as it writes the code for you. You might use this flashing as a signal that the code update is synchronized. If you don't like it, minimize the compiler and the flashing will not occur. There is a special check-box in View > Preferences to start the compiler minimized for this very reason.

Your plug-in code will use the index value 0 (uControlID in the properties dialog) to map to the m_fVolume variable, which is controlled by the slider named "Volume" on the UI.

3.9.4 Volume.h File

Before we add the code, look around the plug-in object files (volume.h and volume.cpp) to get a better understanding of what's going on and what you'll need to modify. First, open the volume.h file and look inside:

```
// RackAFX abstract base class for DSP filters
class CVolume : public CPlugIn
{
public:  // plug-in API Functions
    //
```

```
            // 1. One Time Initialization
            CVolume();

            <SNIP SNIP SNIP>

            // 7. userInterfaceChange() occurs when the user moves a control.
            virtual bool userInterfaceChange(int nControlIndex);

            // Add your code here: ----------------------------------------- //

            // END OF USER CODE -------------------------------------------- //

            // ADDED BY RACKAFX -- DO NOT EDIT THIS CODE!!! ---------------- //
            // **--0x07FD--**
            float m_fVolume;

            // **--0x1A7F--**
            // ------------------------------------------------------------- //
};
```

Aside from the main plug-in functions we discussed in Chapter 2, you will see some more commented areas of code. In the first part marked **// Add your code here:** you can add more variables or function definitions just like you would in any .h file. Try to keep your code in the denoted area to make it easier to find and read. The area below that says:

// ADDED BY RACKAFX—DO NOT EDIT THIS CODE!!!

is very important—you will see your member variable m_fVolume declared in this area. This is the portion of the .h file that RackAFX modifies when you add, edit, or delete controls from your control surface. It is imperative that you let RackAFX maintain this part of the C++ code. There are several other portions of code in the .cpp file that have similar warnings and interesting hex symbols (0x1A7F, etc.); do not edit the code contained between the hex codes or commented areas.

You will see the notation **<SNIP SNIP SNIP>** frequently in the printed code as a reminder that code has been cut out for easier reading.

RackAFX writes C++ code for you! But, you have to be careful not to alter the RackAFX C++ code in any way. You can always tell if the code is RackAFX code because there will be warning comments and strange hex codes surrounding the RackAFX code. The RackAFX code is left for you to see only as a backup to your debugging and should never be altered by anyone except RackAFX.

In this case, check to verify that RackAFX added the float member variable m_fVolume as you anticipated. Next, move on to the volume.cpp file and have a look at it, starting from the top.

3.9.5 Volume.cpp File

Constructor and destructor

The constructor is the One-Time-Init function and is set up to:

- call initUI(): This is where your GUI controls are initialized; m_fVolume is initialized to 0.75 inside this function. It is important to make sure this remains the first line of the constructor so that your GUI variables are always initialized first.
- Set the plug-in name variable: This is what you will see in the user plug-in menu and on the GUI windows.
- Set the plug-in defaults (snipped out here); you will rarely need to change these variables.
- Give you a place to finish any of your own initializations at the end.

The destructor is empty because nothing has been allocated dynamically in this plug-in.

```
CVolume::CVolume()
{
        // Added by RackAFX - DO NOT REMOVE
        //
        // Setup the RackAFX UI Control List and Initialize Variables
        initUI();
        // END InitUI

        // built in initialization
        m_PlugInName = "Volume";

        // Default to Stereo Operation:
        // Change this if you want to support more/less channels

        <SNIP SNIP SNIP>

        // Finish initializations here
}

/* destructor()
        Destroy variables allocated in the contructor()

*/
CVolume::~CVolume(void)
{

}
```

prepareForPlay()

There is nothing to write yet since there are no custom controls or other allocations.

processAudioFrame()

This function is where the signal processing action happens. Above the definition is a comment block as a reminder of how to get the audio data samples into and out of the I/O buffers. Currently, RackAFX only supports mono and stereo plug-ins. The left and right channels are accessed using the normal array-indexed C++ pointer mechanism. Of special note is the reminder that all values in and out are (and should be) between −1.0 and +1.0.

```
/* processAudioFrame

// ALL VALUES IN AND OUT ON THE RANGE OF -1.0 TO + 1.0

LEFT INPUT = pInputBuffer[0];
RIGHT INPUT = pInputBuffer[1]

LEFT OUTPUT = pInputBuffer[0]
LEFT OUTPUT = pOutputBuffer[1]

*/
bool __stdcall CVolume::processAudioFrame(float* pInputBuffer, float* pOutputBuffer,
UINT uNumInputChannels, UINT uNumOutputChannels)
{
        // output = input -- change this for meaningful processing
        //
        // Do LEFT (MONO) Channel; there is always at least one input/one output
        (INSERT Effect)
        pOutputBuffer[0] = pInputBuffer[0];

        // Mono-In, Stereo-Out (AUX Effect)
        if(uNumInputChannels == 1 && uNumOutputChannels == 2)
                pOutputBuffer[1] = pInputBuffer[0];

        // Stereo-In, Stereo-Out (INSERT Effect)
        if(uNumInputChannels == 2 && uNumOutputChannels == 2)
                pOutputBuffer[1] = pInputBuffer[1];

        return true;

}
```

Take a look at the function as RackAFX wrote it for you—it is designed to pass audio through unmodified. In this case, you simply write the output buffer with the data from the input buffer. In your processAudioFrame() function, get used to always processing the first channel then optionally processing the second one. This makes it easy to write mono/stereo capable plug-ins and will also make it easier to extend when RackAFX has more channel options. Because the code is already done, you could compile it right now and try it in RackAFX as a sanity check to make sure your audio hardware is set up properly. In fact, we'll do that right after examining the last few functions in the file (I promise you will write code shortly).

userInterfaceChange()

Perhaps the second most important function is userInterfaceChange(), which is called when the user changes a control on the control surface:

```
/* ADDED BY RACKAFX -- DO NOT EDIT THIS CODE!!! --------------------------- //
        **--0x2983--**

        Variable Name                   Index
        ----------------------------------------------
        m_fVolume                       0


        ----------------------------------------------

        **--0xFFDD--**
// --------------------------------------------------------------------- */
// Add your UI Handler code here ---------------------------------------- //
//
```

As with processAudioFrame(), there is a "hint" comment above the function definition which reminds you how RackAFX has mapped your variable to a control index. In this case, the m_fVolume variable is mapped to index 0.

```
bool __stdcall CVolume::userInterfaceChange(int nControlIndex)
{
        // decode the control index, or delete the switch and use brute force calls
        switch(nControlIndex)
        {
            case 0:
            {
                    break;
            }

            default:
                    break;
        }

        return true;
}
```

userInterfaceChange() implements the first part of a switch/case statement in case you need to decode the control index and do something to the data before finally altering your code to reflect the change. Often, you will have nothing to write here either.

Build the plug-in. Since the default plug-in behavior is to simply pass audio unaffected, you can build the project now and test it in RackAFX to make sure everything is working properly. Rebuild your project from the compiler or from RackAFX's Rebuild button. You should get a clean build with no errors.

> You should *always* build and test your brand-new project first before modifying any code! You want to do this to make sure there are no C++ errors (you might have inadvertently hit a key or changed some code), as well as to make sure your audio system is working and you can hear the audio data.

At this point, you have built a DLL which is designed to serve up CVolume objects when the client requests them. The problem is that RackAFX doesn't yet know your CVolume plug-in is available. During the debugging phase, you need to get used to manually loading and unloading the DLL. You do this with the Load button or the toolbar/menu item. After you hit Load, RackAFX calls the appropriate functions to load your DLL into its address space. You will see the control surface change to reflect that your plug-in is loaded. You will also see the Load button change to read Unload. When it is time to go back to C++, modify, and recompile, you'll need to unload the project first so you can reload it in its later state.

Use Audio > Load Audio File to load a test file. Then use the transport control buttons to play, loop, and pause or stop the file. The volume control should have no effect since we haven't written any code yet. Make sure you get audio at this point before continuing; if you don't, check your audio adapter settings.

In order to make the volume control slider work, we need to wire it into the processing code. The volume slider is directly mapped to the volume coefficient m_fVolume; as the slider moves from 0.0 to 1.0, so does the volume coefficient. So, the algorithm is simple to write: just multiply the input audio sample by the volume coefficient and set the output to that value. Switch to your C++ compiler and find the processAudioFrame() function. Modify it by doing the multiplication described above, which implements the difference equation for the filter.

```
bool __stdcall CVolume::processAudioFrame(float* pInputBuffer, float* pOutputBuffer,
                        UINT uNumInputChannels, UINT uNumOutputChannels)
{
    // output = input -- change this for meaningful processing
    //
    // Do LEFT (MONO) Channel; there is always at least one input/one output
    // (INSERT Effect)
    pOutputBuffer[0] = pInputBuffer[0]*m_fVolume;

    // Mono-In, Stereo-Out (AUX Effect)
    if(uNumInputChannels == 1 && uNumOutputChannels == 2)
        pOutputBuffer[1] = pInputBuffer[0]*m_fVolume;

    // Stereo-In, Stereo-Out (INSERT Effect)
    if(uNumInputChannels == 2 && uNumOutputChannels == 2)
        pOutputBuffer[1] = pInputBuffer[1]*m_fVolume;

    return true;
}
```

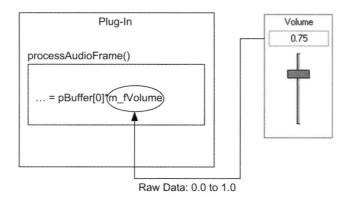

Figure 3.14: Here is the connection between the slider and the variable in the calculation. The processAudioFrame() function is using the raw slider data directly.

There are only three lines to modify, one for the first channel and another two for the other routing combinations. The modification is shown in **bold** where you are scaling the input by the volume coefficient. Can you see how this relates to the difference equation? If not, stop now and go back to figure it out. Now, rebuild the project and reload it into RackAFX. Try the slider and you will see that the volume does indeed change. Congrats on your first plug-in!

What makes this plug-in so easy and quick to develop is that the slider volume control maps *directly* to a variable that is used in the processAudioFrame() function, as depicted in Figure 3.14. This means the data coming from the slider can be used directly in the algorithm. The data coming from the slider and controlling m_fVolume is said to be *raw data*. You use the raw value to affect the signal processing algorithm.

3.10 Design a Volume-in-dB Plug-In

This next example will show you how to *cook* your raw data to be used in the signal processing algorithm. The VolumedB plug-in will also be a volume control, but will operate in dB instead of using a raw multiplier. You may have noticed that your previous volume control didn't seem to do much in the upper half of the throw of the slider. This is because your ears hear logarithmically and so linear deflections of the slider do not correspond to linear changes in perceived loudness. To fix this, we'll design another plug-in that will operate in decibels (dB). The block diagram is identical to the first project, only the control range of values has changed.

- $a_2 = -96$ dB: Mute
- $a_2 = 0$ dB: Max volume

Now, the volume control is specified in dB, so you need a formula to convert the dB value to a scalar multiplier for the algorithm. You should memorize the dB equations now if you haven't already, since they will reappear over and over in audio plug-ins. This is the *cooking formula* that will take the raw data from the slider −96 to 0 dB and cook it into a variable we can use in our scalar multiplication:

$$dB = 20\log(x)$$
$$x = 10^{\frac{dB}{20}} \tag{3.2}$$

You set up the cooking formula by implementing the userInterfaceChange() function which is always called when the user changes anything on the control surface. When the user moves the slider, you cook the data. This is shown conceptually in Figure 3.15.

The cooking function simply converts the dB into a scalar multiplier. In this case, the cooking function is short enough to simply leave inside the userInterfaceChange() function; as the cooking functions become more complex, you will probably want to break them out into separate functions, which are called from userInterfaceChange(). Remember the two RackAFX rules you've learned so far:

> *Coefficients* in a block diagram (or transfer function or algorithm) become *float member variables* in your plug-in code. Each slider or button control on the UI will map to and control a *member variable* in your plug-in.

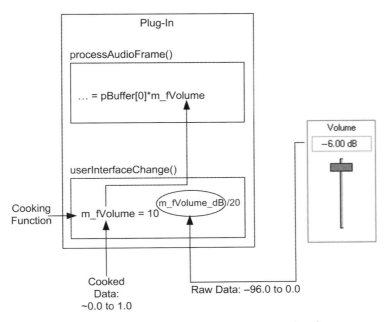

Figure 3.15: The volume-in-dB plug-in will have a single slider that generates values between −96 and 0.0 dB; you need to cook the raw dB values to use in your plug-in.

56 Chapter 3

In the first plug-in, the variable was shared between the slider and the algorithm. Now we need two variables, one for the raw slider data and the other for the cooked algorithm processing. We will name them as follows:

- m_fVolume_dB: The raw slider data
- m_fVolume: The cooked value used in the algorithm

3.10.1 Project: VolumedB

Using the same method as before, create a new project named "VolumedB." As before, you don't have to worry about the advanced options at the bottom of the new project window. Your compiler will automatically start.

3.10.2 VolumedB GUI

We only need a single slider control. It will be adjustable from −96 to 0 dB. Set up the GUI in RackAFX by choosing a slider and right-clicking inside the bounding box. Set the slider according to Table 3.1.

3.10.3 VolumedB.h File

RackAFX has written the code and declared the variable float m_fVolume_dB but we still need to declare our second variable named m_fVolume, which stores the cooked data. Open the VolumedB.h file and declare the variable in the user declarations area:

```
// abstract base class for DSP filters
class CVolumedB : public CPlugIn
{
public:  // plug-in API Functions

        <SNIP SNIP SNIP>

        // Add your code here: ------------------------------------ //

        // our Cooked Volume Multiplier
        float m_fVolume;

        // END OF USER CODE ----------------------------------------- //
```

Table 3.1: The slider properties for the VolumedB project.

Slider Property	Value
Control name	Volume
Units	dB
Variable type	float
Variable name	m_fVolume_dB
Low limit	−96
High limit	0
Initial value	−6

3.10.4 VolumedB.cpp File

Constructor

- Cook and initialize the member variable.
- Use the pow() function.

```
CVolumedB::CVolumedB()
{
        // Added by RackAFX - DO NOT REMOVE
        //
        // Setup the RackAFX UI Control List and Initialize Variables
        initUI();
        // END InitUI

        <SNIP SNIP SNIP>

        // Finish initializations here

        // Cook the raw data:
        m_fVolume = pow(10.0, m_fVolume_dB/20.0);
}
```

prepareForPlay()

There is nothing to do here because the volume variable does not need to be reset on each play event.

userInterfaceChange()

- Decode the control ID value.
- Cook the raw data using the formula.
- When the plug-ins get more complex, you can create separate cooking functions and then share the functions as needed.
- Make sure you check your control ID values from the comments so that they match properly.

```
bool __stdcall CVolumedB::userInterfaceChange(int nControlIndex)
{
        // decode the control index, or delete the switch and use brute force calls
        switch(nControlIndex)
        {
                case 0:
                {
                    // Cook the raw data:
                    m_fVolume = pow(10.0, m_fVolume_dB/20.0);
                }

                default:
                    break;
        }

        return true;
}
```

58 Chapter 3

processAudioFrame()

- Implement the difference equation.

```
bool __stdcall CVolumedB::processAudioFrame(float* pInputBuffer, float* pOutputBuffer,
                            UINT uNumInputChannels, UINT uNumOutputChannels)
{
    // output = input -- change this for meaningful processing
    //
    // Do LEFT (MONO) Channel; there is always at least one input/one output
    // (INSERT Effect)
    pOutputBuffer[0] = pInputBuffer[0]*m_fVolume;

    // Mono-In, Stereo-Out (AUX Effect)
    if(uNumInputChannels == 1 && uNumOutputChannels == 2)
        pOutputBuffer[1] = pInputBuffer[0]*m_fVolume;

    // Stereo-In, Stereo-Out (INSERT Effect)
    if(uNumInputChannels == 2 && uNumOutputChannels == 2)
        pOutputBuffer[1] = pInputBuffer[1]*m_fVolume;

    return true;
}
```

Build and test your plug-in. You should now hear a smooth volume transition as you move the slider. Depending on your sound system and volume levels, you might not hear much below −40 dB.

3.11 Design a High-Frequency Tone Control Plug-In

This example will show you how to implement the last of the digital signal processing (DSP) algorithm building blocks: the delay element (z^{-N}), where N is the number of samples of delay. In this example, $N = 1$, so we are dealing with a one-sample delay, a fundamental building block for many DSP algorithms. Although we won't get to the theory of how this delay creates a tone control until the next chapter, it's worth it to go ahead and build a plug-in that uses it. After this exercise, you will be able to build almost any kind of DSP filter that uses one-sample-delay elements—and that's a lot of DSP filters. The later examples in the book will not be as detailed regarding the operation of RackAFX and your C++ compiler, so make sure you understand how to manipulate the software as needed to complete the project. Figure 3.16 shows the block diagram for the filter.

The design equation is as follows:

$$0.0 \leq a_1 \leq 0.49$$
$$a_0 = a_1 - 1.0$$
(3.3)

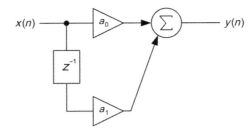

Figure 3.16: The HF tone control block diagram.

You already know that the coefficients a_0 and a_1 will become float member variables in our plug-in. But what about the one-sample-delay element, z^{-1}? In hardware, this would be a register to store the sample for one clock period. In software, it simply becomes another float variable, but it must be able to persist over the lifetime of a processing cycle. Therefore, like the coefficients, it will become a float *member* variable of our plug-in object.

You will do an example using arrays of floats in Chapter 7 when you implement digital delay lines that require long strings of z^{-1} elements. For now, we need to implement the single z^{-1} element in the block diagram. Right away, we need to decide if we are going to support multichannel (stereo) operation.

The last rule is really important and it is easy to get into trouble if you do not follow it. Unless the algorithm specifically deals with multichannel data, you will need to implement a separate algorithm for each channel, which means having separate coefficient and data (z^{-1}) elements for each channel. Even if you might be able to share the coefficients, you can never share the delay elements. We will need to declare variables for the following:

- Left channel a_0 and a_1 variables
- Left channel z^{-1} variable
- Right channel a_0 and a_1 variables
- Right channel z^{-1} variable

> Delay elements will become *float member variables* in your plug-in object. For single-delay elements, you can simply assign separate variables. For multiple-sample-delay elements you may also use float arrays to store the data.

> A DSP filtering algorithm, which is only described in mono or single channel format, that is, one input, $x(n)$, and one output, $y(n)$, cannot share delay elements between multiple channels. This means that you must duplicate your algorithms so that you have one set of variables for the left channel and one for the right channel.

3.11.1 Project: SimpleHPF

This plug-in is going to implement a very primitive HF (high frequency) tone control that behaves like a high-pass filter. It will attenuate low frequencies, leaving only the highest frequencies intact. Using the same method as before, create a new project named "SimpleHPF." Check your compiler to make sure the project was created properly.

3.11.2 SimpleHPF GUI

This simple plug-in will only have one slider that will control the value of the a_1 coefficient. The other coefficient is calculated from it. The specifications show that a_1 varies between 0 and 0.49. Set up the slider according to the properties in Table 3.2.

3.11.3 SimpleHPF.h File

To figure out what the CSimpleHPF object is going to have to do, first write the difference equation. Examine it and figure out which components are going to become coefficients and which are going to be memory locations. Also, figure out any intermediate variables you might need. You can figure out the difference equation by using the rules you learned in Chapter 1 to chart the input and output signals. Make sure you understand how this equation is formed from the block diagram in Figure 3.17.

The difference equation is as follows:

$$y(n) = a_0 x(n) + a_1 x(n - 1) \tag{3.4}$$

Next, figure out which block diagram components become variables in your C++ code. The coefficients a_0 and a_1 will become a float member variable in the code. Even though we might be tempted to share the coefficients, these are separate left and right algorithms that have separate delay elements, so let's implement two sets, one each for the left and right channels.

Table 3.2: The slider properties for the SimpleHPF project.

Slider Property	Value
Control name	a1
Units	
Variable type	float
Variable name	m_fSlider_a1
Low limit	0.0
High limit	0.49
Initial value	0

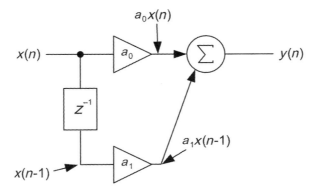

Figure 3.17: The HF tone control block diagram with annotations showing the signal math.

I named mine:

m_f_a0_left
m_f_a1_left
m_f_a0_right
m_f_a1_right

The z^{-1} element will also need to become a member variable and we will definitely need one for each channel because these can never be shared. I named mine

m_f_z1_left
m_f_z1_right

The slider will only modify its own m_fSlider_a1 value. We will calculate the values for the other coefficients using it. We will need to modify the userInterfaceChange() function just like the preceding example to wire the slider into the algorithm. Jump to your C++ compiler and go to the SimpleHPF.h file to add your member variables. Notice the variable that RackAFX added in the code below:

```
// 5. userInterfaceChange() occurs when the user moves a control.
virtual bool userInterfaceChange(int nControlIndex);

// Add your code here: ----------------------------------------------- //
float m_f_a0_left;
float m_f_a1_left;

float m_f_a0_right;
float m_f_a1_right;

float m_f_z1_left;
float m_f_z1_right;

// END OF USER CODE ------------------------------------------------- //

// ADDED BY RACKAFX -- DO NOT EDIT THIS CODE!!! -------------------- //
// **--0x07FD--**
```

```
      float m_fSlider_a1;
      // **--0x1A7F--**
// ------------------------------------------------------------------------ //
```

3.11.4 SimpleHPF.cpp File

Constructor

- Set our coefficient values to match the initialized slider settings.
- Calculate the new a_0 values.
- Clear out the z^{-1} variables.

```
CSimpleHPF::CSimpleHPF()
{
        <SNIP SNIP SNIP>

        // Finish initializations here
        m_f_a1_left = m_fSlider_a1;
        m_f_a1_right = m_fSlider_a1;

        m_f_a0_left = m_f_a1_left - 1.0;
        m_f_a0_right = m_f_a1_right - 1.0;

        m_f_z1_left = 0.0;
        m_f_z1_right = 0.0;
}
```

prepareForPlay()

```
bool __stdcall CSimpleHPF::prepareForPlay()
{
        // Add your code here:
        m_f_z1_left = 0.0;
        m_f_z1_right = 0.0;

        return true;
}
```

processAudioFrame()

The logic for the signal processing of *one channel* will be as follows:

- Read the delayed value $x(n-1)$ out of the z^{-1} element.
- Implement the difference equation.
- Write the current input $x(n)$ into the delay variable; it will be $x(n-1)$ next time around.
- Do this for both channels.

```
bool __stdcall CSimpleHPF::processAudioFrame(float* pInputBuffer, float*
pOutputBuffer, UINT uNumChannels)
{
        // Do LEFT (MONO) Channel
        //
        // Input sample is x(n)
        float xn = pInputBuffer[0];

        // READ: Delay sample is x(n-1)
        float xn_1 = m_f_z1_left;

        // Difference Equation
        float yn = m_f_a0_left*xn + m_f_a1_left*xn_1;

        // WRITE: Delay with current x(n)
        m_f_z1_left = xn;

        // Output sample is y(n)
        pOutputBuffer[0] = yn;
```

OK, now it's your turn to implement the other channel. Give it a try by yourself before proceeding. You should have something like this for the rest of the function:

```
        // Mono-In, Stereo-Out (AUX Effect)
        if(uNumInputChannels == 1 && uNumOutputChannels == 2)
            pOutputBuffer[1] = yn;

        // Stereo-In, Stereo-Out (INSERT Effect)
        if(uNumInputChannels == 2 && uNumOutputChannels == 2)
        {
            // Input sample is x(n)
            float xn = pInputBuffer[1];

            // Delay sample is x(n-1)
            float xn_1 = m_f_z1_right;

            // Difference Equation
            float yn = m_f_a0_right*xn + m_f_a1_right*xn_1;

            // Populate Delay with current x(n)
            m_f_z1_right = xn;

            // Output sample is y(n)
            pOutputBuffer[1] = yn;
        }

        return true;
}
```

> Flush out delay elements in preparation for each play event in the plug-in. You generally do not want old data sitting inside these storage registers. The only exceptions are delay-looping effects where you exploit the old data. This is done in prepareForPlay().

userInterfaceChange()

- Store the new a_1 values.
- Cook the slider data to get the a_0 values.

```
bool __stdcall CSimpleHPF::userInterfaceChange(int nControlIndex)
{
    switch(nControlIndex)
    {
        case 0:
        {
            // save a1
            m_f_a1_left = m_fSlider_a1;
            m_f_a1_right = m_fSlider_a1;

            // calculate a0
            m_f_a0_left = m_f_a1_left - 1.0;
            m_f_a0_right = m_f_a1_right - 1.0;

            break;
        }
        default:
            break;
    }
    return true;
}
```

Build and load the project, open an audio file, and test your plug-in to make sure it's working properly. This plug-in is a simple low-cut filter and with the slider all the way down, you should hear no difference in the music. When you move the slider up, you will lose more and more bass frequencies, allowing only the high frequencies to pass. The cut-off frequency is approximately 11 kHz and the slider controls the amount of low frequency cut. You should easily be able to hear the effect, even over small speakers. Now, to get a little more information about the plug-in's operation, use RackAFX's analysis tools. Stop the music from playing with the transport control. Launch the analyzer window by clicking on the Analyzer button in the toolbar or choose View > Analyzer. The analyzer pops up as shown in Figure 3.18 (yours may look slightly different).

The analyzer is a powerful tool for checking your plug-in's performance. The basic controls are

1. Scope/spectrum analyzer.
2. Basic graphing options.
3. Scope controls.
4. Real-time response buttons, which let you measure the frequency, phase, impulse, and step responses of the plug-in (audio must not be streaming to use these).

Click on the Frequency button and you will get a frequency response plot of the filter. Move the slider all the way down and you should get a flat response, as shown in Figure 3.19. If you move the slider all the way up so that $a_1 = 0.49$, you get the response in Figure 3.20.

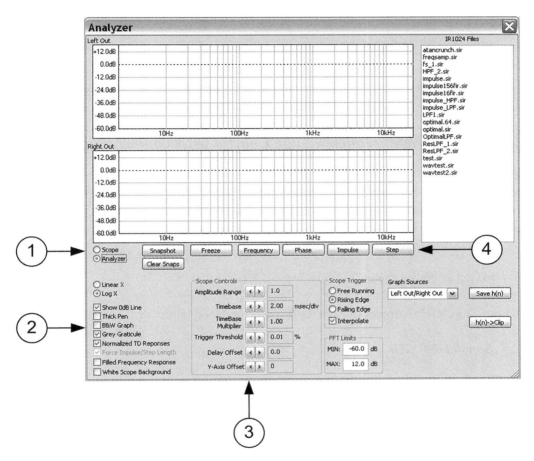

Figure 3.18: The audio analyzer.

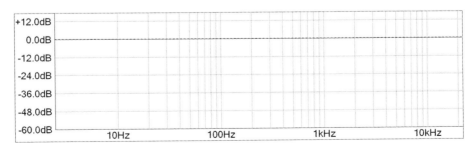

Figure 3.19: A flat frequency response with $a_1 = 0.0$.

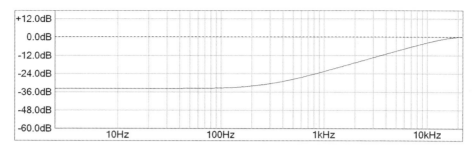

Figure 3.20: A filter that appears to boost high frequencies. You can see that it is really cutting the low frequencies instead; this is something you might only have realized by using the audio analyzer.

3.12 Design a High-Frequency Tone Control with Volume Plug-In

This final example will show you how to deal with more than one slider control by simply adding a volume-in-dB control to the block diagram. The plan is to add another slider to the existing plug-in; the new slider will control the overall volume of the plug-in in dB. You already know how to implement both parts of it, so this exercise is really more about adding new controls to an existing project.

3.12.1 Project: SimpleHPF

Open your SimpleHPF project in RackAFX using the Open button or the menu/toolbar items.

3.12.2 SimpleHPF GUI

Add the new volume slider: Right-click on the second slider group and add a new slider for the volume control, in dB, and use the exact same variable name and settings as in the VolumedB project. You should end up with a GUI like that in Figure 3.21.

3.12.3 SimpleHPF.h File

Add your own m_fVolume variable to handle the cooked volume data, just as before.

```
// Add your code here: ----------------------------------------- //
    float m_f_a0_left;
    float m_f_a1_left;

    float m_f_a0_right;
    float m_f_a1_right;

    float m_f_z1_left;
    float m_f_z1_right;

    float m_fVolume;
```

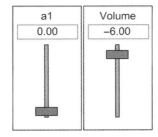

Figure 3.21: The new SimpleHPF GUI with added volume-in-dB control.

```
// END OF USER CODE ------------------------------------------------------ //

// ADDED BY RACKAFX -- DO NOT EDIT THIS CODE!!! ------------------------- //
//    **--0x07FD--**

float m_fSlider_a1;
float m_fVolume_dB;

//    **--0x1A7F--**
// ---------------------------------------------------------------------- //
```

3.12.4 SimpleHPF.cpp File

Constructor

- Cook the volume data after the filter initializations.

```
CSimpleHPF::CSimpleHPF()
{
        <SNIP SNIP SNIP>
        m_f_a0_left = -1.0;
        m_f_a1_left = 0.0;

        m_f_a0_right = -1.0;
        m_f_a1_right = 0.0;

        m_f_z1_left = 0.0;
        m_f_z1_right = 0.0;

        m_fVolume = pow(10.0, m_fVolume_dB/20.0);
}
```

prepareForPlay()

There is nothing to add here because the volume variable does not need to be reset on each play event.

processAudioFrame ()

- Add the volume control scaling *after* the filtering operation.

```
// Do LEFT (MONO) Channel; there is always at least one input/one output
// (INSERT Effect)
// Input sample is x(n)
float xn = pInputBuffer[0];

// READ: Delay sample is x(n-1)
float xn_1 = m_f_z1_left;

// Difference Equation
float yn = m_f_a0_left*xn + m_f_a1_left*xn_1;

// WRITE: Delay with current x(n)
m_f_z1_left = xn;

// Output sample is y(n)
pOutputBuffer[0] = yn*m_fVolume;

// Mono-In, Stereo-Out (AUX Effect)
if(uNumInputChannels == 1 && uNumOutputChannels == 2)
       pOutputBuffer[1] = yn*m_fVolume;

// Stereo-In, Stereo-Out (INSERT Effect)
if(uNumInputChannels == 2 && uNumOutputChannels == 2)
{
       // Input sample is x(n)
       float xn = pInputBuffer[1];

       // Delay sample is x(n-1)
       float xn_1 = m_f_z1_right;

       // Difference Equation
       float yn = m_f_a0_right*xn + m_f_a1_right*xn_1;

       // Populate Delay with current x(n)
       m_f_z1_right = xn;

       // Output sample is y(n)
       pOutputBuffer[1] = yn*m_fVolume;
}
```

userInterfaceChange()

- Cook the volume data.
- Make sure you check your control ID values in case you chose different sliders than I did.

```
bool __stdcall CSimpleHPF::userInterfaceChange(int nControlIndex)
{
    // decode the control index
    switch(nControlIndex)
    {
        case 0:
        {
            m_f_a1_left = m_fSlider_a1;
            m_f_a1_right = m_fSlider_a1;

            m_f_a0_left = m_f_a1_left - 1.0;
            m_f_a0_right = m_f_a1_right - 1.0;

            break;
        }
        case 1:
        {
            m_fVolume = pow(10.0, m_fVolume_dB/20.0);
            break;
        }
        default:
            break;
    }
    return true;
}
```

Build and test your code to make sure the plug-in works as expected. You now understand the basics of writing a plug-in from a block diagram and design equations.

Add some presets

Now that you have a GUI with more than one control, try adding some presets. If you look at the preset list in the toolbar, you will see the first one is named Factory Preset. This preset contains your initial settings for the GUI controls. You cannot delete this preset and it is automatically updated whenever you add, edit, or delete controls. This preset takes you back to your initial state. With your plug-in loaded and (optionally) audio streaming through it, adjust the controls to affect the signal. Hit the Save Preset button on the toolbar or choose it from the plug-in menu. A box will pop up allowing you to name the preset. You can also overwrite an existing preset. You can store up to 32 presets in your plug-in.

3.13 The User Plug-In Menu in RackAFX

As you write more plug-ins, you will notice that they begin to automatically populate the user plug-in menu item in RackAFX. By now, you should have three plug-ins in this menu. There are a few things you need to understand about this menu.

- It allows you to play with the plug-ins without having to open your compiler and manually load and unload the DLLs.
- You can select different plug-ins from this menu while audio is streaming and they will automatically slot in and out, so you can audition or show off your plug-ins quickly.
- It allows RackAFX to behave just like other plug-in clients by loading all the DLLs it finds in its PlugIns folder at once when you start the software. *This can be dangerous!*

That last item poses a problem during the course of development—if you write a DLL that does bad things in the constructor, such as hold pointers with garbage values or try to access memory that hasn't been allocated, it may crash RackAFX when it first starts up. If your DLL behaves really badly, you might even wound the OS too. This is a difficult issue to avoid without complicated rules for commissioning and decommissioning the plug-in. Additionally, you will have the same problem if you are developing a commercial plug-in and you are using a third-party client; most of these are designed to first open all the DLLs in their plug-in folder and check to make sure the plug-ins can be instantiated. If you write a bad DLL, you might also crash these clients and/or the OS. In at least one commercially available client, if your plug-in crashes during startup, it will not be loaded again in future launches. When RackAFX loads your DLL, it does some error checking to try to make sure your plug-in is legal, but it can't check the validity of your construction code.

> If RackAFX crashes each time you open it, remove the last DLL you were working on from the PlugIns folder. Alternatively, you can remove all the DLLs—you will want to copy them and restore them later when you find the bad DLL that caused the crashing. Writing a DLL is challenging and fun, but since you are writing a component, you can wind up with crashes like this.

CHAPTER 4
How DSP Filters Work

During the course of this book you will learn how to implement the following effects:

- EQs/tone controls
- Delay
- Flanger/chorus
- Compressor/limiter/tremolo
- Reverb
- Modulated filters/phaser

The EQ/tone control theory is the most difficult of all the effects to explain in a simple way. These effects are based on DSP filter theory which involves complex algebra, that is, the algebra of complex numbers. Complex numbers contain real and imaginary parts. There are two basic ways to explain basic DSP theory. The first is intuitive and involves no complex math but requires some bookkeeping and can be tedious. The second method uses complex algebra to solve the problem. We'll start with the intuitive method, and then (optionally, if you desire) make the leap into complex math. Don't worry—we don't want to get mired down in theory and forget the fun part, which is making audio effects. If you skip the second part, you will still be able to code EQ and tone control plug-ins, but you will better understand where the equations come from if you know a little theory too. The transfer functions you will learn along the way will reappear in many effects algorithms.

In Figure 4.1 you can see the -6 dB/octave roll-off indicative of a first-order filter. A digital version of the analog filter should have the same shape and roll-off. A key difference is that the digital filter will not operate beyond the Nyquist frequency.

The phase response plot shows the relative phases of different frequencies upon exiting the filter. During the filtering process, the phases of different frequencies get shifted forward or backward. In Figure 4.2, the 1 kHz frequency is shifted by -45 degrees compared to the input. At very high frequencies, the phase shift approaches -90 degrees. This phase shift is not a side effect of the filtering but an integral part of how it works.

To understand the ramifications of the phase shifting, consider a complex waveform entering the filter. Fourier showed that a complex, continuous waveform could be decomposed into a

72　Chapter 4

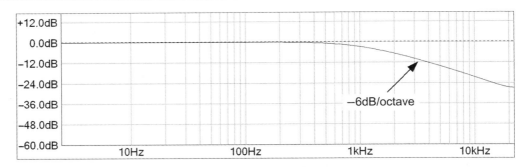

Figure 4.1: The fundamental analysis tool for a DSP filter is its frequency response plot. This graph shows how the filter amplifies or attenuates certain bands of frequencies.

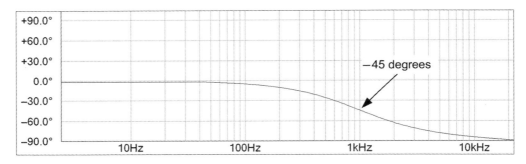

Figure 4.2: The phase response plot of the analog filter in Figure 4.1 shows how the phase of the output is shifted across frequencies.

set of sinusoids with different frequencies and amplitudes. Figure 4.3 shows a filter in action. The input is decomposed into four sinusoids, a fundamental, and three harmonics. The peak-amplitudes of the components are shown as dark bars to the left of the y axes.

We observe several features of the output:

- The composite waveform is smoothed out.
- The amplitude and phase shift of the fundamental are unchanged.
- The three harmonics have decreasing amplitudes and more phase shift as you go higher in frequency.

Figure 4.4 shows the same filter with the information plotted differently; here, the amplitudes and phases are plotted against frequency rather than dealing with a set of time-domain sinusoids. You can see by the output frequency response plot that this filter is a kind of low-pass filter. Its curve is similar to Figure 4.1, the analog equivalent.

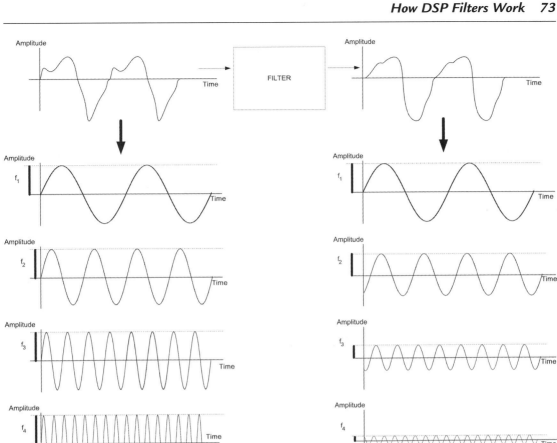

Figure 4.3: A complex waveform is filtered into a smoothed output. The input and output are decomposed into their Fourier-series components.

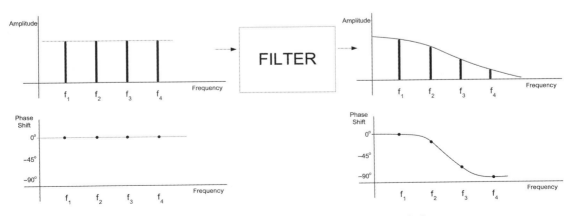

Figure 4.4: The same information is plotted as frequency and phase responses.

4.1 First-Order Feed-Forward Filter

In order to get a grip on the nature of digital filtering, start with the first-order feed-forward filter shown in a block diagram in Figure 4.5. You've already seen a version of it in the HPF tone control you coded in the last chapter.

The difference equation is as follows:

$$y(n) = a_0 x(n) + a_1 x(n-1) \qquad (4.1)$$

You can tell why it's called feed forward—the input branches feed forward into the summer. The signal flows from input to output. There is no feedback from the output back to the input. Now, suppose we let the coefficients a_0 and a_1 both equal 0.5 in Figure 4.6.

In order to analyze this filter we can go the easy but tedious route or the difficult but elegant route. Let's start with the easy way. In order to figure out what this does, you apply the five basic digital test signals you learned in Chapter 1 to the filter and then manually push the values through and see what comes out. You only need a pencil and paper or a simple calculator. The five waveforms we want to test are:

1. DC (0 Hz)
2. Nyquist

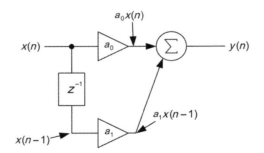

Figure 4.5: The first-order feed-forward filter.

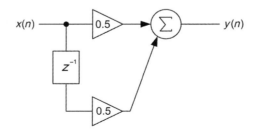

Figure 4.6: What kind of filter is this? What are its frequency and phase responses?

3. ½ Nyquist
4. ¼ Nyquist
5. Impulse

For each audio sample that enters there are two phases to the operation:

1. Read phase: The sample is read in and the output is formed using the difference equation and the previous sample in the delay register.
2. Write phase: The delay element is overwritten with the input value—the sample stored in the single z^{-1} register is effectively lost.

Start with the DC/step input and begin sequentially applying the samples into the filter shown in Figures 4.7 through 4.10.

Now, observe the amplitude and phase shift of the input versus output—the output amplitude eventually settles out to a constant 1.0 or unity gain condition, so at DC or 0 Hz, the output

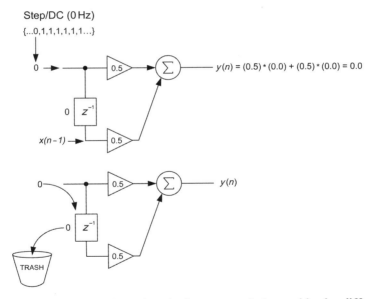

Figure 4.7: On the first iteration the input sample is used in the difference equation to create the output, $x(n) = 0$, $y(n) = 0$ and then the input is shifted into the delay register.

In a feed-forward filter, the amount of time smearing is equal to the maximum delayed path through the feed-forward branches.

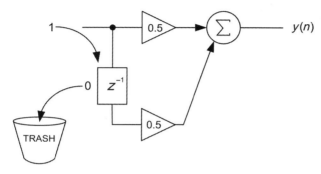

Figure 4.8: The process continues with each sample. Here the input 1.0 is combined with the previous input; the second output $y(n) = 0.5$.

equals the input. However, there is a one sample delay in the response, causing the leading edge of the step-input to be smeared out by one sample interval. This time smearing is a normal consequence of the filtering.

Next, repeat the process for the Nyquist frequency (DC and Nyquist are the easiest, so we'll do them first). The filter behaves in an entirely different way when presented with Nyquist (Figures 4.11 through 4.14).

Now, make your observations about amplitude and phase. The amplitude at Nyquist eventually becomes zero after the one-sample-delay time. The phase is hard to tell because the signal has vanished. Why did the amplitude drop all the way to zero at Nyquist? The answer is one of the keys to understanding digital filter theory: the one-sample delay introduced

> Delay elements create phase shifts in the signal. The amount of phase shift depends on the amount of delay as well as the frequency in question.

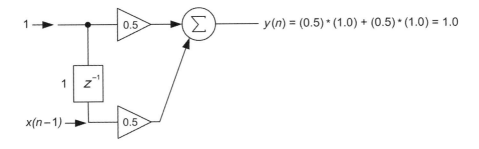

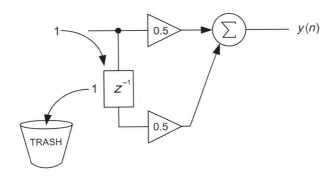

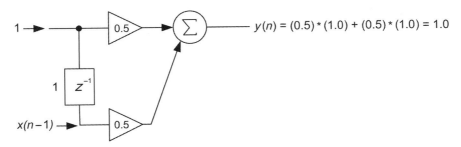

Figure 4.9: The sequence continues until we observe a repeating pattern; 1.0 is repeating here.

exactly 180 degrees of phase shift at the Nyquist frequency and caused it to cancel out when recombined with the input branch through a_0.

In the case of Nyquist, the one-sample delay is exactly enough to cancel out the original signal when they are added together in equal ratios. What about other frequencies like

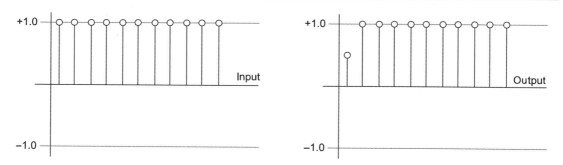

Figure 4.10: The input and output sequences for the filter in Figure 4.6 at DC or 0 Hz.

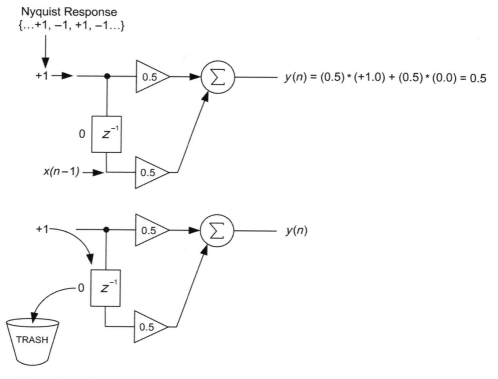

Figure 4.11: The Nyquist sequence is applied to the filter. Notice how the delay element has been zeroed out. The output for the first iteration is $y(n) = 0.5$.

½ and ¼ Nyquist? They are a bit more laborious to work through but worth the effort. By now you can see how the data moves through the filter, so let's use a table (Table 4.1) and move the data through *it* instead. The ½ Nyquist sequence is $x(n) = \{\ldots 0, +1.0, 0.0, -1.0, 0.0, +1.0, 0.0, \ldots\}$.

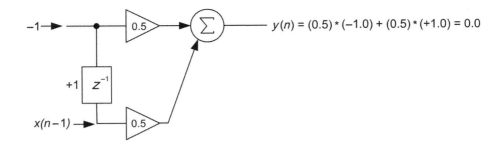

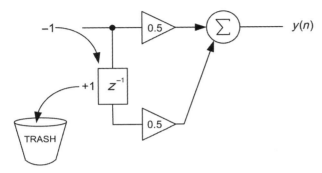

Figure 4.12: The second iteration at Nyquist produces an output $y(n) = 0$.

Table 4.1: The manual labor continues as we work through the ½ Nyquist frequency.

$x(n)$	$x(n-1)$	$y(n) = 0.5x(n) + 0.5x(n-1)$
0	0	0
1	0	0.5
0	1	0.5
-1	0	-0.5
0	-1	-0.5
1	0	0.5
0	1	0.5
-1	0	-0.5
0	-1	-0.5

Can you see how $x(n)$ becomes $x(n-1)$ for the next row? The $x(n-1)$ column holds a one-sample-delayed version of the input $x(n)$. The output is $y(n) = \{0, +0.5, +0.5, -0.5, -0.5, +0.5, +0.5\}$.

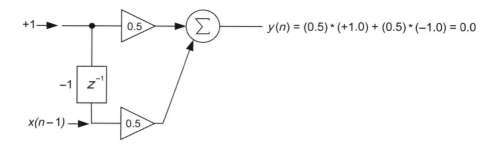

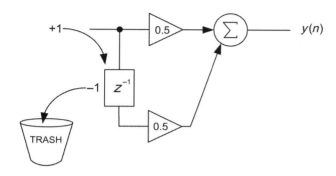

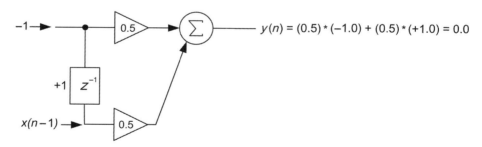

Figure 4.13: Continuing the operation at Nyquist, we see that eventually the output settles to a repeating 0, 0, 0, 0 sequence.

Next we observe the amplitude and phase relationship from input to output in Figure 4.15. At first it might seem difficult to figure out the sequence {... −0.5, −0.5, +0.5, +0.5, ...}. ½ Nyquist is also encoded with a repeating sequence of four values (0, 1, 0, −1).

Work through ¼ Nyquist the same way (Table 4.2). The ¼ Nyquist frequency sequence is $x(n) = \{0, 0.707, 1, 0.707, 0, -0.707, -1, -0.707, 0, ...\}$.

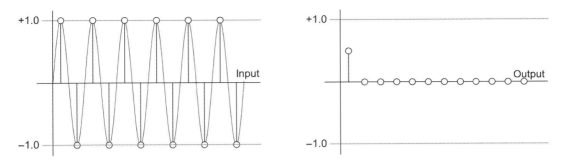

Figure 4.14: The input and output sequences for the filter in Figure 4.6 at Nyquist.

Table 4.2: ¼ Nyquist input/output.

$x(n)$	$x(n-1)$	$y(n) = 0.5x(n) + 0.5x(n-1)$
0	0	0
0.707	0	+0.354
1	0.707	+0.854
0.707	1	+0.854
0	0.707	+0.354
−0.707	0	−0.354
−1	−0.707	−0.854
−0.707	−1	−0.854
0	−0.707	−0.354

The output is $y(n) = \{\ldots +0.354, +0.854, +0.854, +0.354, -0.354, -0.854, -0.854, -0.354, +0.354, \ldots\}$. Analysis of the output sequence reveals the phase-shifted and slightly attenuated output signal at ¼ Nyquist. Both the phase shift and the attenuation are smaller than ½ Nyquist. As you can see in Figure 4.16 there is also one sample of time smearing at the start of the signal.

Finally, apply the impulse sequence and find the impulse response. The impulse response is the third analysis tool. The impulse response defines the filter in the time domain like the frequency response defines it in the frequency domain. The basic idea is that if you know how the filter reacts to a single impulse you can predict how it will react to a series of impulses of varying amplitudes. Take a Fast Fourier Transform (FFT) of the impulse response and you get the frequency response. An inverse FFT converts the frequency response back into the impulse response. For this filter the impulse response is simple (Table 4.3).

82   Chapter 4

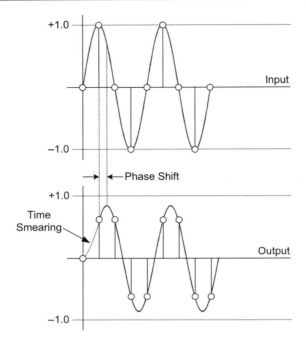

Figure 4.15: The input/output relationship in time at ½ Nyquist. The ½ Nyquist frequency is attenuated almost by one-half. The output is also phase shifted by 45 degrees. The leading edge of the first cycle is smeared out by one sample's worth of time.

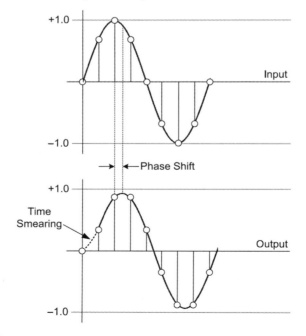

Figure 4.16: The input/output relationship at ¼ Nyquist.

Table 4.3: The impulse response input/output relationship.

x(n)	x(n − 1)	y(n) = 0.5x(n) + 0.5x(n − 1)
0	0	0
1	0	0.5
0	1	0.5
0	0	0
0	0	0
0	0	0

Here you can see that the impulse is flattened and smeared out. It is actually two points on a sin(x)/(x)-like curve, as shown in Figure 4.17. Now, you can combine all the frequency amplitude and phase values into one big graph, as shown in Figure 4.18.

We observe that this digital filter is a low-pass variety with a typical low pass filter (LPF) magnitude response. However the phase response is quite interesting—it is linear instead of nonlinear like the analog example at the beginning of the chapter. In fact, this simple filter is a *linear phase filter*.

Figures 4.19 and 4.20 show the measured frequency and phase responses for this filter. Compare it with our estimation. Notice the linear phase produces a straight line only when plotted on the linear frequency axis. What makes this filter a low-pass filter? It is a combination of the coefficients and the filter topology (first-order feed-forward). There are three basic topologies: feed forward (FF), feed back (FB), and a combination of FF/FB. Once the topology has been chosen, it's really the *coefficients* that determine what the filter will do.

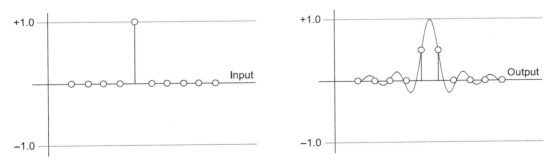

Figure 4.17: The time domain plots of the impulse response input/output.

A feed-forward filter will be a linear phase filter if its coefficients are symmetrical about their center. In this case (0.5, 0.5) is symmetrical. Another example would be (−0.25, 0, −0.25).

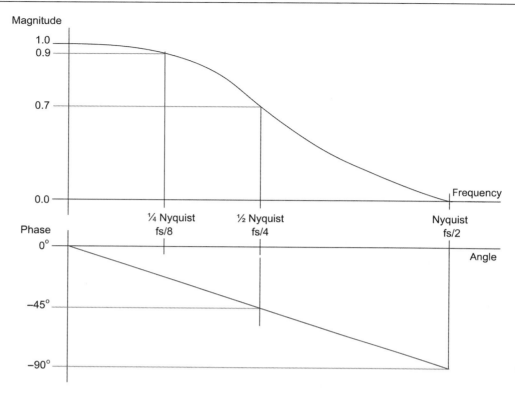

Figure 4.18: Final frequency and phase response plots for the digital filter in Figure 4.6. Notice that this is a linear frequency plot since ½ Nyquist is halfway across the x-axis. The phase at Nyquist pops back up to 0 degrees since there is no Nyquist component present (output = 0).

4.2 Design a General First-Order Feed-Forward Filter

To illustrate this and crystallize it as a concept, modify your SimpleHPF filter so that two sliders control the a_0 and a_1 coefficients directly. Also, alter the range of values they can control to (-1.0 to $+1.0$). Then, experiment with the two values and watch what happens in the analyzer. How to do this is described next.

Open your SimpleHPF project and modify the user interface (UI). First, change the values for the a_1 slider to match the new low and high limits. As usual, you right-click inside the slider's bounding box and alter the limits and initial value (shown in bold) as in Table 4.4.

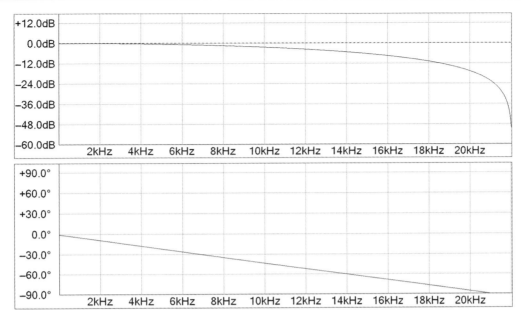

Figure 4.19: Measured frequency and phase response plots for the filter you just analyzed by hand. These are plotted with a linear frequency base.

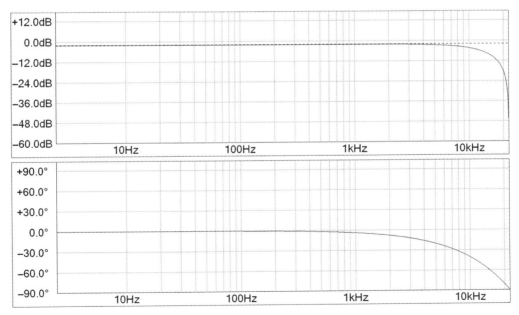

Figure 4.20: Measured frequency and phase response plots with log frequency base.

Table 4.4: The altered a_1 slider properties.

Slider Property	Value
Control Name	a1
Units	
Variable Type	float
Variable Name	m_fSlider_a1
Low Limit	−1.0
High Limit	+1.0
Initial Value	0.0

Now add a new slider for the a_0 coefficient just below the a_1 slider (Table 4.5).

Table 4.5: The new a_0 slider properties.

Slider Property	Value
Control Name	a0
Units	
Variable Type	float
Variable Name	m_fSlider_a0
Low Limit	−1.0
High Limit	+1.0
Initial Value	1.0

Change the userInterfaceChange() function to directly map the slider values to the coefficient values. The new slider has a control ID of 10; always check your nControlIndex value since it might be different depending on your UI.

```
        Variable Name                          Index
--------------------------------------------------------
        m_fSlider_a1                             0
        m_fVolume_dB                             1
        m_fSlider_a0                            10

--------------------------------------------------------
bool CSimpleHPF::userInterfaceChange(int nControlIndex)

{
        switch(nControlIndex)
        {
```

```
case 0:
        // direct map to the a1 Slider
        m_f_a1_left = m_fSlider_a1;
        m_f_a1_right = m_fSlider_a1;
        break;

case 1:
        // cook the Volume Slider
        m_fVolume = pow(10.0, m_fVolume_dB/20.0);
        break;

case 10:
        // direct map to the a0 Slider
        m_f_a0_left = m_fSlider_a0;
        m_f_a0_right = m_fSlider_a0;
        break;

default:
        ; // do nothing
    }
    return true;
}
```

Rebuild the DLL and load it. Place the volume control at 0 decibels (dB) (maximum) and set a_0 and a_1 to 1.0 (maxima); then open the analyzer and hit the Frequency button—play with the different values for a_0 and a_1. Figure 4.21 shows the response with various coefficient values.

After you play around with the controls, there are several things to note from this experiment:

- You can get a low-pass or high-pass response, or anything between, including a flat response ($a_0 = 1.0$, $a_1 = 0.0$).

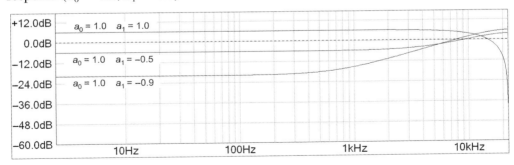

Figure 4.21: Three different combinations of coefficient settings yield three different filters.

> The topology of the filter determines its difference equation. The coefficients (a_N) of a filter determine its filter frequency and phase response and therefore its type (HPF, LPF, etc.) and its sonic qualities. Your plug-in implements the difference equation in processAudioFrame(). Your plug-in calculates the coefficients in userInterfaceChange() in response to the user making changes to the control surface.

- Inverted settings give the identical frequency response but the phase responses will be inverted (use the Phase button and have a look for yourself).
- You can get gain out of the filter.
- You can also get attenuation out of the filter.

4.3 First-Order Feed-Back Filter

A first-order feed-back filter is shown in Figure 4.22.

The difference equation is as follows:

$$y(n) = a_0 x(n) - b_1 y(n-1) \qquad (4.2)$$

You can see the feed-back nature of the filter; the output $y(n)$ is fed back into the summer through a one-sample-delay z^{-1} element. Notice that the feedback coefficient has a negative sign in front of it and the difference equation reflects this with the $-b_1$ term. The negative sign is for mathematical convenience and will make more sense in the next chapter when we analyze the difference equations in more detail. Also notice that there is no b_0 coefficient— there will not be a b_0 coefficient in any of the feed-back filters. To analyze this without math requires going through the same procedure as before, analyzing the amplitude and phase of the basic test signals. Another option would be to code it as a plug-in in RackAFX and map the coefficients directly to the sliders as you did in the previous example. Then, you can experiment with the coefficients and see how they affect the filter's frequency, phase, DC/step, and impulse responses.

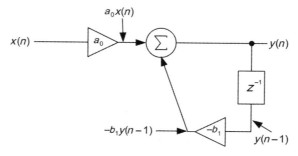

Figure 4.22: First-order feed-back filter block diagram.

4.4 Design a General First-Order Feed-Back Filter

4.4.1 Project FeedBackFilter

Create a new RackAFX project called "FeedBackFilter." The plan is for this filter to let you directly control the coefficients with UI sliders and use the analyzer to check the resulting filters.

4.4.2 FeedBackFilter GUI

The GUI will consist of two sliders, one for the a_0 coefficient and the other for the b_1 coefficient. You can use any two sliders you like, but make sure you keep track of the control index values later on. Figure 4.23 shows my version of the GUI. Right-click near the sliders you want to use and set them up according to Table 4.6.

4.4.3 FeedBackFilter.h File

Add the z^{-1} elements and a_0, b_1 coefficient variables for the right and left channels in your .h file:

```
// Add your code here: ---------------------------------------------- //
float m_f_a0_left;
float m_f_b1_left;
```

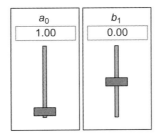

Figure 4.23 The GUI for the FeedBackFilter project.

Table 4.6: The a_0 slider properties.

Slider Property	Value	Slider Property	Value
Control Name	b1	Control Name	a0
Units		Units	
Variable Type	float	Variable Type	float
Variable Name	m_fSlider_b1	Variable Name	m_fSlider_a0
Low Limit	−1.0	Low Limit	−1.0
High Limit	+1.0	High Limit	+1.0
Initial Value	0.0	Initial Value	1.0

Chapter 4

```
float m_f_a0_right;
float m_f_b1_right;

float m_f_z1_left;
float m_f_z1_right;
// END OF USER CODE ----------------------------------------------------- //
```

4.4.4 FeedBackFilter.cpp File

Constructor

- Initialize the internal a_0 and b_1 variables to match our GUI variables.
- Zero out the delay line elements.

```
CFeedBackFilter::CFeedBackFilter()
{
        // Added by RackAFX - DO NOT REMOVE
        //
        // Setup the RackAFX UI Control List and Initialize Variables
        initUI();
        // END InitUI

        <SNIP SNIP SNIP>

        // Finish initializations here
        //
        // setup our coefficients
        m_f_a0_left = m_fSlider_a0;
        m_f_b1_left = m_fSlider_b1;

        m_f_a0_right = m_fSlider_a0;
        m_f_b1_right = m_fSlider_b1;

        // flush the memory registers
        m_f_z1_left = 0.0;
        m_f_z1_right = 0.0;
}
```

prepareForPlay()

- Flush the z^{-1} storage registers.

```
bool __stdcall CFeedBackFilter::prepareForPlay()
{
        // Add your code here:
        m_f_z1_left = 0.0;
```

```
            m_f_z1_right = 0.0;

        return true;
    }
```

processAudioFrame()

- Implement the difference equation in processAudioFrame(); notice the (−) sign in the difference equation too.

```
bool __stdcall CFeedBackFilter::processAudioFrame(float* pInputBuffer, float*
pOutputBuffer, UINT uNumInputChannels, UINT uNumOutputChannels)
{
        // Do LEFT (MONO) Channel; there is always at least one input/one output
        // (INSERT effect)
        // Input sample is x(n)
        float xn = pInputBuffer[0];

        // Delay sample is y(n-1)
        float yn_1 = m_f_z1_left;

        // Difference Equation
        float yn = m_f_a0_left*xn - m_f_b1_left*yn_1;

        // Populate Delay with current y(n)
        m_f_z1_left = yn;

        // Output sample is y(n)
        pOutputBuffer[0] = yn;

        // Mono-In, Stereo-Out (AUX effect) -- COPY for now
        if(uNumInputChannels == 1 && uNumOutputChannels == 2)
                pOutputBuffer[1] = yn;

        // Stereo-In, Stereo-Out (INSERT effect)
        if(uNumInputChannels == 2 && uNumOutputChannels == 2)
        {
                // Input sample is x(n)
                float xn = pInputBuffer[1];

                // Delay sample is x(n-1)
                float yn_1 = m_f_z1_right;
```

```
            // Difference Equation
            float yn = m_f_a0_right*xn - m_f_b1_right*yn_1;

            // Populate Delay with current y(n)
            m_f_z1_right = yn;

            // Output sample is y(n)
            pOutputBuffer[1] = yn;
    }
    return true;
}
```

userInterfaceChange()

- Update the coefficients when the user changes a slider.
- Check your nControlIndex values to make sure you are processing the correct control.

```
bool CFeedBackFilter::userInterfaceChange(int nControlIndex)
{
    // add your code here
    switch(nControlIndex)
    {
    case 0:
            // map the a0 Slider
            m_f_a0_left = m_fSlider_a0;
            m_f_a0_right = m_fSlider_a0;
            break;
    case 1:
            // map the b1 Slider
            m_f_b1_left = m_fSlider_b1;
            m_f_b1_right = m_fSlider_b1;
            break;

    default:
            ; // do nothing
    }

    return true;
}
```

Build and load the plug-in into RackAFX. Open the analyzer and use the Frequency, Phase, Impulse, and Step buttons to analyze the output. Play with the controls. Notice first that the a_0 slider only controls the gain and phase of the signal; in the frequency response this just moves the curve up or down and it disappears when $a_0 = 0.0$, which makes sense. The real action is with the b_1 control; try the examples in Figures 4.24 through 4.26.

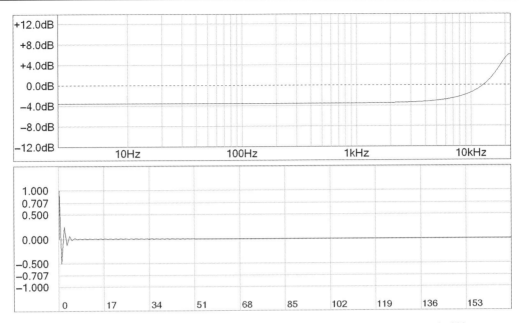

Figure 4.24: $a_0 = 1.0$ and $b_1 = 0.5$; the frequency response is high-pass/low-shelf in nature and has gain above 11 kHz, while the impulse response shows slight ringing.

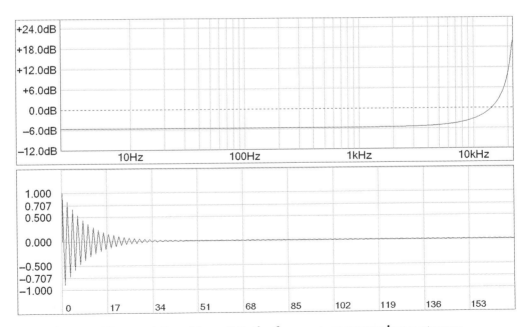

Figure 4.25: $a_0 = 1.0$ and $b_1 = 0.9$; the frequency response has a steeper high-pass response than before and has gain of +20 dB at Nyquist, while the impulse response shows considerable ringing.

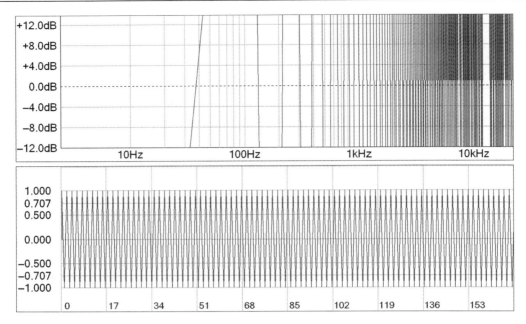

Figure 4.26: $a_0 = 1.0$ and $b_1 = 1.0$; the frequency response has blown up while the impulse response rings forever.

What happened in that last filter? Why was there no frequency response? The filter became unstable and *blew up*. It blew up because the b_1 coefficient was 1.0, which introduced 100% positive feedback into the loop. The output recirculated through the feedback loop forever causing the infinite ringing seen in the step and impulse responses. Also notice that as the b_1 variable was increased, the gain at Nyquist also increased. With $b_1 = 1.0$, the gain at Nyquist is actually infinite.

4.5 Observations

In doing these exercises, you have made a lot of progress—you know how to implement both feed-forward and feed-back filters in a plug-in. You also have a good intuitive knowledge about how the coefficients control the filter type. Plus you got to blow up a filter, so that is pretty cool. Here are some observations.

4.5.1 General

- Feed-forward and feed-back topologies can both make high-pass or low-pass-shaped filter responses.
- The coefficients of the filter ultimately determine the kind of filter and what it sounds like.
- The phase shift introduced by the delay element(s) is responsible for the filtering operation.

4.5.2 Feed-Forward Filters

- Operate by making some frequencies go to zero; in the case of $a_0 = 1.0$ and $a_1 = 1.0$, the Nyquist frequency went to zero; this is called a *zero of transmission* or a *zero frequency* or just a *zero*.
- The step and impulse responses show smearing. The amount of smearing is exactly equal to the total amount of delay in the feed-forward branches.
- Don't blow up.
- Are called *finite impulse response* (FIR) filters because their impulse responses, though they may be smeared, are always finite in length.

4.5.3 Feed-Back Filters

- Operate by making some frequencies go to infinity; in the case of $a_0 = 1.0$ and $b_1 = 1.0$, the Nyquist frequency went to infinity and with $a_0 = 1.0$ and $b_1 = -1.0$, DC or 0 Hz went to infinity; this is called a *pole of transmission* or a *pole frequency* or just a *pole*.
- The step and impulse responses show overshoot and ringing or smearing depending on the coefficients. The amount of ringing or smearing is proportional to the amount of feedback.
- Can blow up (or go unstable) under some conditions.
- Are called *infinite impulse response* (IIR) filters because their impulse responses can become infinite.

The problem now is that we want to be able to specify the filter in a way that makes sense in audio—a low-pass filter with a cut-off of 100 Hz or a band-pass filter with a Q of 10 and center frequency of 1.2 kHz. What we need is a better way to analyze the filtering than just randomly trying coefficients, and we need methods to come up with filter coefficients based on how we specify the filter. In the next chapter, we'll work on the basic DSP theory that will make this happen.

Bibliography

Dodge, C. and Jerse, T. 1997. *Computer Music Synthesis, Composition and Performance*, Chapters 3 and 6. New York: Schirmer.

Steiglitz, K. 1996. *A DSP Primer with Applications to Digital Audio and Computer Music*, Chapter 4. Menlo Park, CA: Addison-Wesley.

CHAPTER 5
Basic DSP Theory

You want to get a grip on the underlying digital signal processing (DSP) theory of filters for several reasons. It helps to understand the anatomy of the filter because you have to implement it in code; a deeper understanding of the theory can only help your coding strategy. Also, the same DSP filter analysis and mathematical models can be applied to other effects including delay, chorusing, reverb, and compression. In order to intuitively understand the foundation of DSP theory, you need to review some math and engineering concepts.

5.1 The Complex Sinusoid

The analysis and design of digital filters uses the sinusoid as its basic stimulus function. Since Fourier showed that a signal can be decomposed into sinusoids, if you know how the system reacts to a sinusoidal stimulus at a bunch of different frequencies, you can plot the frequency and phase responses like you did by hand in the last chapter. This is akin to the impulse response—since the input signal is a train of impulses of varying amplitudes, if you know how the filter responds to a single impulse, you can figure out how it will respond to multiple impulses. You also did this by hand when you took the impulse response of the low-pass filter.

Everyone is familiar with the sine and cosine functions—sine and cosine are related by an offset of 90 degrees and the sine function starts at zero, whereas cosine starts at 1.0. In Figure 5.1 you can identify the sine and cosine waveforms by their starting position. But what about the sine-like waveform that starts at an arbitrary time in the lower plot? Is it a sine that has been phase shifted backwards or a cosine that has been phase shifted forward? You have to be careful how you answer because sine and cosine have different mathematical properties; their derivatives are not the same and it usually becomes difficult when you try to multiply them or combine them in complex ways. Add a phase offset to the argument of the sin() or cos() function and then it really turns into a mess when you do algebra and calculus with them. What you need is a function that behaves in a sinusoidal manner, encapsulates both sine and cosine functions, and is easy to deal with mathematically. Such a function exists, and it's called the complex sinusoid:

$$\text{Complex sinusoid} = e^{j\omega t} \tag{5.1}$$

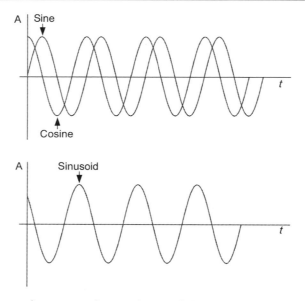

Figure 5.1: Sine, cosine, and sinusoid signals.

Euler's equation is shown below:

$$e^{j\theta} = \cos(\theta) + j\sin(\theta)$$
$$j = \sqrt{-1} \tag{5.2}$$

You can see that it includes both sine and cosine functions. The j term is the imaginary number, the square root of -1 (mathematicians call it i but since that represents current in engineering, we rename it j instead). The j is known as the phase rotation operator; multiplying a function by j rotates the phase by 90 degrees.

Suppose you want to shift the phase of a waveform by 180°, thereby inverting it. Mathematically, you can do this by multiplying the waveform by -1, inverting the values of all its points. Suppose you wanted to invert the waveform again (which would bring it back to its original shape)—you could do that by multiplying by -1 again. But suppose that you only wanted to shift the phase by 90°? Is there a number (η) you could use as a multiplier to shift by 90°? In other words,

$$\text{90-degree shifted waveform} = (\text{original waveform})\,(\eta)$$

You don't know what η is yet, but you can figure it out. Suppose you then wanted to shift the waveform by another 90 degrees, which would be the same as shifting the original waveform by 180 degrees. You would multiply it by η again. You'd then have the following:

$$\begin{aligned}\text{180-degree shifted waveform} &= (\text{original waveform})\,(\eta)(\eta) \\ &= (\text{original waveform})\,(\eta)^2 \\ &= (\text{original waveform})\,(-1)\end{aligned}$$

This leads to Equation 5.3:

$$\eta^2 = -1$$
$$\eta = \sqrt{-1} = j \qquad (5.3)$$

So, you can perform a conceptual 90-degree phase shift by multiplying a waveform by j. A -90-degree phase shift is accomplished by multiplying by $-j$. Some other useful relationships with j are

$$j^2 = -1$$
$$\frac{1}{j} = -j \qquad (5.4)$$

Euler's equation is complex and contains a real part (cos) and imaginary part (sin), and the plus sign (+) in the equation is not a literal addition—you can't add real and imaginary numbers together. In a complex number of the format $A + jB$, the two parts coexist as part of one complex number.

For our purposes, we replace the θ with ωt instead, where ω is the frequency in radians/second and t is the time variable. Plug in various values for t and you get the plot in Figure 5.1 when you plot the sine and cosine in the same plane. So, we will reject using the sin() and cos() functions independently and adopt the complex sinusoid as a prepackaged mixture of the two. The reason is partly mathematical—as it turns out, e is simple to deal with mathematically. You only need to learn four rules (Equation 5.5) in addition to Euler's equation.

$$\text{Euler's equation: } e^{j\omega t} = \cos(\omega t) + j\sin(\omega t)$$

The four rules:

$$e^a e^b = e^{(a+b)} \text{ or } e^{at} e^{bt} = e^{(a+b)t}$$
$$\frac{e^a}{e^b} = e^{(a-b)} \text{ or } \frac{e^{at}}{e^{bt}} = e^{(a-b)t}$$
$$\frac{d(e^{at})}{dt} = ae^{at} \qquad (5.5)$$
$$\int e^{at} dt = \frac{1}{a}e^{at}$$

> So, what Euler's equation is really describing is a sine and cosine pair of functions, coexisting in two planes that are 90 degrees apart. The word *orthogonal* is the engineering term for *90 degrees apart*.

The two equations in Equation 5.5 demonstrate that e behaves like a polynomial ($x^2 * x^5 = x^7$) even when the argument is a function of time, t. Equation 5.5 also shows how simply it behaves in calculus—multiple derivatives or integrations are done by simply multiplying the argument's constant (a or 1/a) over and over. Before we leave this topic, make sure you remember how to deal with complex numbers—you'll need it to understand where the frequency and phase responses come from.

5.2 Complex Math Review

Because a complex number has both real and imaginary parts, it cannot be plotted on a single axis. In order to plot a complex number, you need two axes: a real and imaginary axis. The two axes are aligned at right angles, just like the x- and y-axes in two-dimensional geometry. They are orthogonal. The x-dimension is the real axis (Re), and the y-dimension is the imaginary axis (Im). So, a complex number is plotted as a point on this x-y plane, also called the *complex plane*. Complex numbers are usually written in the form A + jB where A is the real part and B is the imaginary part. Notice that the notation always stays in the form A + jB even when common sense might contradict it. For the point in the second quadrant, you still write $-2 + j2$ even though $j2 - 2$ might look nicer. Also notice that you write $2 - j1$ instead of just $2 - j$. The points are plotted the way you would plot (x, y) pairs in a plane (Figure 5.2).

This example plots points as $(x + jy)$ pairs; this is called the Cartesian coordinate system. You can also use polar notation to identify the same complex numbers. In polar notation, you specify the radius (R) and angle (θ) for the vector that results from drawing a line from the origin of the axes out to the point in question. For example, consider the point $2 + j3$ above. You could plot this point in polar form as shown in Figure 5.3.

We are leaving the Cartesian form $(2 + j3)$ for reference. Normally, all you would see is R and θ. Most engineers prefer polar notation for dealing with complex numbers. The polar number is often written as $R < \theta$. Fortunately for us, the conversion from Cartesian to polar

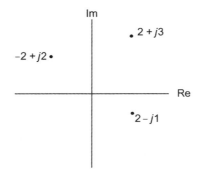

Figure 5.2: Several points plotted in the complex plane using the Cartesian (x, y) coordinate system.

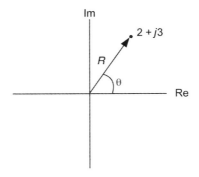

Figure 5.3: Plotting 2 + j3 in polar form.

notation is simple. Starting with a complex number in the form $A + jB$, you can find the resulting radius and angle from Equations 5.6 and 5.7.

$$R = \sqrt{A^2 + B^2} \qquad (5.6)$$

$$\theta = \tan^{-1}\left[\frac{B}{A}\right] \qquad (5.7)$$

The radius (R) is sometimes called the *magnitude* and the angle θ is called the *argument*. You sometimes only care about the square of the magnitude (called the "magnitude-squared" or $|R|^2$) (Equation 5.8).

$$|R|^2 = A^2 + B^2 \qquad (5.8)$$

Equation 5.9 shows how to extract the magnitude and phase from a transfer function.

$$H = \frac{\text{num}}{\text{denom}}$$

then

$$|H| = \frac{|\text{num}|}{|\text{denom}|} \qquad (5.9)$$

and

$$\text{Arg}(H) = \text{Arg}(\text{num}) - \text{Arg}(\text{denom})$$

The frequency response plots of filters are actually magnitude responses of a complex function called the *transfer function* of the filter. The phase response plots are actually argument responses of this function. The transfer function is complex because it contains complex numbers; many transfer functions are actually quite simple. We use the letter H to denote a transfer function.

5.3 Time Delay as a Math Operator

The next piece of DSP theory you need to understand is the concept of time delay as a mathematical operator. This will be pretty easy since we are going to exploit the simple mathematical behavior of the $e^{j\omega t}$ function. First, consider a complex sinusoid and a delayed complex sinusoid (Figure 5.4).

How does the delay of n seconds change the complex sinusoid equation? Since positive time goes in the positive x direction, a delay of n seconds is a shifting of $-n$ seconds. In the complex sinusoid equation, you would then replace t with $t-n$. In other words, any point on the delayed curve is the same as the nondelayed curve minus n seconds. Therefore, the delayed sinusoid is

$$\text{Delayed sinusoid} = e^{j\omega(t-n)} \tag{5.10}$$

But, by using the polynomial behavior of e and running Equation 5.5 in reverse, you can rewrite it as shown in Equation 5.11:

$$e^{j\omega(t-n)} = e^{j\omega t}e^{-j\omega n} \tag{5.11}$$

It's a subtle mathematical equation but it says a lot: if you want to delay a complex sinusoid by n seconds, multiply it by $e^{-j\omega n}$—this allows us to express time delay as a mathematical operator.

In the last two sections you've learned that phase rotation and time delay can both be expressed as mathematical operators.

> Time delay can be expressed as a mathematical operator by multiplying the signal to be delayed n seconds by $e^{-j\omega n}$. This is useful because $e^{-j\omega n}$ is not dependent on the time variable. In discrete-time systems, the n refers to samples rather than seconds.

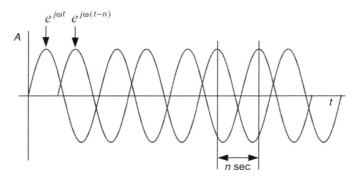

Figure 5.4: A complex sinusoid $e^{j\omega t}$ and another one delayed by n seconds.

5.4 First-Order Feed-Forward Filter Revisited

Being able to express delay as the mathematical operation of multiplication by $e^{-j\omega n}$ means you can take the block diagram and difference equation for a DSP filter and apply a sinusoid to the input in the form of $e^{j\omega t}$ rather than having to plug in sequences of samples as you did in Chapter 4. Then, you can see what comes out of the filter as a mathematical expression and evaluate it for different values of ω (where $\omega = 2\pi f$ with f in Hz) to find the frequency and phase responses directly rather than having to wait and see what comes out then try to guesstimate the amplitude and phase offsets. Consider the first-order feed-forward filter from the last chapter but with $e^{j\omega t}$ applied as the input signal, shown in a block diagram in Figure 5.5.

The difference equation is as follows:

$$y(t) = a_0 e^{j\omega t} + a_1 [e^{j\omega t} e^{-j\omega 1}] \quad (5.12)$$

Figure 5.5 shows the familiar block diagram but this time with the input $x(t)$ and output $y(t)$ instead of $x(n)$ and $y(n)$. Notice the delay element has been replaced by $e^{-j\omega 1}$ since there is a one-sample delay. When you apply the complex sinusoid $e^{j\omega t}$ to the input, the difference equation uses the delay-as-multiplication operation to produce the output. With a little math you can arrive at Equation 5.13.

$$y(t) = a_0 e^{j\omega t} + a_1 [e^{j\omega t} e^{-j\omega 1}]$$
$$= e^{j\omega t}(a_0 + a_1 e^{-j\omega 1})$$

and the input $x(t) = e^{j\omega t}$ so

$$y(t) = x(t)(a_0 + a_1 e^{-j\omega 1}) \quad (5.13)$$

The transfer function is defined as the ratio of output to input, therefore

$$\frac{y(t)}{x(t)} = a_0 + a_1 e^{-j\omega 1}$$

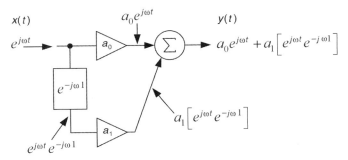

Figure 5.5: Block diagram of a first-order feed-forward filter with signal analysis.

> The *transfer function* of the filter is the ratio of output to input. The *frequency response* of the filter is the magnitude of the transfer function evaluated at different frequencies across its spectrum. The *phase response* of the filter is the argument (or angle) of the transfer function evaluated at different frequencies across its spectrum.
>
> To produce the frequency and phase response graphs, you evaluate the function for various values of ω then find the magnitude and argument at each frequency. The evaluation uses Euler's equation to replace the *e* term and produce the real and imaginary components.

What is so significant about this is that the transfer function is not dependent on time even though the input and output signals are functions of time. The transfer function (Equation 5.14) is only dependent on frequency ω, so we call it $H(\omega)$.

$$H(\omega) = a_0 + a_1 e^{-j\omega 1} \tag{5.14}$$

Notice that the transfer function is complex.

But what values of ω are to be used in the evaluation? We know that $\omega = 2\pi f$, but do we really care about the frequency in Hz? In Chapter 4 when you analyzed the same filter, you applied DC, Nyquist, ½ Nyquist, and ¼ Nyquist without thinking about the actual sampling frequency. This is called *normalized frequency* and is usually the way you want to proceed in analyzing DSP filters. The actual sampling rate determines Nyquist but the overall frequency range (0 Hz to Nyquist) is what we care about. To normalize the frequency, you let $f = 1$ Hz in $\omega = 2\pi f$, then ω varies from 0 to 2π or across a 2π range. There is also one detail we have to be aware of: negative frequencies.

5.4.1 Negative Frequencies

You may have never thought a frequency could be negative, but it can. When you first learned about the concept of a waveform's frequency, you were taught that the frequency is $1/T$, where T is the period, as shown in Figure 5.6.

The reason the frequencies came out as positive numbers is because the period is defined as $t_2 - t_1$, which makes it a positive number. But, there's no reason you couldn't define the period to be the other way around: $T = t_1 - t_2$, except that it implies that time is running backwards. Mathematically, time can run backwards. This means that for every positive frequency that exists, there is also a negative "twin" frequency. When you look at a frequency response plot you generally only look at the positive side. Figure 5.7 shows a low-pass response up to the highest frequency in the system, Nyquist.

However, in reality, the filter also operates on the negative frequencies just the same in a mirror image. In this case, as the negative frequencies get higher and higher, they are attenuated just like their positive counterparts (Figure 5.8). And it makes sense too. If

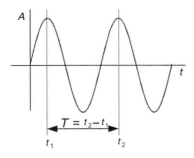

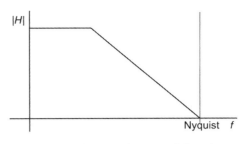

Figure 5.6: The classic way of defining the period, T.

Figure 5.7: The classic way of showing a frequency response plot only shows the positive portion.

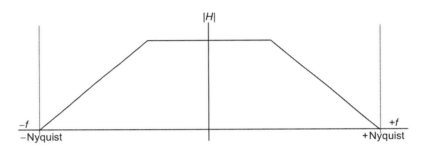

Figure 5.8: The more complete frequency response plot contains both positive and negative sides.

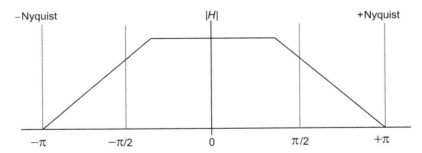

Figure 5.9: One way to divide the 2π range of frequencies includes both positive and negative frequencies.

you take an audio file and reverse it in time, then run it through a low-pass filter, the same frequency filtering still occurs.

For filter evaluation, ω varies on a 0 to 2π radians/second range and one way to think about this 2π range is to split it up into the range from $-\pi$ to $+\pi$ corresponding to $-$Nyquist to $+$Nyquist (Figure 5.9).

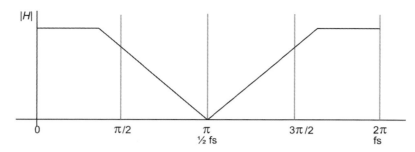

Figure 5.10: Mapping the 0 to 2π range of frequencies across the 0 to fs range.

> To evaluate the transfer function, let ω vary from 0 to π and get the first half of the response. The other half is a mirror image of the data.

5.4.2 Frequencies Above and Below ±Nyquist

The sampling theorem sets up the Nyquist criteria with regards to completely recovering the original, band-limited signal without aliasing. However, frequencies above Nyquist and all the way up to the sampling frequency are also allowed mathematically. And in theory, any frequency could enter the system and you could sample it without limiting Nyquist. For a frequency or phase response plot, the frequencies from Nyquist up to the sampling frequency are a mirror image about Nyquist. This is another way to divide up the 2π range by going from 0 Hz to the sampling frequency (Figure 5.10).

Notice that in either method the same information is conveyed as we get both halves of the curves, and in both cases, Nyquist maps to π and 0 Hz to 0 and positive frequencies map to the range 0 to π.

5.5 Evaluating the Transfer Function $H(\omega)$

DSP filter transfer functions will contain $e^{-j\omega n}$ terms that need to be evaluated over the range of 0 to π; the way to do this is by using Euler's equation to decompose the sinusoid into its real (cos) and imaginary (sin) components. Then, evaluate the cos and sin terms at the frequency in question. In the last chapter you manually calculated the input/output relationship of a filter by cranking through the filter operation, one step at a time. In this improved method, you only need to solve the transfer function equation. Start with the block diagram in Figure 5.11.

The transfer function is as follows:

$$H(\omega) = a_0 + a_1 e^{-j\omega 1} \tag{5.15}$$

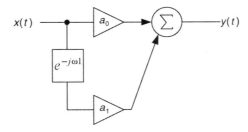

Figure 5.11: First-order feed-forward block diagram.

Table 5.1: Sine and cosine function evaluations at DC, ¼ Nyquist, ½ Nyquist, ¾ Nyquist, and Nyquist.

Frequency ω	$\cos(\omega)$	$\sin(\omega)$
0	1.0	0.0
$\pi/4$	0.707	0.707
$\pi/2$	0.0	1.0
$3\pi/4$	0.707	−0.707
π	−1.0	0.0

Use the filter coefficients $a_0 = 0.5$, $a_1 = 0.5$. You can use Table 5.1 to help with the evaluation. Evaluate at the following frequencies:

- DC: 0
- Nyquist: π
- ½ Nyquist: $\pi/2$
- ¼ Nyquist: $\pi/4$

Evaluation is a two-step process for each frequency:

1. Use Euler's equation to convert the e terms into real and imaginary components.
2. Find the magnitude and argument of the complex equation.

5.5.1 DC (0 Hz)

$$\begin{aligned}
H(\omega) &= 0.5 + 0.5e^{-j\omega 1} \\
&= 0.5 + 0.5(\cos(\omega) - j\sin(\omega)) \\
&= 0.5 + 0.5(\cos(0) - j\sin(0)) \\
&= 0.5 + 0.5(1 - j0) \\
&= 1.0 + j0
\end{aligned} \quad (5.16)$$

Find the magnitude and phase at this frequency:

$$\begin{aligned}|H(\omega)| &= \sqrt{(a+jb)(a-jb)} \\ &= \sqrt{(1+j0)(1-j0)} \\ &= 1.0 \\ \text{Arg}(H) &= \tan^{-1}(b/a) \\ &= \tan^{-1}(0/1) \\ &= 0.0°\end{aligned} \quad (5.17)$$

Compare these mathematical results (Equations 5.16 and 5.17) with the graphical ones from the last chapter (Figure 5.12).

5.5.2 Nyquist (π)

$$\begin{aligned}H(\omega) &= 0.5 + 0.5e^{-j\omega 1} \\ &= 0.5 + 0.5(\cos(\omega) - j\sin(\omega)) \\ &= 0.5 + 0.5(\cos(\pi) - j\sin(\pi)) \\ &= 0.5 + 0.5(-1 - j0) \\ &= 0 + j0\end{aligned} \quad (5.18)$$

$$\begin{aligned}|H(\omega)| &= \sqrt{(a+jb)(a-jb)} \\ &= \sqrt{(0+j0)(0-j0)} \\ &= 0.0 \\ \text{Arg}(H) &= \tan^{-1}(b/a) \\ &= \tan^{-1}(0/0) \\ &= 0°\end{aligned} \quad (5.19)$$

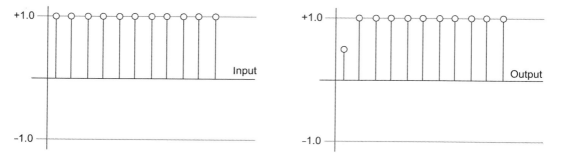

Figure 5.12: The graphical results show the same information. The magnitude is 1.0 and the phase shift is 0.

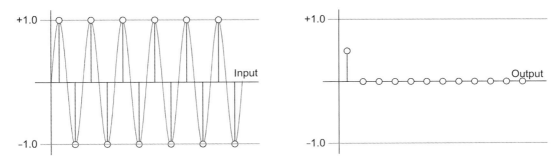

Figure 5.13: The graphical results show the same information at Nyquist—the magnitude is 0 and there is no phase shift since there is nothing there to shift.

The inverse tangent argument is 0/0 and the phase or Arg(H) is defined to be 0 under this condition. The C++ function you use is *arctan2(im,re)*, which performs the inverse tangent function; it will also evaluate to 0 in this case. Now, compare our results to the last chapter's graphical results (Figure 5.13).

5.5.3 ½ Nyquist (π/2)

$$\begin{aligned} H(\omega) &= 0.5 + 0.5e^{-j\omega 1} \\ &= 0.5 + 0.5(\cos(\omega) - j\sin(\omega)) \\ &= 0.5 + 0.5(\cos(\pi/2) - j\sin(\pi/2)) \\ &= 0.5 + 0.5(0 - j1) \\ &= 0.5 - j0.5 \end{aligned} \quad (5.20)$$

$$\begin{aligned} |H(\omega)| &= \sqrt{(a+jb)(a-jb)} \\ &= \sqrt{(0.5+j0.5)(0.5-j0.5)} \\ &= \sqrt{0.25 + 0.25} = \sqrt{0.5} \\ &= 0.707 \end{aligned} \quad (5.21)$$

$$\begin{aligned} \text{Arg}(H) &= \tan^{-1}(b/a) \\ &= \tan^{-1}(-0.5/0.5) \\ &= -45° \end{aligned}$$

Compare this to the last chapter's graphical results (Figure 5.14); the magnitude is 0.707 with a phase shift of −45 degrees, and the results agree.

5.5.4 1/4 Nyquist (π/4)

$$\begin{aligned} H(\omega) &= 0.5 + 0.5e^{-j\omega 1} \\ &= 0.5 + 0.5(\cos(\omega) - j\sin(\omega)) \end{aligned} \quad (5.22)$$

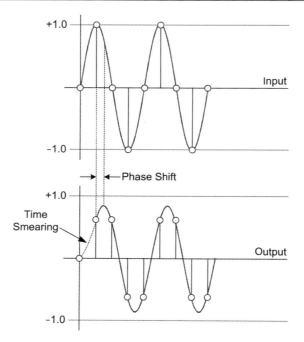

Figure 5.14: Graphical results from the last chapter at ½ Nyquist.

$$= 0.5 + 0.5(\cos(\pi/4) - j\sin(\pi/4))$$
$$= 0.5 + 0.5(0.707 - j0.707)$$
$$= 0.853 - j0.353$$

$$|H(\omega)| = \sqrt{(a+jb)(a-jb)}$$
$$= \sqrt{(0.853 + j0.353)(0.853 - j0.353)}$$
$$= \sqrt{0.728 + 0.125} = \sqrt{0.853}$$
$$= 0.923 \tag{5.23}$$

$$\text{Arg}(H) = \tan^{-1}(b/a)$$
$$= \tan^{-1}(-0.353/0.853)$$
$$= -22.5°$$

Compare to the last chapter's graphical results (Figure 5.15); you can see how much more accuracy we get with the mathematical calculation. The magnitude and phase shift look about right when compared to the graphs.

Now, you can combine all the evaluations together and sketch out the frequency response of the filter (Figure 5.16).

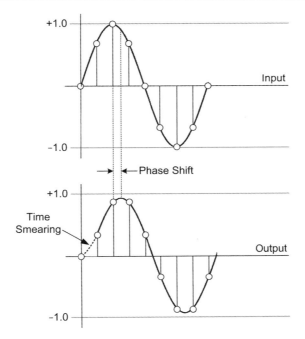

Figure 5.15: Graphical results from the last chapter at ¼ Nyquist.

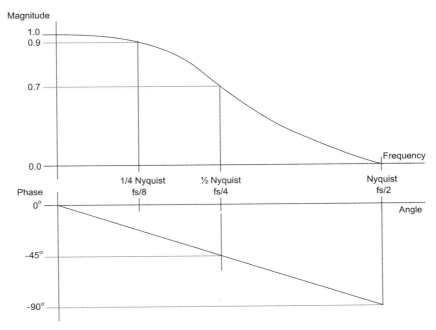

Figure 5.16: The final composite frequency and phase response plots show the same results as the last chapter, but with a lot less work.

Hopefully, this quick example has convinced you that it is better to do a little complex math than have to analyze and design these filters by brute force analysis of time domain input sequences.

5.6 Evaluating $e^{j\omega}$

In the evaluation of the transfer function, you had to substitute values of ω from 0 to π into the $e^{j\omega}$ terms of the equation. But what would the plot look like if you evaluate a single $e^{j\omega}$ term? You saw that the use of Euler's equation produced the real and imaginary components of the term and now it's time to plot them over the range of 0 to π.

If you evaluate $e^{j\omega}$ over more frequencies and plot the resulting values in the complex plane, you get an interesting result. The frequencies in Table 5.2 from 0 to $+\pi$ map to an arc that is the top half of a circle with radius = 1.0, shown in Figure 5.17. Remember, the magnitude is the radius and the argument is the angle when using polar notation, which simplifies the

Table 5.2: The magnitude and angle of $e^{j\omega}$ from DC to Nyquist.

Frequency π	$e^{j\omega} = \cos(\omega) + j\sin(\omega)$	$|e^{j\omega}|$	$\mathrm{Arg}(e^{j\omega})$
DC (0Hz)	$1 + j0$	1.0	0
¼ Nyquist	$0.707 + j0.707$	1.0	$\pi/4$
½ Nyquist	$0 + j1$	1.0	$\pi/2$
Nyquist	$-1 + j0$	1.0	π

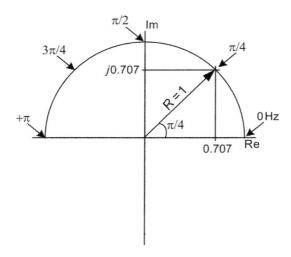

Figure 5.17: The positive frequencies map to the upper half of the unit circle.

analysis. You don't have to keep track of the real and imaginary parts. The evaluation at $\omega = \pi/4$ is plotted on the curve. The circle this arc is laying over would have a radius of 1.0 and is called the *unit circle*. If you evaluate $e^{j\omega}$ over the negative frequencies that correspond to 0 to $-\pi$, you get a similar but inverted table (Table 5.3).

This table translates to a mapping across the lower half of the same unit circle (Figure 5.18). The negative frequencies increase as you move clockwise from 0 Hz, the radius stays 1.0 during the entire arc.

Why bother to evaluate $e^{j\omega}$? It will be useful very soon when we start picking apart the transfer functions in an effort to figure out how to design filters. It also shows the limited "frequency space" of the digital domain. All the frequencies that could exist from $-$Nyquist to $+$Nyquist map to outline of a simple unit circle. In contrast the analog domain has an infinitely long frequency axis and an infinite frequency space.

Table 5.3: The magnitude and angle of $e^{j\omega}$ from DC to $-$Nyquist.

Frequency ω	$e^{j\omega} = \cos(\omega) + j\sin(\omega)$	$\lvert e^{j\omega} \rvert$	$\text{Arg}(e^{j\omega})$
DC (0Hz)	$1 + j0$	1.0	0
$-¼$ Nyquist	$0.707 - j0.707$	1.0	$-\pi/4$
$-½$ Nyquist	$0 - j1$	1.0	$-\pi/2$
$-$Nyquist	$-1 + j0$	1.0	$-\pi$

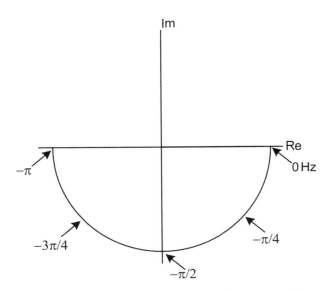

Figure 5.18: The negative frequencies map to the lower half of the unit circle.

5.7 The z Substitution

It's going to get messy having to write $e^{j\omega}$ so often and we know $e^{j\omega}$ behaves like a polynomial mathematically. So, we can simplify the equations by making a simple substitution of Equation 5.24.

$$z = e^{j\omega} \tag{5.24}$$

This is just a substitution right now and nothing else. Making the substitution in Equation 5.24 and noting the resulting transfer function is now a function of z, not ω, we can write it like Equation 5.25:

$$H(z) = a_0 + a_1 z^{-1} \tag{5.25}$$

The reason this is useful is that it turns the transfer function into an easily manipulated polynomial in z. In this case, the polynomial is a first-order polynomial (the highest exponent absolute value is 1) and this is the real reason the filter is named a first-order filter—it's the polynomial order of the transfer function.

5.8 The z Transform

The z substitution does a good job at simplifying the underlying polynomial behavior of $e^{j\omega}$ and it lets us use polynomial math to solve DSP problems. But, there is an interesting application of $z = e^{j\omega}$ that simplifies the design and analysis of digital filters. In the graph from Chapter 1 (Figure 5.19) you can see how the indexing of the samples determines their place in time. The future samples have positive indices and the past (delayed) samples have negative indices.

We've seen that $e^{j\omega}$ is the delay operator, so $e^{-j\omega 1}$ means one sample of delay, or one sample *behind* the current one. Likewise, $e^{-j\omega 2}$ would be two samples behind and $e^{+j\omega 4}$ indicates 4 samples *ahead* or in the future. That means that $e^{j\omega}$ could also be used as a book-keeping device since it can relate the position in time of something (a sample, for us).

The rules for implementing the z transform on a discrete signal or difference equation are easy and can often be done by inspection. The current sample $x(n)$ or $y(n)$ transforms into the signal $X(z)$ or $Y(z)$. Instead of thinking of the *sample $x(n)$* you think of the *signal $X(z)$*, where $X(z)$ is the whole signal—past, present, and future.

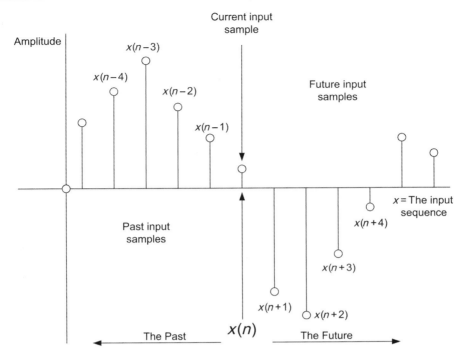

Figure 5.19: Our book-keeping rules shown graphically.

> The *z transform* changes a sequence of samples in *n* to a sequence of samples in *z* by replacing the indices ...n−1, n, n+1... with ...z^{-1}, z, z^{+1}... This works because multiplication by $z = e^{j\omega}$ represents the operation of delay or time shift. The resulting transformed sequence now is a function of the complex frequency $e^{j\omega}$, therefore it transforms things from the *time domain* into the *complex frequency domain*.

- $x(n) \rightarrow X(z)$
- $y(n) \rightarrow Y(z)$

Instead of thinking of the sample $x(n-1)$ as being delayed by one sample, you think of the signal $X(z)$ delayed by one sample, z^{-1}. The delayed signals are the result of the whole signal multiplied by the z^{-N} terms:

- $x(n-1) \rightarrow X(z)\, z^{-1}$
- $y(n-1) \rightarrow Y(z)\, z^{-1}$
- $x(n-2) \rightarrow X(z)\, z^{-2}$
- $y(n+6) \rightarrow Y(z)\, z^{+6}$

You can see that this concept relies on the ability to express delay as a mathematical operator. It not only allows us to express an *algorithm* based on z, it also lets us express a *signal* based on z. Mathematically, the z transform is

$$X(z) = \sum_{n=-\infty}^{n=+\infty} x(n) z^{-n} \tag{5.26}$$

5.9 The z Transform of Signals

Remember, $x(n)$ is the sequence of samples just like the ones you used for analysis. Figure 5.20 shows an example. This simple, finite length signal consists of five samples. The remaining zero samples don't need to be counted.

$$x(n) = \{0, 0.25, 0.5, 0.75, 1.0\}$$

The sequence $x(n)$ could also be written $x(n) = \{x(0), x(1), x(2), x(3), x(4)\}$, so using Equation 5.26 we transform $x(n)$ into $X(z)$ and write Equation 5.27:

$$X(z) = 0z^0 + 0.25z^{-1} + 0.5z^{-2} + 0.75z^{-3} + 1.0z^{-4} \tag{5.27}$$

You could read Equation 5.27 as follows: "The whole signal $X(z)$ consists of a sample with an amplitude of 0 at time 0 followed by a sample with an amplitude of 0.25 one sample later and a sample with an amplitude of 0.5 two samples later and a sample with an amplitude of 0.75 three samples later and . . ." This should shed light on the fact that the transform really involves the book-keeping of sample locations in time and that the result is a polynomial. You can multiply and divide this signal with other signals by using polynomial math. You can mix two signals by linearly combining the polynomials.

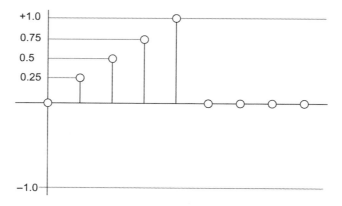

Figure 5.20: A simple signal for analysis.

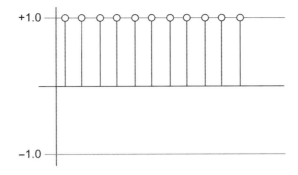

Figure 5.21: The DC signal is infinite in length.

Let's do one more example regarding the transformation of an input signal. This time, let's choose the DC signal—it goes on forever. Figure 5.21 shows the DC signal with the first sample at 1.0 and all preceding samples at 0.0 with a sequence of $x(n) = \{1, 1, 1, 1, 1\ldots\}$.

Using Equation 5.26 you can directly write the z transform as (remember, $1z^0 = 1$) in Equation 5.28:

$$X(z) = 1 + z^{-1} + z^{-2} + z^{-3} + \ldots + z^{-\infty} \qquad (5.28)$$

While that looks like an ugly, infinitely long equation, it can also be represented in a *closed form*. In fact, a closed form representation exists for this polynomial using a polynomial series expansion, as shown in Equation 5.29:

$$\begin{aligned} X(z) &= 1 + z^{-1} + z^{-2} + z^{-3} + \ldots + z^{-\infty} \\ &= \frac{1}{1 - z^{-1}} \\ &= \frac{z}{z - 1} \end{aligned} \qquad (5.29)$$

5.10 The z Transform of Difference Equations

The z transform of signals is interesting, but something fascinating happens when you take the z transform of a difference equation, converting it into a transfer function in z all at once. And, the same easy rules apply. Let's do that with the basic first-order feed-forward filter. The difference equation is

$$y(n) = a_0 x(n) + a_1 x(n - 1) \qquad (5.30)$$

Taking the z transform:

$$Y(z) = a_0 X(z) + a_1 X(z) z^{-1}$$
$$Y(z) = X(z)[a_0 + a_1 z^{-1}]$$
$$H(z) = \frac{Y(z)}{X(z)} = a_0 + a_1 z^{-1}$$

This is a really useful result—you got the final transfer function in just a handful of steps, using the simple z transform rules. Let's recap what you had to do before you learned how to take the z transform of a difference equation:

- Redraw the block diagram with $e^{-j\omega n}$ operators in the n-sample delay elements.
- Apply the complex sinusoid $e^{-j\omega t}$ to the input.
- Find out what comes out, $y(t)$, and formulate the transfer function $H(\omega)$.
- Apply the z substitution to the transfer function.

Taking the z transform does all these steps at once and we're left with the same simple polynomial in z. If we evaluate the transfer function for different values of $z = e^{j\omega}$, we can find the frequency and phase plots. You'll get more practice taking the z transforms of more difference equations soon.

5.11 The z Transform of an Impulse Response

The z transform of a difference equation results in the transfer function. But what if you don't have the difference equation? Suppose you only have a black box that performs some kind of DSP algorithm and you'd like to figure out the transfer function, evaluate it, and plot the frequency and phase responses. It can be done without knowing the algorithm or any details about it by taking the impulse response of the system. You apply the input sequence $x(n) = \{1, 0, 0, 0, 0...\}$ and capture what comes out, which we'll call $h(n)$. If you take the z transform of the impulse response, you get the transfer function expanded out into a series form.

In fact, this is exactly what the RackAFX software's audio analyzer does—it takes the impulse response of the filter and then runs a z transform on it to make the magnitude and phase plots you see on the screen. Mathematically, Equation 5.31 is identical to Equation 5.26 except we've changed the signal from X to H:

> The z transform of the impulse response $h(n)$ is the transfer function $H(z)$ as a series expansion. Evaluate the transfer function to plot the frequency and phase responses.

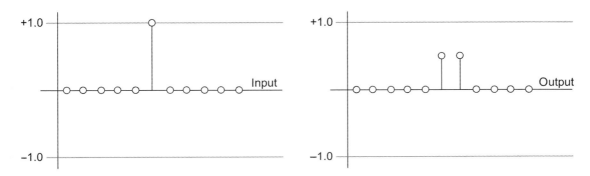

Figure 5.22: The impulse response from the first-order feed-forward filter.

$$H(z) = \sum_{n=-\infty}^{n=+\infty} h(n)z^{-n} \tag{5.31}$$

Try this on the first-order feed-forward filter we've been working on; you already have the impulse response "captured" from the last chapter (Figure 5.22).

The impulse response is $h(n) = \{0.5, 0.5\}$. Applying the z transform yields Equation 5.32:

$$\begin{aligned} H(z) &= 0.5z^0 + 0.5z^{-1} \\ &= 0.5 + 0.5z^{-1} \end{aligned} \tag{5.32}$$

Notice that this is the identical result as taking the z transform of the difference equation and the filter coefficients (0.5, 0.5) *are* the impulse response $\{0.5, 0.5\}$.

5.12 The Zeros of the Transfer Function

When we used the coefficients $a_0 = a_1 = 0.5$ we wound up with a filter that completely destroys the Nyquist frequency and you saw how its output became 0 in both the manual and complex sinusoid evaluations. We noted that feed-forward filters have *zeros of transmission* or *zero frequencies* or just *zeros* when their output becomes zero. In both the manual and complex sinusoid evaluations, we just got lucky when we stumbled upon this value as Nyquist happened to be one of the signals we were testing or evaluating. There's a way to precisely find these critical frequencies by using the polynomial result of the z transform. You probably remember factoring polynomials in high school or college. When you did that, you set the polynomial equal to 0 and then you factored to find the roots of the polynomial. What you were really doing was finding the zeros of the polynomial, that is, the values of the dependent variable that make the polynomial become zero. You can do the same thing with the transfer function by setting it equal to zero and then factoring the polynomial. Suppose $a_0 = a_1 = 0.5$ and we factor the transfer function in Equation 5.32 to get Equation 5.33:

$$H(z) = 0.5 + 0.5z^{-1}$$
$$= 0.5 + \frac{0.5}{z} \qquad (5.33)$$

You can find the zero by inspection—it's the value of z that forces $H(z)$ to be 0 and in this case there is a zero at $z = -1.0$. But what does it mean to have a zero at -1.0? This is where the concept of evaluating $e^{j\omega}$ comes into play. When you did that and plotted the various points, noting they were making a unit circle in the complex plane, you were actually working in the z-plane, that is, the plane of $e^{j\omega}$. The location of the zero at $z = -1.0$ is really at the location $z = -1.0 + j0$ purely on the real axis and at Nyquist. In Figure 5.23 the zero is shown as a small circle sitting at the location $z = -1.0$.

There are several reasons to plot the zero frequencies. First, you can design a filter directly in the z-plane by deciding where you want to place the zero frequencies first, then figuring out the transfer function that will give you those zeros. Secondly, plotting the zeros gives you a quick way to sketch the frequency response without having to evaluate the transfer function directly. You can estimate a phase plot too, but it is a bit more involved. So, you have two really good reasons for wanting to plot the zeros; one for design, the other for analysis.

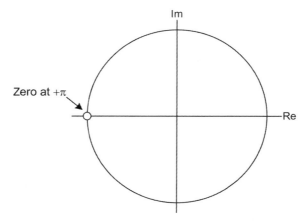

Figure 5.23: Zero is plotted in the z-plane at its location on the real axis $z = -1 + j0$.

5.13 Estimating the Frequency Response: Zeros

An interesting property of the z-plane and z transform is that you can measure the frequency response graphically on the z-plane. In the simplest case of only one zero, the method is as follows:

- Locate each evaluation frequency on the outer rim of the unit circle.
- Draw a line from the point on the circle to the zero and measure the length of this vector. Do it for each evaluation frequency.
- The *lengths* of the lines will be the *magnitudes* at each frequency in the frequency response.

In Figure 5.24 you can see the complete operation, first drawing and measuring the lines—you can use graph paper and a ruler, if you want—then building the frequency response plot from them. Notice also that the magnitude of a line drawn from Nyquist to the zero at -1 has a length of zero. The lengths of the vectors are the mathematical definition of magnitude and you are evaluating the whole filter at once. These z-plane plots are going to be useful for filter design. You can also derive the phase response, which involves measuring the angles of incidence of each vector on the zero. With multiple zeros, it becomes cumbersome. But, estimating the frequency response is pretty simple, even for more complex filters.

You might notice that even though this frequency response looks like the one we produced earlier, the gain values are not the same. In this filter, the gain is 2.0 at DC, and in ours, it's half of that. In fact, this filter's magnitudes at the evaluation frequencies are all twice what ours are.

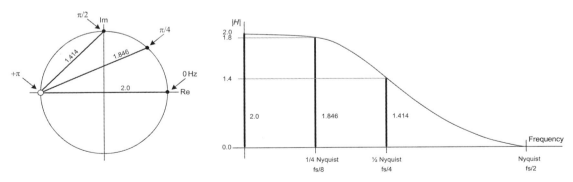

Figure 5.24: The geometric interpretation shows how the length of each vector from the evaluation frequency to the zero is really a magnitude in the response.

5.14 Filter Gain Control

The last thing you need to do is remove the overall gain factor from the transfer function so that overall filter gain (or attenuation) can be controlled by just one variable. This is actually pretty simple to do, but requires re-working the transfer function a bit. The idea is to pull out the a_0 variable as a multiplier for the whole function. This way, it behaves like a volume knob, gaining the whole filter up or down. The way you do it is to normalize the filter by a_0 (Equation 5.34):

$$H(z) = a_0 + a_1 z^{-1}$$
$$= \frac{a_0}{a_0} + \frac{a_1}{a_0} z^{-1}$$
$$= a_0 \left[1 + \frac{a_1}{a_0} z^{-1} \right] \quad (5.34)$$
$$\text{Let } \alpha_1 = \frac{a_1}{a_0}$$
$$H(z) = a_0 [1 + \alpha_1 z^{-1}]$$

By normalizing by a_0 and using the α_1 variable you can produce a transfer function that looks basically the same in the polynomial but pulls a_0 out as a scalar multiplier—a gain control. Where is the zero of the new transfer function in Equation 5.35?

$$H(z) = a_0 [1 + \alpha_1 z^{-1}]$$
$$= a_0 \left[1 + \frac{\alpha_1}{z} \right] \quad (5.35)$$

Once again, by inspection we can tell that if $z = -\alpha_1$ then the function will become 0 regardless of the value of a_0. This transfer function has a zero at $-\alpha_1$. If you plug our values of $a_0 = a_1 = 0.5$, you still get the same zero at -1.0. The difference is in the gain of the filter.

After extracting the magnitude response from the z-plane plot, scale it by your a_0 value. This makes the response in Figure 5.24 match our others because everything gets multiplied by $a_0 = 0.5$. The idea of controlling the gain independently of the magnitude response is useful in audio filters, so we will keep it and use it in all future analyses.

At this point, you have all the DSP theory tools you need to understand the rest of the classical DSP filters (first-order feed back, second-order feed forward and feed back) as well as many filter design techniques. The rest of the chapter will be devoted to applying these same fundamentals to the other classical DSP filters but we will move much more quickly,

---

The *graphical interpretation* method of evaluating a filter in the z-plane assumes the filter is normalized so that $a_0 = 1.0$.

applying each analysis technique to the other algorithms. For example, we will dispense with the evaluation of $e^{j\omega}$ terms and start off directly in the z transform of the difference equations.

5.15 First-Order Feed-Back Filter Revisited

Now let's go through the same analysis technique on the first-order feed-back filter from the last chapter. We can move much more quickly now that we have the basic DSP theory down. There will be many similarities but also several key differences when dealing with feed-back designs. You already saw that the feed-back topology can blow up or ring forever and that the feed-forward design cannot. We will find a way to figure out if this is going to happen and how to prevent it. Start with the original first-order feed-back filter (block diagram in Figure 5.25) and its difference equation.

The difference equation is as follows:

$$y(n) = a_0 x(n) - b_1 y(n-1) \tag{5.36}$$

Step 1: Take the z transform of the difference equation

This can be done by inspection, using the rules from Section 5.8 (Figure 5.26). Therefore, the z transform is shown in Equation 5.37.

$$Y(z) = a_0 X(z) - b_1 Y(z) z^{-1} \tag{5.37}$$

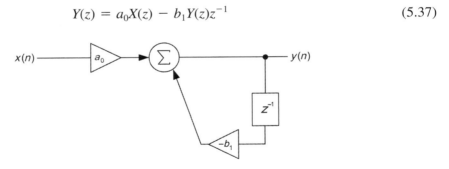

Figure 5.25: The first-order feed-back filter.

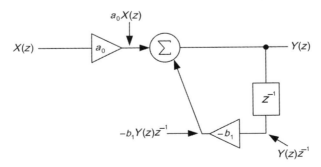

Figure 5.26: Pushing the input X(z) through the algorithm produces the z transform.

Step 2: Fashion the difference equation into a transfer function

Now apply some algebra to convert the transformed difference equation to $H(z)$. The process is always the same: separate the $X(z)$ and $Y(z)$ variables, then form their quotient (Equation 5.38).

$$Y(z) = a_0 X(z) - b_1 Y(z) z^{-1}$$

Separate variables:

$$Y(z) + b_1 Y(z) z^{-1} = a_0 X(z)$$
$$Y(z)[1 + b_1 z^{-1}] = a_0 X(z) \quad (5.38)$$
$$Y(z) = \frac{a_0 X(z)}{1 + b_1 z^{-1}}$$

From $H(z)$:

$$H(z) = \frac{Y(z)}{X(z)} = \frac{a_0}{1 + b_1 z^{-1}}$$

Step 3: Factor out a_0 as the scalar gain coefficient

In this case, this step is simple since pulling a_0 out is trivial, as in Equation 5.39. However, in more complex filters this requires making substitutions as you did in the last section.

$$H(z) = \frac{a_0}{1 + b_1 z^{-1}}$$
$$= a_0 \frac{1}{1 + b_1 z^{-1}} \quad (5.39)$$

5.16 The Poles of the Transfer Function

The next step in the sequence is to do a quick estimation of the frequency response using the graphical interpretation method in the z-plane. The pure feed-forward filter you analyzed produced zeros of transmission or zeros at frequencies where its output becomes zero. A pure feed-back filter produces *poles* at frequencies where its output becomes infinite. We were able to make this happen by applying 100% feed-back in the last chapter. For the simple first-order case, finding the poles is done by inspection.

> When the denominator of the transfer function is zero, the output is infinite. The complex frequency where this occurs is the pole frequency or pole.

Examining the transfer function, we can find the single pole in Equation 5.40:

$$H(z) = a_0 \frac{1}{1 + b_1 z^{-1}}$$
$$= a_0 \frac{1}{1 + \frac{b_1}{z}} \quad (5.40)$$
$$= a_0 \frac{z}{z + b_1}$$

By rearranging the transfer function, you can see that the denominator will be zero when $z = -b_1$ and so there is a pole at $z = -b_1$. You might also notice something interesting about this transfer function—it has a z in the numerator. If $z = 0$, then this transfer function has a zero at $z = 0$. This zero is called a *trivial zero* because it has no impact on the filter's frequency response. Thus, you can ignore the zero at $z = 0$. In fact, you can also ignore poles at $z = 0$ for the same reason.

And, if you look back at the transfer function in the feed-forward filter in Equation 5.35 you can see that it also had a pole in Equation 5.41:

$$H(z) = a_0[1 + \alpha_1 z^{-1}]$$
$$= a_0\left[1 + \frac{\alpha_1}{z}\right] \quad (5.41)$$

Pole at $z = 0$

The poles are plotted in the z-plane in the same manner as the zeros but you use an **x** to indicate the pole frequency. In Equation 5.40, the pole is at $-b_1 + j0$ and so it is a real pole located on the real axis in the z-plane. For this filter let's analyze it with $a_0 = 1.0$ and $b_1 = 0.9$. You wrote a plug-in and implemented this filter in the last chapter. The results are shown in Figure 5.27.

Let's see how this is estimated first, and then we can do a direct evaluation as before.

Step 4: Estimate the frequency response

The single pole is plotted on the real axis at $z = -0.9 + j0$ and a trivial zero at $z = 0 + j0$ (Figure 5.28). In the future, we will ignore the trivial zeros or poles.

> A pole or zero at $z = 0$ is trivial and can be ignored for the sake of analysis since it has no effect on the frequency response.

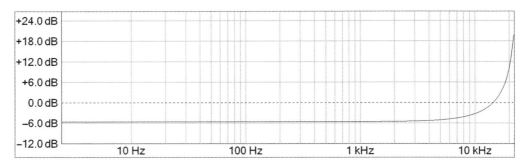

Figure 5.27: The frequency response of the first-order feed-back filter with these coefficients.

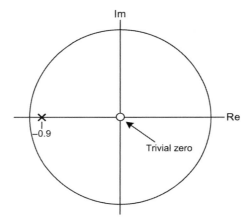

Figure 5.28: The pole is plotted in the z-plane along with the trivial zero.

In the simplest case of only one pole, the method for estimating the frequency response is as follows:

- Locate each evaluation frequency on the outer rim of the unit circle.
- Draw a line from the point on the circle to the pole and measure the length of this vector. Do it for each evaluation frequency.
- The *inverse of the lengths* of the lines will be the *magnitudes* at each frequency in the frequency response in Equation 5.42.

$$\text{Magnitude} = \frac{1}{L_p} \qquad (5.42)$$

L_p = length from evaluation frequency to pole

Thus, the mechanism is the same as for the one-zero case, except you take the inverse of the length of the vector. This means that as you near the pole, the vector becomes shorter, but the amplitude becomes larger—exactly opposite of the zero case. You can see from Figure 5.29 that our estimate is pretty close to what we saw in the real world when we coded the filter and tested it. Note the gain at Nyquist is 10.0; convert that to dB and you get +20 dB of gain, which is what we expect.

Step 5: Direct evaluation of frequency response

Now you can evaluate the filter the same way as before using Euler's equation to separate the real and imaginary components from the transfer function. Evaluate at the following frequencies:

- DC: 0
- Nyquist: π
- ½ Nyquist: $\pi/2$
- ¼ Nyquist: $\pi/4$

First, get the transfer function in a form from Equation 5.43 to use for all the evaluation frequencies:

$$H(z) = a_0 \frac{1}{1 + b_1 z^{-1}}$$
$$a_0 = 1.0 \quad b_1 = 0.9$$
$$H(z) = \frac{1}{1 + 0.9 z^{-1}}$$
$$z = e^{j\omega}$$

(5.43)

Next, let's make a simplification in the magnitude response equation and use the reduced form in Equation 5.44. Next, evaluate.

$$|a + jb| = \sqrt{(a + jb)(a - jb)}$$
$$= \sqrt{a^2 + b^2}$$

(5.44)

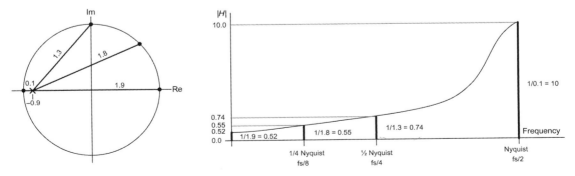

Figure 5.29: Estimating the frequency response of the first-order feed-back design.

5.16.1 DC (0 Hz)

$$H(\omega) = \frac{1}{1 + 0.9e^{-j\omega 1}}$$

$$= \frac{1}{1 + 0.9[\cos(\omega) - j\sin(\omega)]}$$

$$= \frac{1}{1 + 0.9[\cos(0) - j\sin(0)]} \quad (5.45)$$

$$= \frac{1}{1 + 0.9[1 - j0]}$$

$$= \frac{1}{1.9}$$

$$= 0.526 + j0$$

$$|H(\omega)| = \sqrt{a^2 + b^2}$$

$$= \sqrt{0.526^2 + 0^2}$$

$$= 0.526$$

$$\text{Arg}(H) = \tan^{-1}(b/a) \quad (5.46)$$

$$= \tan^{-1}(0/0.526)$$

$$= 0.0°$$

5.16.2 Nyquist (π)

$$H(\omega) = \frac{1}{1 + 0.9e^{-j\omega 1}}$$

$$= \frac{1}{1 + 0.9[\cos(\omega) - j\sin(\omega)]}$$

$$= \frac{1}{1 + 0.9[\cos(\pi) - j\sin(\pi)]} \quad (5.47)$$

$$= \frac{1}{1 + 0.9[-1 - j0]}$$

$$= \frac{1}{0.1}$$

$$= 10 + j0$$

$$|H(\omega)| = \sqrt{a^2 + b^2}$$
$$= \sqrt{10.0^2 + 0^2}$$
$$= 10.0$$
$$\mathrm{Arg}(H) = \tan^{-1}(b/a) \tag{5.48}$$
$$= \tan^{-1}(0/10.0)$$
$$= 0.0°$$

5.16.3 ½ Nyquist (π/2)

$$H(\omega) = \frac{1}{1 + 0.9e^{-j\omega 1}}$$
$$= \frac{1}{1 + 0.9[\cos(\omega) - j\sin(\omega)]}$$
$$= \frac{1}{1 + 0.9[\cos(\pi/2) - j\sin(\pi/2)]} \tag{5.49}$$
$$= \frac{1}{1 + 0.9[0 - j1]}$$
$$= \frac{1}{1 - j0.9}$$

$$H(\omega) = \frac{1}{1 - j0.9}$$
$$|H(\omega)| = \frac{|1|}{|1 - j0.9|}$$
$$= \frac{1}{\sqrt{a^2 + b^2}}$$
$$= \frac{1}{\sqrt{1 + 0.81}} \tag{5.50}$$
$$= 0.743$$
$$\mathrm{Arg}(H) = \mathrm{Arg(Num)} - \mathrm{Arg(Denom)}$$
$$= \tan^{-1}(0/1) - \tan^{-1}(-0.9/1)$$
$$= +42°$$

5.16.4 ¼ Nyquist ($\pi/4$)

$$H(\omega) = \frac{1}{1 + 0.9e^{-j\omega 1}}$$

$$= \frac{1}{1 + 0.9[\cos(\omega) - j\sin(\omega)]}$$

$$= \frac{1}{1 + 0.9[\cos(\pi/4) - j\sin(\pi/4)]} \quad (5.51)$$

$$= \frac{1}{1 + 0.636 - j0.636}$$

$$= \frac{1}{1.636 - j0.636}$$

$$H(\omega) = \frac{1}{1.636 - j0.636}$$

$$|H(\omega)| = \frac{|1|}{|1.636 - j0.636|}$$

$$= \frac{1}{\sqrt{1.636^2 + 0.636^2}} \quad (5.52)$$

$$= 0.57$$

$$\text{Arg}(H) = \text{Arg}(\text{Num}) - \text{Arg}(\text{Denom})$$

$$= \tan^{-1}(0/1) - \tan^{-1}(-0.636/1.636)$$

$$= +21°$$

Make a special note about how we have to handle the magnitude of a fraction with numerator and denominator. You need to use the two equations in Equation 5.9 to deal with this. The main issue is that the phase is the difference of the Arg(numerator) − Arg(denominator). If the numerator was a complex number instead of 1.0, you would need to take the magnitude of it separately then divide. The final composite frequency/phase response plot is shown in Figure 5.30. You can see that the phase behaves linearly until it gets near the pole, and then it behaves nonlinearly. This is not a linear phase filter.

Step 6: z transform of impulse response

From Chapter 4 you will recall that the impulse response rings for this particular set of coefficients (Figure 5.31). Finding the impulse response by hand is going to be tedious. There are a lot of points to capture and it could take many lines of math before the impulse settles out. Fortunately, you can use RackAFX to do the work for you. The frequency and phase response plots are made using a z transform of the impulse response.

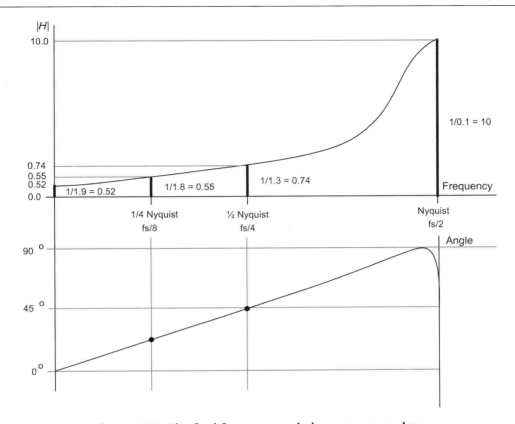

Figure 5.30: The final frequency and phase response plots.

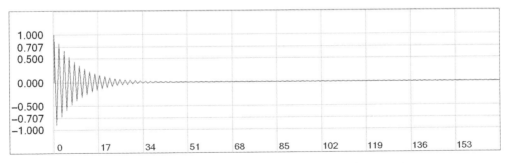

Figure 5.31: The impulse response of the filter in question.

In Figure 5.32 we observe excellent agreement with our evaluation; the response is down −6 dB at DC (0.52) and +20 dB at Nyquist (10) and the phase is 45 degrees at $\pi/2$. You followed six steps in evaluation of this filter:

1. Take the z transform of the difference equation.
2. Fashion the difference equation into a transfer function.
3. Factor out a_0 as the scalar gain coefficient.

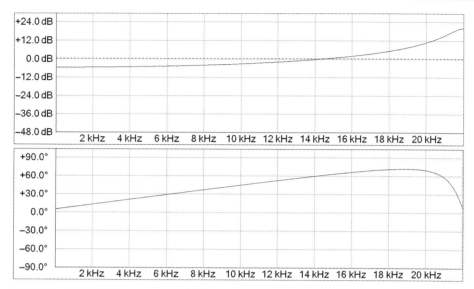

Figure 5.32: RackAFX's frequency and phase responses are taken from the z transform of the impulse response. Compare these measured results to the response we predicted using direct evaluation in Figure 5.30—notice these are plotted in dB rather than raw magnitudes.

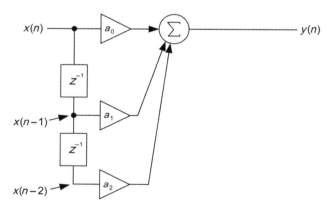

Figure 5.33: Second-order feed-forward filter.

4. Estimate the frequency response.
5. Direct evaluation of frequency response.
6. z transform of impulse response as a final check.

5.17 Second-Order Feed-Forward Filter

Analysis of the second-order feed-forward filter proceeds much like the first-order filters you've seen so far, but there's a bit more math we have to deal with. The topology of a second-order feed-forward filter is shown in the block diagram in Figure 5.33.

The difference equation is as follows:

$$y(n) = a_0 x(n) + a_1 x(n-1) + a_2 x(n-2) \tag{5.53}$$

Steps 1 & 2: Take the z transform of the difference equation and fashion it into a transfer function

We can combine steps to save time. The z transform can be taken by inspection, using the rules from Section 5.8, and then you need to get it into the form $Y(z)/X(z)$ for the transfer function in Equation 5.54.

$$\begin{aligned} y(n) &= a_0 x(n) + a_1 x(n-1) + a_2 x(n-2) \\ Y(z) &= a_0 X(z) + a_1 X(z) z^{-1} + a_2 X(z) z^{-2} \\ &= X(z)[a_0 + a_1 z^{-1} + a_2 z^{-2}] \end{aligned} \tag{5.54}$$

Form the transfer function $H(z)$:

$$H(z) = \frac{\text{output}}{\text{input}} = \frac{Y(z)}{X(z)} = a_0 + a_1 z^{-1} + a_2 z^{-2}$$

Step 3: Factor out a_0 as the scalar gain coefficient

We'll need to make some substitutions to get this in the form we are used to, shown in Equation 5.55:

$$\begin{aligned} H(z) &= a_0 + a_1 z^{-1} + a_2 z^{-2} \\ \text{Let } \alpha_1 &= \frac{a_1}{a_0} \quad \alpha_2 = \frac{a_2}{a_0} \\ H(z) &= a_0(1 + \alpha_1 z^{-1} + \alpha_2 z^{-2}) \end{aligned} \tag{5.55}$$

Step 4: Estimate the frequency response

First, this is a pure feed-forward filter, so you know there will only be nontrivial zeros; there are no poles to deal with. This transfer function is a second-order function because of the z^{-2} term. In fact, this is a quadratic equation. In order to find the poles or zeros, you need to first factor this equation and find the roots. The problem is that this is a complex equation, and the roots could be real, imaginary, or a combination of both. The mathematical break that we get is that our coefficients α_1 and α_2 are *real numbers*. The only way that could work out is if the locations of the zeros are complex conjugates of one another. When you multiply complex conjugates together the imaginary component disappears. So, with some algebra, you can arrive at the deduction shown in Equation 5.56:

$$H(z) = 1 + \alpha_1 z^{-1} + \alpha_2 z^{-2} \tag{5.56}$$

can be factored as

$$H(z) = (1 - Z_1 z^{-1})(1 - Z_2 z^{-1})$$

where

$$Z_1 = Re^{j\theta} = a + jb$$
$$Z_2 = Re^{-j\theta} = a - jb$$

This analysis results in two zeros, Z_1 and Z_2, located at complex conjugate positions in the z-plane. Figure 5.34 shows an arbitrary conjugate pair of zeros plotted in the z-plane. You can see how they are at complementary angles to one another with the same radii. Remember, any arbitrary point in the z-plane is located at $Re^{j\theta}$ and the outer rim of the circle is evaluated for $R = 1$ and θ between $-\pi$ and $+\pi$ (or 0 to 2π).

But, how do the complex conjugate pair of zeros at $Re^{j\theta}$ and $Re^{-j\theta}$ relate to the coefficients α_1 and α_2? The answer is to just multiply everything out, use Euler's equation, and compare functions as shown in Equation 5.57:

$$H(z) = a_0(1 + \alpha_1 z^{-1} + \alpha_2 z^{-2}) \qquad (5.57)$$
$$= a_0(1 - Z_1 z^{-1})(1 - Z_2 z^{-1})$$

where

$$Z_1 = Re^{j\theta}$$
$$Z_2 = Re^{-j\theta}$$

$$(1 - Z_1 z^{-1})(1 - Z_2 z^{-1}) = (1 - Re^{j\theta} z^{-1})(1 - Re^{-j\theta} z^{-1})$$
$$= 1 - Re^{j\theta} z^{-1} - Re^{-j\theta} z^{-1} + R^2(e^{j\theta} e^{-j\theta}) z^{-2}$$

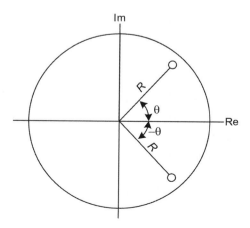

Figure 5.34: A complementary pair of zeros in the z-plane.

noting that $(e^{j\theta}e^{-j\theta}) = e^{j\theta-j\theta} = e^0 = 1$

$$= 1 - (Re^{j\theta} + Re^{-j\theta})z^{-1} + R^2z^{-2}$$
$$= 1 - R(\cos(\theta) + j\sin(\theta) + \cos(\theta) - j\sin(\theta))z^{-1} + R^2z^{-2}$$
$$= 1 - 2R\cos(\theta)z^{-1} + R^2z^{-2}$$

compare functions:

$$H(z) = a_0(1 + \alpha_1 z^{-1} + \alpha_2 z^{-2})$$
$$= a_0(1 - 2R\cos(\theta)z^{-1} + R^2 z^{-2})$$

then

$$\alpha_1 = -2R\cos(\theta)$$
$$\alpha_2 = R^2$$

Equation 5.57 shows how the coefficients α_1 and α_2 create the zeros at the locations $Re^{j\theta}$ and $Re^{-j\theta}$. Once again you see that the coefficients *are* the filter—they determine the locations of the zeros, and these determine the frequency and phase responses of the filter. To estimate, we'll need some coefficients to test with. Use the following: $a_0 = 1.0$, $a_1 = -1.27$, $a_2 = 0.81$. Now, calculate the location of the zeros from Equation 5.35; since $a_0 = 1.0$, then $\alpha_1 = -1.27$ and $\alpha_2 = 0.81$. Start with α_2 then solve for α_1 as shown in Equation 5.58. The zeros are plotted in Figure 5.35.

$$R^2 = \alpha_2 = 0.81 \tag{5.58}$$
$$R = \sqrt{0.81} = 0.9$$

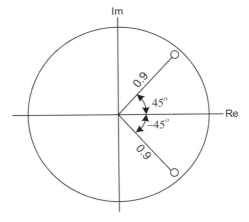

Figure 5.35: The complementary pair of zeros in the z-plane at radii 0.9 and angles ±45 degrees.

then

$$-2R\cos(\theta) = -1.27$$
$$2(0.9)\cos(\theta) = 1.27$$
$$\cos(\theta) = \frac{1.27}{2(0.9)}$$
$$\theta = \arccos(0.705)$$
$$\theta = 45°$$

Evaluating the frequency response of the complex pair is similar to before, but with an extra step. When estimating the frequency response with more than one zero:

- Locate each evaluation frequency on the outer rim of the unit circle.
- Draw a line from the point on the circle to *each* zero and measure the length of these vectors. Do it for each evaluation frequency.
- For each evaluation frequency, the magnitude of the transfer function is the *product of the two vectors* to each zero pair.

Mathematically, this last rule looks like Equation 5.59:

$$|H(e^{j\omega})|_\omega = a_0 \prod_{i=1}^{N} U_i \qquad (5.59)$$

where

N = filter order

U_i = geometric distance from the point ω on the unit circle to the ith pole

Follow the progression in Figures 5.36 through 5.39 through the four evaluation frequencies, starting at DC (0 Hz). Finally, put it all together in one plot and sketch the magnitude response in Figure 5.40.

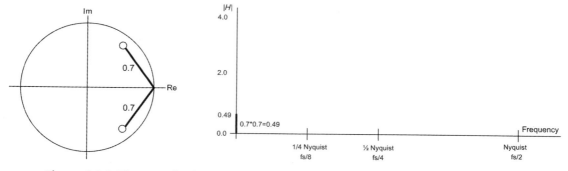

Figure 5.36: The magnitude response at 0 Hz (DC) is the product of the two vectors drawn to each zero, or 0.49.

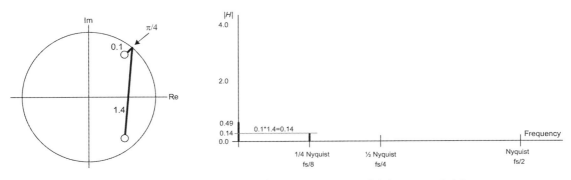

Figure 5.37: At ¼ Nyquist, the two vectors multiply out to 0.14.

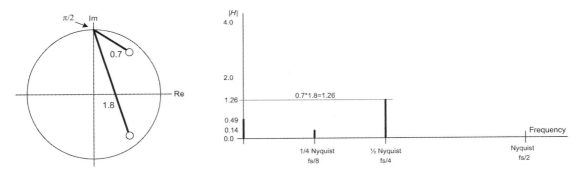

Figure 5.38: At ½ Nyquist, the response reaches 1.26 as the vectors begin to stretch out again.

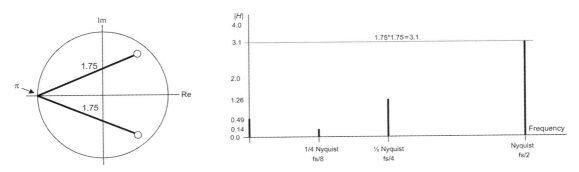

Figure 5.39: At Nyquist, the response reaches a maximum of 3.7 as the vectors stretch out to their longest possible lengths.

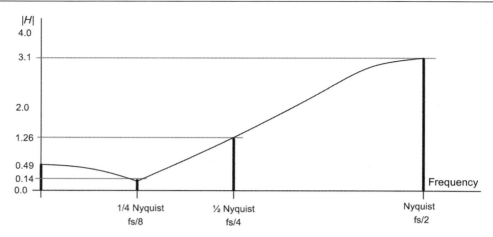

Figure 5.40: The combined response reveals a band-stop (notch) type of filter. The minimum amplitude occurs at the zero frequency, where the vector product is the lowest; this is where the smallest vector is obtained when evaluating on the positive frequency arc.

Step 5: Direct evaluation

Now you can evaluate the filter the same way as before using Euler's equation to separate the real and imaginary components from the transfer function. Evaluate at the following frequencies:

- DC: 0
- Nyquist: π
- ½ Nyquist: $\pi/2$
- ¼ Nyquist: $\pi/4$

First, get the transfer function in the form of Equation 5.60 to use for all the evaluation frequencies:

$$H(z) = a_0(1 + \alpha_1 z^{-1} + \alpha_2 z^{-2})$$

$$\text{Where: } \alpha_1 = \frac{a_1}{a_0} \quad \alpha_2 = \frac{a_2}{a_0}$$

$$H(z) = 1 - 1.27z^{-1} + 0.81z^{-2}$$

Let: $z = e^{j\omega}$ \hfill (5.60)

$$H(\omega) = 1 - 1.27e^{-j1\omega} + 0.81e^{-j2\omega}$$

Apply Euler's Equation:
$$H(\omega) = 1 - 1.27[\cos(\omega) - j\sin(\omega)] + 0.81[\cos(2\omega) - j\sin(2\omega)]$$

Now evaluate for each of our four frequencies in Equations 5.61 through 5.68.

5.17.1 DC (0 Hz)

$$\begin{aligned}
H(\omega) &= 1 - 1.27[\cos(\omega) - j\sin(\omega)] + 0.81[\cos(2\omega) - j\sin(2\omega)] \\
&= 1 - 1.27[\cos(0) - j\sin(0)] + 0.81[\cos(2*0) - j\sin(2*0)] \\
&= 1 - 1.27[1 - j0] + 0.81[1 - j0] \\
&= 1 - 1.27 + 0.81 \\
&= 0.54 + j0
\end{aligned}$$
(5.61)

$$\begin{aligned}
|H(\omega)| &= \sqrt{a^2 + b^2} \\
&= \sqrt{0.54^2 + 0^2} \\
&= 0.54 \\
\text{Arg}(H) &= \tan^{-1}(b/a) \\
&= \tan^{-1}(0/0.54) \\
&= 0.0°
\end{aligned}$$
(5.62)

The exact magnitude is 0.54, which is pretty close to our estimated value of 0.49 using the z-plane graphical method.

5.17.2 Nyquist (π)

$$\begin{aligned}
H(\omega) &= 1 - 1.27[\cos(\omega) - j\sin(\omega)] + 0.81[\cos(2\omega) - j\sin(2\omega)] \\
&= 1 - 1.27[\cos(\pi) - j\sin(\pi)] + 0.81[\cos(2\pi) - j\sin(2\pi)] \\
&= 1 - 1.27[-1 - j0] + 0.81[1 - j0] \\
&= 1 + 1.27 + 0.81 \\
&= 3.08
\end{aligned}$$
(5.63)

$$\begin{aligned}
|H(\omega)| &= \sqrt{a^2 + b^2} \\
&= \sqrt{3.08^2 + 0^2} \\
&= 3.08 \\
\text{Arg}(H) &= \tan^{-1}(b/a) \\
&= \tan^{-1}(0/3.08) \\
&= 0.0°
\end{aligned}$$
(5.64)

The exact magnitude is 3.08, which is pretty close to our estimated value of 3.1 using the z-plane graphical method.

5.17.3 ½ Nyquist (π/2)

$$\begin{aligned}
H(\omega) &= 1 - 1.27[\cos(\omega) - j\sin(\omega)] + 0.81[\cos(2\omega) - j\sin(2\omega)] \\
&= 1 - 1.27[\cos(\pi/2) - j\sin(\pi/2)] + 0.81[\cos(2\pi/2) - j\sin(2\pi/2)] \\
&= 1 - 1.27[\cos(\pi/2) - j\sin(\pi/2)] + 0.81[\cos(\pi) - j\sin(\pi)] \\
&= 1 - 1.27[0 - j1] + 0.81[-1 - j0] \\
&= 1 + j1.27 - 0.81 \\
&= 0.19 + j1.27
\end{aligned} \quad (5.65)$$

$$\begin{aligned}
|H(\omega)| &= \sqrt{a^2 + b^2} \\
&= \sqrt{0.19^2 + 1.27^2} \\
&= 1.28 \\
\mathrm{Arg}(H) &= \tan^{-1}(b/a) \\
&= \tan^{-1}(1.27/0.19) \\
&= 82°
\end{aligned} \quad (5.66)$$

The exact magnitude is 1.28, which is pretty close to our estimated value of 1.26 using the z-plane graphical method.

5.17.4 ¼ Nyquist (π/4)

$$\begin{aligned}
H(\omega) &= 1 - 1.27[\cos(\omega) - j\sin(\omega)] + 0.81[\cos(2\omega) - j\sin(2\omega)] \\
&= 1 - 1.27[\cos(\pi/4) - j\sin(\pi/4)] + 0.81[\cos(2\pi/4) - j\sin(2\pi/4)] \\
&= 1 - 1.27[\cos(\pi/4) - j\sin(\pi/4)] + 0.81[\cos(\pi/2) - j\sin(\pi/2)] \\
&= 1 - 1.27[0.707 - j0.707] + 0.81[-j1] \\
&= 0.11 + j0.08
\end{aligned} \quad (5.67)$$

$$\begin{aligned}
|H(\omega)| &= \sqrt{a^2 + b^2} \\
&= \sqrt{0.11^2 + 0.08^2} \\
&= 0.136 \\
\mathrm{Arg}(H) &= \tan^{-1}(b/a) \\
&= \tan^{-1}(0.08/0.11) \\
&= 36°
\end{aligned} \quad (5.68)$$

The exact magnitude is 0.136, which is pretty close to our estimated value of 0.14 using the z-plane graphical method.

Step 6: z transform of impulse response

This second-order feed-forward filter is actually pretty easy to examine for its impulse response. For an impulse stimulus, the impulse response $h(n)$ is $h(n) = \{1.0, -1.27, 0.81\}$. Taking the z transform of the impulse response is easy, as shown in Equation 5.69:

$$\begin{aligned} H(z) &= \sum_{n=-\infty}^{n=+\infty} h(n)z^{-n} \\ &= 1.0z^0 - 1.27z^{-1} + 0.81z^{-2} \\ &= 1 - 1.27z^{-1} + 0.81z^{-2} \end{aligned} \quad (5.69)$$

This is exactly what we expect. This should help you understand two more very important details about pure feed-forward filters.

> In a pure feed-forward filter:
> - The coefficients $\{a_0, a_1, a_2,...\}$ are the impulse response, $h(n)$.
> - The transfer function is the z transform of the coefficients.

Finally, I'll use RackAFX to verify the frequency and phase response from our analysis by using a plug-in I wrote for second-order feed-forward filters. Figure 5.41 shows the frequency and phase response plots.

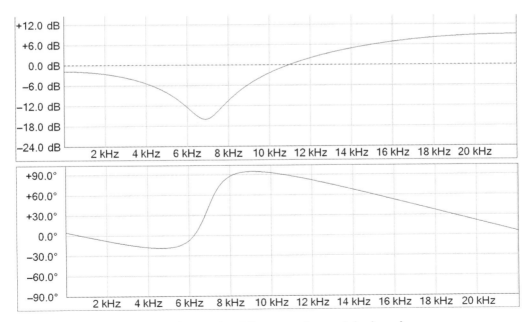

Figure 5.41: Plots using RackAFX's z-transform of the impulse response.

5.18 Second-Order Feed-Back Filter

Analysis of the second-order feed-back filter starts with the block diagram and difference equation. Figure 5.42 shows the topology of a second-order feed-back filter.

The difference equation is as follows:

$$y(n) = a_0 x(n) - b_1 y(n-1) - b_2 y(n-2) \tag{5.70}$$

Steps 1 to 3: Take the z transform of the difference equation to get the transfer function, then factor out a_0 as the scalar gain coefficient

We'll continue to combine steps. Once again, the z transform can be taken by inspection using the rules from Section 5.8, and then you need to get it into the form $Y(z)/X(z)$ for the transfer function in Equation 5.71.

$$y(n) = a_0 x(n) - b_1 y(n-1) - b_2 y(n-2)$$
$$Y(z) = a_0 X(z) - b_1 Y(z) z^{-1} - b_2 Y(z) z^{-2}$$

Separate variables:

$$Y(z) + b_1 Y(z) z^{-1} + b_2 Y(z) z^{-2} = a_0 X(z)$$
$$Y(z)[1 + b_1 z^{-1} + b_2 z^{-2}] = a_0 X(z) \tag{5.71}$$

Form transfer function:

$$H(z) = \frac{Y(z)}{X(z)} = \frac{a_0}{1 + b_1 z^{-1} + b_2 z^{-2}}$$

Factor out a_0:

$$H(z) = a_0 \frac{1}{1 + b_1 z^{-1} + b_2 z^{-2}}$$

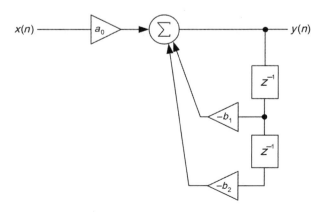

Figure 5.42: Second-order feed-back filter.

Step 4: Estimate the frequency response

Now you can see why the feed-back filter block diagrams have all the b coefficients negated. It puts the quadratic denominator in the final transfer function in Equation 5.71 in the same polynomial form as the numerator of the feed-forward transfer function. Thus, you can use the same logic to find the poles of the filter; since the coefficients b_1 and b_2 are real values, the poles must be complex conjugates of each other, as shown in Equation 5.72:

$$H(z) = a_0 \frac{1}{1 + b_1 z^{-1} + b_2 z^{-2}}$$

can be factored as

$$H(z) = a_0 \frac{1}{(1 - P_1 z^{-1})(1 - P_2 z^{-1})}$$

where

$$P_1 = Re^{j\theta}$$
$$P_2 = Re^{-j\theta}$$
$$(1 - P_1 z^{-1})(1 - P_1 z^{-1}) = 1 - 2R\cos(\theta)z^{-1} + R^2 z^{-2}$$

(5.72)

therefore

$$H(z) = a_0 \frac{1}{1 + b_1 z^{-1} + b_2 z^{-2}}$$
$$= a_0 \frac{1}{(1 - 2R\cos(\theta)z^{-1} + R^2 z^{-2})}$$

and

$$b_1 = -2R\cos(\theta)$$
$$b_2 = R^2$$

This results in two poles, P_1 and P_2, located at complex conjugate positions in the z-plane.

Figure 5.43 shows an arbitrary conjugate pair of poles plotted in the z-plane. You can see how they are at complementary angles to one another with the same radii.

To estimate we'll need some coefficients to test with. Use the following: $a_0 = 1.0$, $b_1 = -1.34$, $b_2 = 0.902$. Now, calculate the location of the poles from Equation 5.72:

$$R^2 = b_2 = 0.902 \qquad (5.73)$$
$$R = \sqrt{0.902} = 0.95$$

then

$$-2R\cos(\theta) = -1.34$$
$$2(0.95)\cos(\theta) = 1.34$$

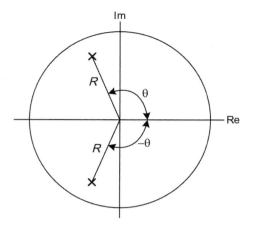

Figure 5.43: A complementary pair of poles in the z-plane.

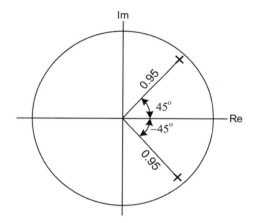

Figure 5.44: The poles of the filter.

$$\cos(\theta) = \frac{1.34}{2(0.95)}$$

$$\theta = \arccos(0.707)$$

$$\theta = 45°$$

Figure 5.44 shows the complex conjugate pair of poles plotted in the z-plane at angles $\pm 45°$ and radii of 0.95. Evaluating the frequency response of the complex pair is similar to before, but with an extra step. When estimating the frequency response with more than one pole:

- Locate each evaluation frequency on the outer rim of the unit circle.
- Draw a line from the point on the circle to *each* pole and measure the length of these vectors. Do it for each evaluation frequency.
- For each evaluation frequency, the magnitude of the transfer function is the *product of the inverse lengths* of the two vectors to each pole pair.

Mathematically, this last rule looks like Equation 5.74:

$$|H(e^{j\omega})|_\omega = a_0 \frac{1}{\prod_{i=1}^{N} V_i} \qquad (5.74)$$

where

N = the filter order
V_i = the geometric length from the point (ω) on the unit circle to the ith pole

Thus, the process is the same as with the zeros, except that you take the inverse of the length to the pole.

For feed-forward filters:

- The closer the frequency is to the zero, the more attenuation it receives.
- If the zero is *on* the unit circle, the magnitude would go to zero at that point.

For feed-back filters:

- The closer the evaluation frequency is to the pole, the more gain it receives.
- If a pole is *on* the unit circle, the magnitude would theoretically go to infinity, and it would produce an oscillator, ringing forever at the pole frequency.

You blew up the first-order feed-back filter as an exercise in Chapter 4. All feed-back filters are prone to blowing up when their poles go outside the unit circle. We can now continue with the estimation process for our standard four evaluation frequencies. This time, we'll convert the raw magnitude values into dB. The reason for doing this is that there will be a very wide range of values that will be difficult to sketch if we don't use dB. Follow the evaluation sequence in Figures 5.45 through 5.48. Finally, you can put it all together to form the frequency response plot in Figure 5.49.

In a digital filter:
- Zeros may be located anywhere in the z-plane, inside, on, or outside the unit circle since the filter is always stable; it's output can't go lower than 0.0.
- Poles must be located inside the unit circle.
- If a pole is on the unit circle, it produces an oscillator.
- If a pole is outside the unit circle, the filter blows up as the output goes to infinity.

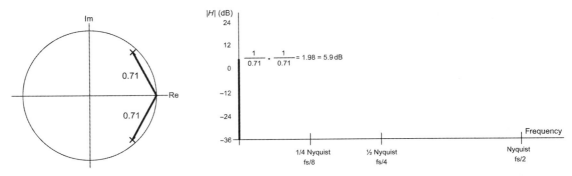

Figure 5.45: The magnitude response at 0 Hz (DC) is the product of the inverse of the two vectors drawn to each zero, or $(1/0.71)(1/0.71) = 5.9$ dB.

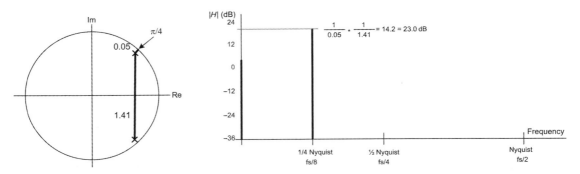

Figure 5.46: The magnitude response at 1/4 Nyquist is a whopping +23 dB since the inverse of 0.05 is a large number.

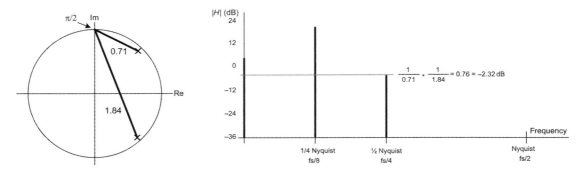

Figure 5.47: The magnitude response at $\pi/2$ is -2.98 dB.

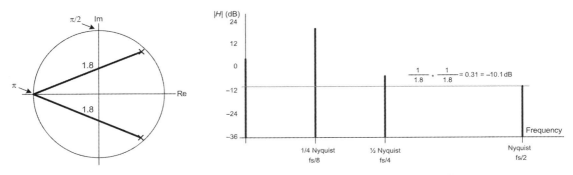

Figure 5.48: The magnitude response at Nyquist is −10.1 dB.

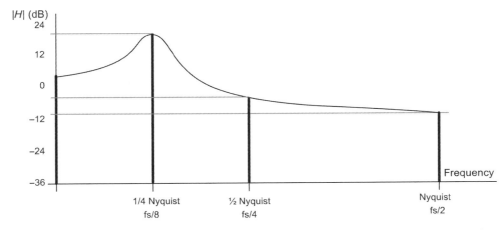

Figure 5.49: The composite magnitude response of the filter shows that it is a resonant low-pass filter; the resonant peak occurs at the pole frequency.

Step 5: Direct evaluation

Now you can evaluate the filter the same way as before using Euler's equation to separate the real and imaginary components from the transfer function. Evaluate at the following frequencies:

- DC: 0
- Nyquist: π
- ½ Nyquist: $\pi/2$
- ¼ Nyquist: $\pi/4$

First, get the transfer function in a form to use for all the evaluation frequencies:

$$H(z) = a_0 \frac{1}{1 + b_1 z^{-1} + b_2 z^{-2}} \tag{5.75}$$

$$= \frac{1}{1 - 1.34 z^{-1} + 0.902 z^{-2}}$$

Let $z = e^{j\omega}$.

$$H(\omega) = \frac{1}{1 - 1.34e^{-j1\omega} + 0.902e^{-j2\omega}}$$

Apply Euler's equation:

$$H(\omega) = \frac{1}{1 - 1.34e^{-j1\omega} + 0.902e^{-j2\omega}}$$

$$H(\omega) = \frac{1}{1 - 1.34[\cos(\omega) - j\sin(\omega)] + 0.902[\cos(2\omega) - j\sin(2\omega)]}$$

Now evaluate for each of our four frequencies starting with DC.

5.18.1 DC (0 Hz)

$$H(\omega) = \frac{1}{1 - 1.34[\cos(\omega) - 1 - j0\, j\sin(\omega)] + 0.902[\cos(2\omega) - j\sin(2\omega)]}$$

$$= \frac{1}{1 - 1.34[\cos(0) - j\sin(0)] + 0.902[\cos(2*0) - j\sin(2*0)]}$$

$$= \frac{1}{1 - 1.34[1 - j0] + 0.902} \quad (5.76)$$

$$= \frac{1}{1 - 1.34 + 0.902}$$

$$= \frac{1}{0.562 + j0}$$

$$H(\omega) = \frac{1}{0.562 + j0}$$

$$|H(\omega)| = \frac{|1|}{|0.562 + j0|}$$

$$= \frac{1}{\sqrt{a^2 + b^2}} \quad (5.77)$$

$$= \frac{1}{\sqrt{0.562^2}} = 1.78 = 5.00 \text{ dB}$$

$$\text{Arg}(H) = \text{Arg}(\text{Num}) - \text{Arg}(\text{Denom})$$
$$= \tan^{-1}(0/1) - \tan^{-1}(0/0.562)$$
$$= 0°$$

Remember that for magnitudes of fractions, you need to take the magnitude of the numerator and denominator separately; also for phase, the total is the difference of the Arg(num) and Arg(denom). The direct evaluation yields 5.0 dB and shows our sketch evaluation was a little off at 5.9 dB.

Table 5.4: Challenge answers.

Frequency (ω)	$|H(\omega)|$	Arg(H)
Nyquist (π)	−10.2 dB	0.0°
½ Nyquist ($\pi/2$)	−2.56 dB	−85.8°
¼ Nyquist ($\pi/4$)	+23.15 dB	−40.3°

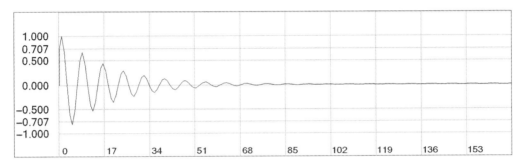

Figure 5.50: Impulse response of the filter from RackAFX.

5.18.2 Challenge

Finish the rest of the direct evaluation calculations on your own. The answers are in Table 5.4. The exact evaluation once again produces results pretty close to our estimated version. You can get the complete answers at the RackAFX websites.

Step 6: z transform of impulse response

The impulse response for this filter is shown in Figure 5.50. Once again, finding the impulse response by hand is going to be tedious. Instead, I will use RackAFX to do the analysis so we can compare our estimated and direct evaluation results. The measured responses are shown in Figure 5.51.

5.19 First-Order Pole-Zero Filter: The Shelving Filter

The first-order pole-zero filter consists of a first-order pole and first-order zero in the same algorithm. The topology in the block diagram in Figure 5.52 is a combination feed forward and feed back since it contains both paths.

The difference equation is as follows:

$$y(n) = a_0 x(n) + a_1 x(n - 1) - b_1 y(n - 1) \qquad (5.78)$$

Steps 1 to 3: Take the z transform of the difference equation to get the transfer function, then factor out a_0 as the scalar gain coefficient

$$Y(z) = a_0 X(z) + a_1 X(z)z^{-1} - b_1 Y(z)z^{-1} \qquad (5.79)$$

Separate variables:

$$Y(z) + b_1Y(z)z^{-1} = a_0X(z) + a_1X(z)z^{-1}$$
$$Y(z)[1 + b_1z^{-1}] = X(z)[a_0 + a_1z^{-1}]$$

Form the transfer function:

$$H(z) = \frac{Y(z)}{X(z)} = \frac{a_0 + a_1z^{-1}}{1 + b_1z^{-1}}$$

Factor out a_0:

$$H(z) = a_0\frac{1 + \alpha_1z^{-1}}{1 + b_1z^{-1}}$$

where

$$\alpha_1 = \frac{a_1}{a_0}$$

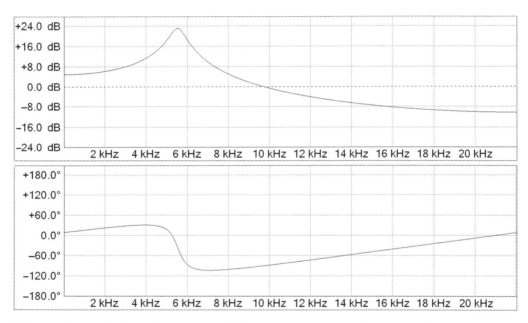

Figure 5.51: RackAFX's frequency and phase responses are taken from the z-transform of the impulse response.

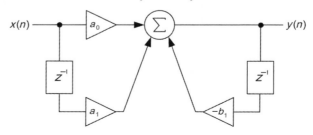

Figure 5.52: First-order pole-zero filter.

Step 4: Estimate the frequency response

This transfer function has one pole and one zero and both are first order. Like the other first-order cases, we can find the pole and zero by inspection of the transfer function:

$$H(z) = a_0 \frac{1 + \alpha_1 z^{-1}}{1 + b_1 z^{-1}} \qquad (5.80)$$

$$= a_0 \frac{1 + \dfrac{\alpha_1}{z}}{1 + \dfrac{b_1}{z}}$$

In the numerator, you can see that if $z = -\alpha_1$ the numerator will go to zero and the transfer function will go to zero. In the denominator, you can see that if $z = -b_1$ the denominator will go to zero and the transfer function will go to infinity. Therefore we have a zero at $z = -\alpha_1$ and a pole at $z = -b_1$. For this example, use the following values for the coefficients: $a_0 = 1.0$, $a_1 = -0.92$, $b_1 = -0.71$. Then, $\alpha_1 = -0.92$, and so we now have a zero at $z = -\alpha_1 = 0.92 + j0$ and a pole at $z = -b_1 = 0.71 + j0$. The pole/zero pair are plotted in Figure 5.53.

Evaluating the frequency response when you have mixed poles and zeros is the same as before, but you have to implement both magnitude steps.

- Locate each evaluation frequency on the outer rim of the unit circle.
- Draw a line from the point on the circle to *each* zero and measure the length of these vectors. Do it for each evaluation frequency.
- Draw a line from the point on the circle to *each* pole and measure the length of these vectors. Do it for each evaluation frequency.

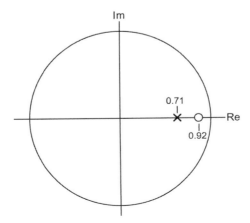

Figure 5.53: The pole and zero are both purely real and plotted on the real axis in the z-plane.

- Multiply all the zero magnitudes together.
- Multiply all the *inverse* pole magnitudes together.
- Divide the zero magnitude by the pole magnitude for the final result at that frequency.

Mathematically, this last rule looks like Equation 5.81:

$$|H(e^{j\omega})|_\omega = a_0 \frac{\prod_{i=1}^{N} U_i}{\prod_{i=1}^{N} V_i} \qquad (5.81)$$

where

N = the filter order
U_i = the geometric length from the point (ω) on the unit circle to the ith zero
V_i = the geometric length from the point (ω) on the unit circle to the ith pole

Equation 5.81 is the final, generalized magnitude response equation for the geometric interpretation of the pole/zero plots in the z-plane. For completeness, here's the equation for calculating the phase response of a digital filter using the geometric method:

$$\text{Arg}(H(e^{j\omega}))|_\omega = \sum_{i=1}^{N} \theta_i - \sum_{i=1}^{N} \phi_i \qquad (5.82)$$

where

N = the filter order
θ_i = the angle between the ith zero and the vector U_i
ϕ_i = the angle between the ith pole and the vector V_i

Equations 5.81 and 5.82 together complete the DSP theory for pole/zero interpretation for estimating the frequency and phase responses of any filter. So, let's do the analysis for this filter—by going through the math, you can see the tug of war going on between the pole and zero. Follow the analysis sequence in Figures 5.54 through 5.58.

Step 5: Direct evaluation

You can evaluate the filter the same way as before using Euler's equation to separate the real and imaginary components from the transfer function. Evaluate at the following frequencies:

- DC: 0
- Nyquist: π
- ½ Nyquist: $\pi/2$
- ¼ Nyquist: $\pi/4$

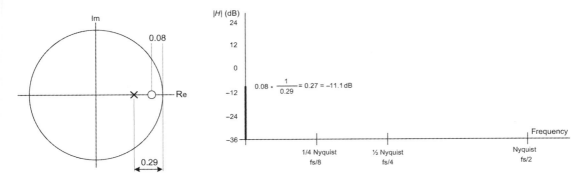

Figure 5.54: The magnitude response at DC is −11.1 dB. Look at the equation and you can see the zero value bringing down the total while the pole value is trying to push it back up. In this case, the zero wins and the response is down almost 12 dB. Geometrically, you can see this is because the zero is closer to the evaluation point and so it has more effect on the outcome.

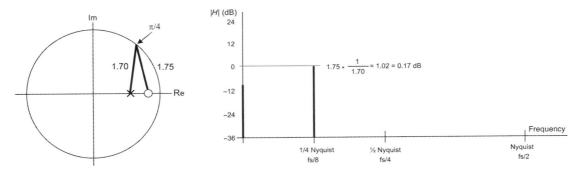

Figure 5.55: The magnitude response at π is almost unity gain because the pole and zero distances are almost the same. The tug of war ends in stalemate here at 0.17 dB of gain.

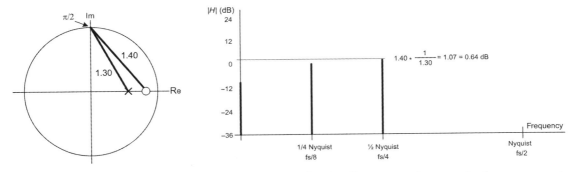

Figure 5.56: With the pole slightly closer to the evaluation frequency, the magnitude response at $\pi/2$ picks up a bit to +0.64 dB.

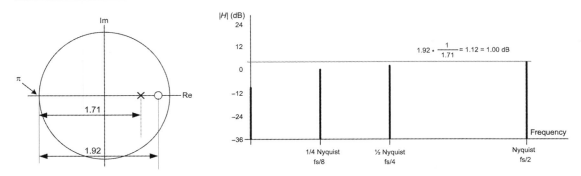

Figure 5.57: At $\pi/4$ the pole/zero ratio favors the pole and the response perks up to 1.0 dB; notice that this is the frequency where the pole is clearly dominating, but just barely.

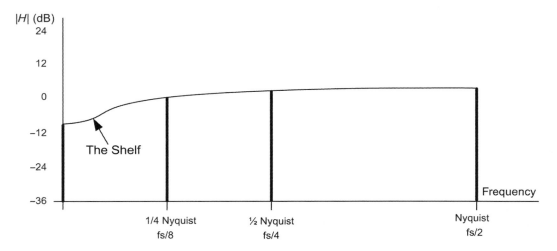

Figure 5.58: The composite frequency response plot shows a -12 dB low shelving filter response, a useful filter in audio.

First, get the transfer function in a form to use for all the evaluation frequencies (Equation 5.83). Then evaluate at our four frequencies.

$$H(z) = \frac{a_0 + a_1 z^{-1}}{1 + b_1 z^{-1}} \tag{5.83}$$

$$= \frac{1 - 0.92 z^{-1}}{1 - 0.71 z^{-1}}$$

Let $z = e^{j\omega}$

$$H(\omega) = \frac{1 - 0.92 e^{-j1\omega}}{1 - 0.71 e^{-j1\omega}}$$

Apply Euler's equation:

$$H(\omega) = \frac{1 - 0.92e^{-j1\omega}}{1 - 0.71e^{-j1\omega}}$$

$$H(\omega) = \frac{1 - 0.92[\cos(\omega) - j\sin(\omega)]}{1 - 0.71[\cos(\omega) - j\sin(\omega)]}$$

5.19.1 DC (0Hz)

$$\begin{aligned}
H(\omega) &= \frac{1 - 0.92[\cos(\omega) - j\sin(\omega)]}{1 - 0.71[\cos(\omega) - j\sin(\omega)]} \\
&= \frac{1 - 0.92[\cos(0) - j\sin(0)]}{1 - 0.71[\cos(0) - j\sin(0)]} \\
&= \frac{1 - 0.92[1 - j0]}{1 - 0.71[1 - j0]} \\
&= \frac{0.08 + j0}{0.29 + j0}
\end{aligned} \quad (5.84)$$

$$\begin{aligned}
|H(\omega)| &= \frac{|0.08 + j0|}{|0.29 + j0|} \\
&= \frac{\sqrt{a^2 + b^2}}{\sqrt{a^2 + b^2}} \\
&= \frac{\sqrt{0.08^2}}{\sqrt{0.29^2}} = 0.276 = -11.2 \text{ dB}
\end{aligned} \quad (5.85)$$

$$\begin{aligned}
\text{Arg}(H) &= \text{Arg(Num)} - \text{Arg(Denom)} \\
&= \tan^{-1}(0/0.08) - \tan^{-1}(0/0.29) \\
&= 0°
\end{aligned}$$

By now this should be getting very familiar; the only difference in this situation is that we have to evaluate the numerator and denominator, but the method is the same.

15.19.2 Challenge

Finish the rest of the direct evaluation calculations on your own. The answers are in Table 5.5.

Table 5.5: Challenge answers.

Frequency (ω)	$|H(\omega)|$	Arg(H)
Nyquist (π)	1.00 dB	0.0°
½ Nyquist ($\pi/2$)	0.82 dB	7.23°
¼ Nyquist ($\pi/4$)	0.375 dB	16.60°

Thus, once again the direct evaluation backs up the estimation from the z-plane. Because we have a feedback path, extracting the impulse response will be tedious but we can use RackAFX's pole/zero filter module to analyze the impulse response. Figure 5.59 shows the measured impulse response, while Figure 5.60 shows the frequency and phase responses.

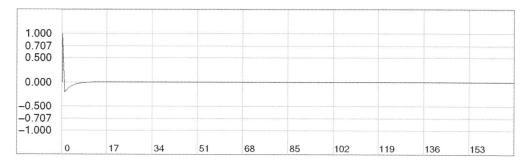

Figure 5.59: Impulse response of the first-order shelving filter.

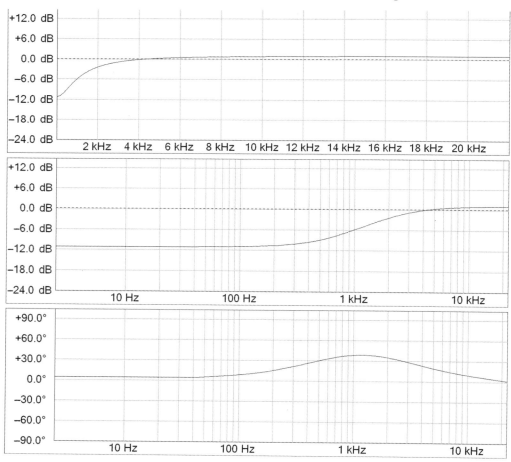

Figure 5.60: Frequency and phase responses of the first-order shelving filter.

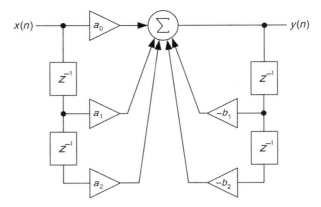

Figure 5.61: The bi-quad.

A first-order shelving filter is a pole-zero design. The shelf will be located in the region between where the zero dominates and where the pole dominates. When neither really dominates, the response is approximately unity gain. The RackAFX frequency and phase plots are shown in Figure 5.60; the log frequency plot has been included to reveal the textbook shelving filter curve.

5.20 The Bi-Quadratic Filter

The last filter topology to study is the bi-quadratic (or bi-quad) filter. The bi-quad consists of two second-order components: a second-order feed-forward and a second-order feed-back filter combined together as shown in the block diagram in Figure 5.61. The resulting transfer function will have two quadratic equations, thus the name.

The difference equation is as follows:

$$y(n) = a_0 x(n) + a_1 x(n-1) + a_2 x(n-2) - b_1 y(n-1) - b_2 y(n-2) \quad (5.86)$$

Steps 1 to 3: Take the z transform of the difference equation to get the transfer function, then factor out a_0 as the scalar gain coefficient

$$y(n) = a_0 x(n) + a_1 x(n-1) + a_2 x(n-2) - b_1 y(n-1) - b_2 y(n-2)$$
$$Y(z) = a_0 X(z) + a_1 X(z) z^{-1} + a_2 X(z) z^{-2} - b_1 Y(z) z^{-1} - b_2 Y(z) z^{-2}$$

Separate variables:

$$Y(z) + b_1 Y(z) z^{-1} + b_2 Y(z) z^{-2} = a_0 X(z) + a_1 X(z) z^{-1} + a_2 X(z) z^{-2}$$
$$Y(z)[1 + b_1 z^{-1} + b_2 z^{-2}] = X(z)[a_0 + a_1 z^{-1} + a_2 z^{-2}]$$

Form transfer function

$$H(z) = \frac{Y(z)}{X(z)} = \frac{a_0 + a_1 z^{-1} + a_2 z^{-2}}{1 + b_1 z^{-1} + b_2 z^{-2}}$$

Factor out a_0

$$H(z) = a_0 \frac{1 + \alpha_1 z^{-1} + \alpha_2 z^{-2}}{1 + b_1 z^{-1} + b_2 z^{-2}}$$

where (5.87)

$$\alpha_1 = \frac{a_1}{a_0}$$

$$\alpha_2 = \frac{a_2}{a_0}$$

Step 4: Plot the poles and zeros of the transfer function

The bi-quad will produce a conjugate pair of zeros and conjugate pair of poles from the numerator and denominator respectively. Calculating these locations is the same as in the pure second-order feed forward and feed back topologies. All you need to do is plot them in the same unit circle. The transfer function becomes (by simple substitution from previous sections):

$$H(z) = a_0 \frac{1 - 2R_z \cos(\theta) z^{-1} + R_z^2 z^{-2}}{1 - 2R_p \cos(\phi) z^{-1} + R_p^2 z^{-2}} \tag{5.88}$$

Figure 5.62 shows a pair of poles and a pair of zeros plotted together. Each has its own radius, R_z and R_p, and angle, θ and φ. The same kind of pole/zero tug of war goes on with the bi-quad, only now there are more competing entities.

Estimating the frequency response is complicated by the additional poles and zeros, but the rules are still the same:

- Locate each evaluation frequency on the outer rim of the unit circle.
- Draw a line from the point on the circle to *each* zero and measure the length of these vectors. Do it for each evaluation frequency.

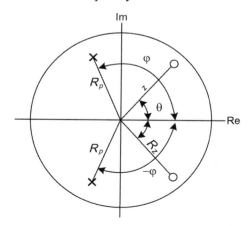

Figure 5.62: Second-order poles and zeros.

- Draw a line from the point on the circle to *each* pole and measure the length of these vectors. Do it for each evaluation frequency.
- Multiply all the zero magnitudes together.
- Multiply all the *inverse* pole magnitudes together.
- Divide the zero magnitude by the pole magnitude for the final result at that frequency.

Using the following coefficients, $a_0 = 1.0$, $a_1 = 0.73$, $a_2 = 1.00$, $b_1 = -0.78$, $b_2 = 0.88$, we can directly find the pole and zero locations from Equation 5.88 (note that because a_0 is 1.0, you don't have to calculate the α terms). The pairs of poles and zeros are plotted in Figure 5.63.

Zeros are calculated as follows:

$$a_2 = R_z^2 = 1.00$$
$$R_z = \sqrt{1.00} = 1.00$$

and

$$a_1 = -2R\cos(\theta) = 0.73$$ (5.89)
$$2(1.00)\cos(\theta) = 0.73$$
$$\cos(\theta) = 0.365$$
$$\theta = \arccos(0.365)$$
$$\theta = 68.6°$$

Poles are calculated as follows:

$$b_2 = R_p^2 = 0.88$$
$$R_p = \sqrt{0.88} = 0.94$$

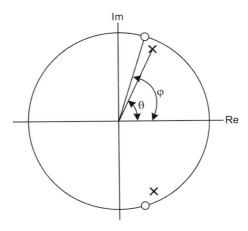

Figure 5.63: Pole/zero plot for the example bi-quad filter.

and

$$b_1 = -2R\cos(\phi) = -0.78$$
$$2(0.94)\cos(\phi) = 0.78$$
$$\cos(\phi) = \frac{0.78}{2(0.94)} \quad (5.90)$$
$$\phi = \arccos(0.414)$$
$$\phi = 65.5°$$

The poles and zeros are in close proximity to each other. The zero is directly on the unit circle ($R_z = 1.0$), so we expect a notch to occur there. The pole is near the unit circle but not touching it, so we expect a resonance there.

We are not going to go through the full response estimation or direct evaluation since it's just repetitive algebra at this point. But, we can use RackAFX's bi-quad module to set up the filter and evaluate it (Figure 5.64). The frequency response clearly shows the resonant peaking due to the pole, then the sudden notch due to the nearby zero.

In Figure 5.64, it's easy to locate the places where the pole or zero dominates. In the low frequencies, the pole dominates and at high frequencies the zero dominates. This is an example of a direct z-plane design where we place a pole and zero pair directly in the z-plane, then calculate the coefficients. In the next chapter, we will examine some basic Infinite Impulse Response (IIR) design techniques and the direct z-plane method will be the first.

Final Review Notes

In this chapter you learned the basics of DSP theory; specifically, you learned the sequence:

1. Take the z transform of the difference equation.
2. Fashion the difference equation into a transfer function.
3. Factor out a_0 as the scalar gain coefficient.

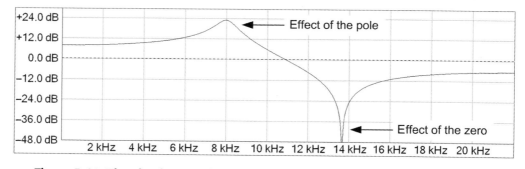

Figure 5.64: The plot from RackAFX shows the resonant peak and notch formed by the pole and zero.

4. Estimate the frequency response.
5. Direct evaluation of frequency response.
6. z transform of impulse response as a final check.

For geometric estimation, the frequency response and phase responses of a filter can be found with Equations 5.91 and 5.92.

$$|H(e^{j\omega})|_\omega = a_0 \frac{\prod_{i=1}^{N} U_i}{\prod_{i=1}^{N} V_i} \qquad (5.91)$$

where

N = the filter order
U_i = the geometric length from the point (ω) on the unit circle to the ith zero
V_i = the geometric length from the point (ω) on the unit circle to the ith pole

$$\mathrm{Arg}(H(e^{j\omega}))|_\omega = \sum_{i=1}^{N} \theta_i - \sum_{i=1}^{N} \phi_i$$

where (5.92)

N = the filter order
θ_i = the angle between the ith zero and the vector U_i
ϕ_i = the angle between the ith pole and the vector V_i

For direct evaluation, you simply plug in various values for frequency and crank through the algebra. We applied this same routine to feed-forward, feed-back, and combination types of algorithms, and then used RackAFX to check our results. We also classified the filters into IIR and Finite Impulse Response (FIR) types.

IIR Filters

- Any filter with a feed-back path is IIR in nature, even if it has feed-forward branches as well.
- The feed-back paths in IIR filters produce poles in the z-plane and the poles cause gain to occur in the magnitude response.
- An IIR filter can blow up when its output steadily increases toward infinity, which occurs when the poles are located outside the unit circle.
- If the IIR filter also has feed-forward branches it will produce zeros as well as poles.
- IIR filters can have fast transient responses but may ring.

FIR Filters

- The FIR filter only has feed-forward paths.
- It only produces zeros in the z-plane.

- The FIR filter is unconditionally stable since its output can only go to zero in the worst case.
- FIR filters will have slower transient responses because of the time smearing they do on the impulse response.
- The more delay elements in the FIR, the poorer the transient response becomes.

Bibliography

Ifeachor, E. C. and Jervis, B. W. 1993. *Digital Signal Processing: A Practical Approach*, Chapter 3. Menlo Park, CA: Addison-Wesley.

Kwakernaak, H. and Sivan, R. 1991. *Modern Signals and Systems*, Chapter 3. Englewood Cliffs, NJ: Prentice-Hall.

Moore, R. 1990. *Elements of Computer Music*, Chapter 2. Englewood Cliffs, NJ: Prentice-Hall.

Oppenheim, A. V. and Schafer, R. W. 1999. *Discrete-Time Signal Processing* (2nd ed.), Chapter 3. Englewood Cliffs, NJ: Prentice-Hall.

Orfanidis, S. 1996. *Introduction to Signal Processing*, Chapters 10–11. Englewood Cliffs, NJ: Prentice-Hall.

Steiglitz, K. 1996. *A DSP Primer with Applications to Digital Audio and Computer Music*, Chapters 4–5. Menlo Park, CA: Addison-Wesley.

CHAPTER 6
Audio Filter Designs: IIR Filters

It's time to put the theory into practice and make some audio filters and equalizers (EQs). You know that the coefficients of a filter determine its frequency response and other characteristics. But how do you find the coefficients? There are two fundamental ways to find the coefficients of the infinite impulse response (IIR) filter:

- Direct z-plane design
- Analog filter to digital filter conversion

This chapter uses the following filter naming conventions:

- LPF: Low-pass filter
- HPF: High-pass filter
- BPF: Band-pass filter
- BSF: Band-stop filter

6.1 Direct z-Plane Design

In this first category of design techniques, you manipulate the poles and zeros directly in the z-plane to create the response you want. You take advantage of the simple equations that relate the coefficients to the pole/zero locations. Consider the bi-quad. Equation 6.1 shows the numerator. Equation 6.2 shows the denominator.

$$H(z) = a_0(1 + \alpha_1 z^{-1} + \alpha_2 z^{-2}) \\ = a_0(1 - 2R\cos(\theta)z^{-1} + R^2 z^{-2}) \tag{6.1}$$

then

$$\alpha_1 = -2R\cos(\theta)$$
$$\alpha_2 = R^2$$

$$H(z) = a_0 \left[\frac{1}{1 + b_1 z^{-1} + b_2 z^{-2}} \right]$$
$$= a_0 \left[\frac{1}{(1 - 2R\cos(\theta)z^{-1} + R^2 z^{-2})} \right] \quad (6.2)$$

then
$$b_1 = -2R\cos(\theta)$$
$$b_2 = R^2$$

For the numerator or denominator, the a_1 or b_1 coefficients are in direct control over the angles of the zeros or poles. The distance R to the zeros or poles is determined by both a_1, a_2 or b_1, b_2. For first-order filters, the coefficients only control the location of the pole and zero on the real axis. There are no conjugate pairs. However, careful placement of the pole and zero can still result in useful audio filters.

6.2 Single Pole Filters

A block diagram of a single pole filter is shown in Figure 6.1.

The difference equation is as follows:

$$y(n) = a_0 x(n) - b_1 y(n-1) \quad (6.3)$$

6.2.1 First-Order LPF and HPF

Specify:

- f_c, the corner frequency

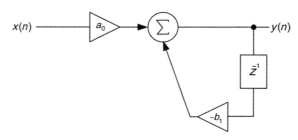

Figure 6.1: The first-order feed-back filter and difference equation.

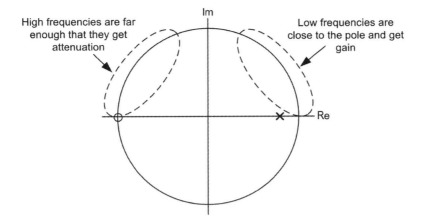

Figure 6.2: The first-order LPF has a single pole and zero on the real axis.

The design equations are as follows:

$$\begin{array}{ll} \text{LPF} & \text{HPF} \\ \theta_c = 2\pi f_c/f_s & \theta_c = 2\pi f_c/f_s \\ \gamma = 2 - \cos(\theta_c) & \gamma = 2 + \cos(\theta_c) \\ b_1 = \sqrt{\gamma^2 - 1} - \gamma & b_1 = \gamma - \sqrt{\gamma^2 - 1} \\ a_0 = 1 + b_1 & a_0 = 1 - b_1 \end{array} \qquad (6.4)$$

These simple first-order filters work by moving the pole back and forth across the real axis while holding a fixed zero at Nyquist. When the pole is on the right side of the unit circle, low frequencies are gained up due to their close proximity to the pole. High frequencies are close to the zero and get attenuated (Figure 6.2). This produces the low-pass response in Figure 6.3.

When the pole/zero pair are reversed, the opposite effect happens and the high frequencies are now boosted with low frequencies attenuated. There is also a zero at 0 Hz. This produces the high-pass response. These are simple but very useful designs and are found in many effects including delay and reverb algorithms.

6.3 Resonators

A resonator is a band-pass filter that can be made to have a very narrow peak. The simple version uses a second-order feed-back topology.

6.3.1 Simple Resonator

A block diagram of a simple resonator is shown in Figure 6.4.

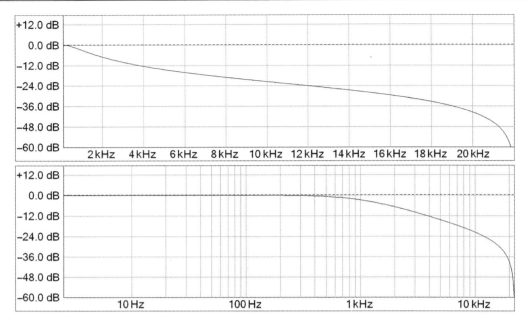

Figure 6.3: Linear and log frequency responses of the one-pole low-pass filter with f_c = 1 kHz.

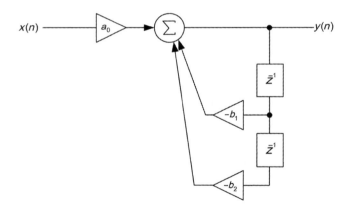

Figure 6.4: Second-order feed-back filter block diagram.

The difference equation is as follows:

$$y(n) = a_0 x(n) - b_1 y(n-1) - b_2 y(n-2) \qquad (6.5)$$

Specify:

- f_c, center frequency
- BW, 3 dB bandwidth; or Q, quality factor

The design equations are as follows:

$$\theta_c = 2\pi f_c/f_s$$
$$W = f_c/Q$$
$$b_2 = e^{\left(-2\pi \frac{BW}{f_s}\right)}$$
$$b_1 = \frac{-4b_2}{1+b_2}\cos(\theta_c)$$
$$a_0 = (1-b_2)\sqrt{1 - \frac{b_1^2}{4b_2}}$$

(6.6)

The resonator works by simple manipulation of the conjugate poles formed with the second-order feed-back network. The b_2 term controls the distance out to the pole which makes the resonant peak sharper (when the pole is close to the unit circle) or wider (when the pole is farther away from the unit circle). The b_1 and b_2 terms control the angle of the pole, which controls the center frequency, shown in Figure 6.5. The a_0 term is used to scale the filter so its peak output is always normalized to unity gain or 0 dB.

The drawback to this design is that it is only symmetrical in frequency response at one frequency, $\pi/2$, when the low-frequency and high-frequency magnitude vectors are symmetrical. At all other frequencies, the response is shifted asymmetrically. When the pole is in the first quadrant, it is nearer to the low frequencies than high frequencies and so the

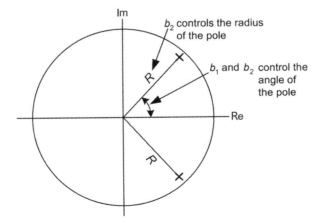

Figure 6.5: The location of the resonator's conjugate poles are determined by the coefficients.

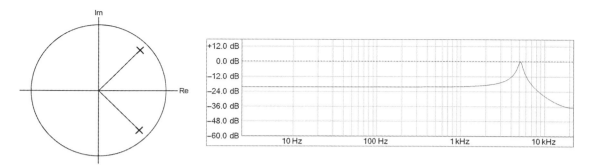

Figure 6.6: The asymmetrical response shape when the pole is in the first quadrant; notice the difference in gain at DC versus Nyquist. This filter has f_s = 44.1 kHz, $f_c = \pi/4$ = 5.5125 kHz, and Q = 10.

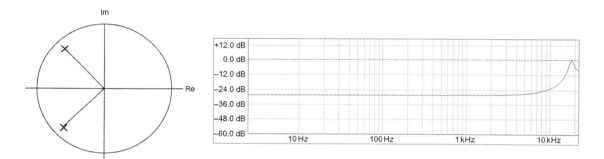

Figure 6.7: The asymmetrical response shape when the pole is in the second quadrant; notice the difference in gain at DC versus Nyquist. This filter has f_s = 44.1 kHz, $f_c = 3\pi/4$ = 16.5375 kHz, and Q = 10.

low end is boosted (Figure 6.6). The opposite happens (Figure 6.7) when the pole moves to the second quadrant and the high frequencies get the boost. The solution to the asymmetry problem is a really useful filter, especially for making filter banks of extremely narrow bandpass filters.

6.3.2 Smith-Angell Improved Resonator

A block diagram of a Smith-Angell improved resonator is shown in Figure 6.8.

The difference equation is as follows:

$$y(n) = a_0 x(n) + a_2 x(n-2) - b_1 y(n-1) - b_2 y(n-2) \tag{6.7}$$

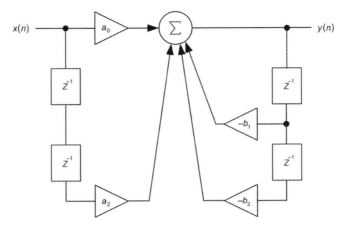

Figure 6.8: Block diagram for the Smith-Angell improved resonator.

Specify:

- f_c, center frequency
- BW, 3 dB bandwidth; or Q, quality factor

The design equations are as follows:

$$\theta_c = 2\pi f_c / f_s$$
$$BW = f_c / Q$$
$$b_2 = e^{\left(-2\pi \frac{BW}{f_s}\right)}$$
$$b_1 = \frac{-4 b_2}{1 + b_2} \cos(\theta_c) \quad (6.8)$$
$$a_0 = 1 - \sqrt{b_2}$$
$$a_2 = -a_0$$

This design is also gain normalized with a_0 and as before, the radius is set with b_2 first, then b_1 is calculated using the b_2 coefficient and the desired pole frequency. The filter is not truly normalized to 0.0 dB; there is actually a slight fluctuation of less than 1 dB.

The improved resonator is also named the Smith-Angell resonator, after its inventors. They improved the design by dropping two zeros into the filter, one at $z = 1$ and one at $z = -1$ in order to pin down DC and Nyquist with a response of zero. This forces the filter to become somewhat more symmetric (but not entirely) and has the advantage of making the band pass even more selective in nature (Figure 6.9).

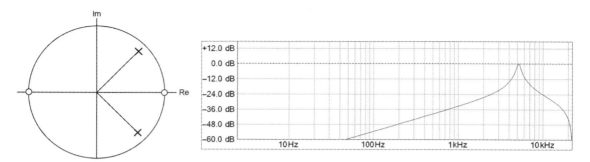

Figure 6.9: The effect on the resonator shape with the added zeros to hold down DC and Nyquist. This filter has f_s = 44.1 kHz, f_c = $\pi/4$ = 5.5125 kHz, and Q = 20 (notice how close the poles are to the unit circle).

6.4 Analog Filter to Digital Filter Conversion

A more widely used approach to filter design is to first start with the classic analog designs and then convert them into digital versions. There's good reason to do this because there are many excellent analog designs already done for us. We just need a way to make them work in our digital world. While analog filter design is outside the scope of the book, there are many similarities between the two design worlds. For example, both analog and digital filter design involve a transfer function that you manipulate to produce poles and zeros. They also both use a transform to get from the time domain to the frequency domain. A fundamental difference is that in analog, there is no Nyquist restriction and all frequencies from $-\infty$ to $+\infty$ are included. Also, in analog, reactive components or circuits like inductors, capacitors, and gyrators (a circuit that simulates an inductor) are used to create the phase shift that is at the heart of the filtering operation. Table 6.1 summarizes the similarities and differences.

Table 6.1: The differences between analog and digital filter design technique.

Digital	Analog
• Uses a transfer function to relate I/O • Delay elements create phase shift • Uses the z transform (sampled time to frequency) • Poles and zeros in the z-plane • Nyquist limitation • Poles must be inside the unit circle for stable operation	• Uses a transfer function to relate I/O • Reactive components create phase shift • Uses the Laplace transform (continuous time to frequency) • Poles and zeros in the s-plane • All frequencies from $-\infty$ to $+\infty$ allowed • Poles must be in the left-hand part of the s-plane for stable operation

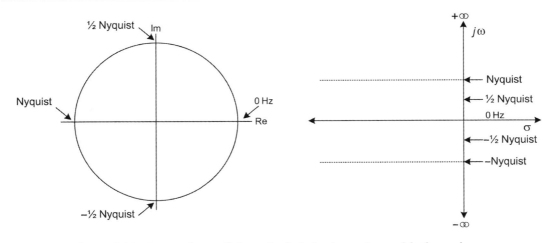

Figure 6.10: Comparison of the unit circle in the z-plane with the s-plane.

In Figure 6.10 you can see the *s*-plane on the right—it is also a complex plane. The real axis is named the σ axis and the imaginary axis is the $j\omega$ axis. The $j\omega$ axis is the frequency axis and it spans $-\infty$ to $+\infty$ rad/sec. The unit circle maps to the portion on the $j\omega$ axis between $-$Nyquist and $+$Nyquist. In order to transform an analog filter design into a digital filter design, we need a mapping system to get the poles and zeros from the *s*-plane to become poles and zeros in the *z*-plane. Once we have that, the coefficients that produce those poles and zeros in the digital locations can be calculated. In the analog world, poles that are close to the $j\omega$ axis will result in poles that are close to the unit circle in the digital world. The problem is that poles and zeros can exist anywhere along the $j\omega$ axis, even at frequencies outside the digital Nyquist zone. It is common for analog designs to have a pole or a zero at $-\infty$ and/or $+\infty$ on either the σ and/or $j\omega$ axes.

In Figure 6.11 you can see the problem: the pair of zeros close to the $j\omega$ axis at ½ Nyquist will map to locations close to the unit circle at the ½ Nyquist angles. But what about the pair of poles that are outside the Nyquist frequency in the *s*-plane? The other way to think about the issue is that in the analog *s*-plane, the entire left-hand side, including all the infinite frequencies, must map to the *interior* of the unit circle in the *z*-plane. Also, the entire right-hand plane must map to the *exterior* of the unit circle, as shown in Figure 6.12.

So, what we need is a mathematical device to make the transformation from the *s*-plane to the *z*-plane. It must map an infinitely large area into the unit circle. Fortunately, many of these mathematical transforms already exist. The one best suited for us is called the bilinear transform, and we can use it to transform an analog transfer function, $H(s)$ into a digital one, $H(z)$.

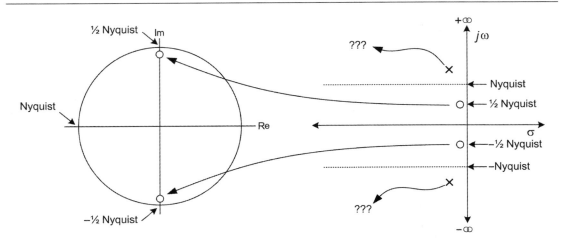

Figure 6.11: Mapping the poles and zeros from the analog *s*-plane to the digital *z*-plane.

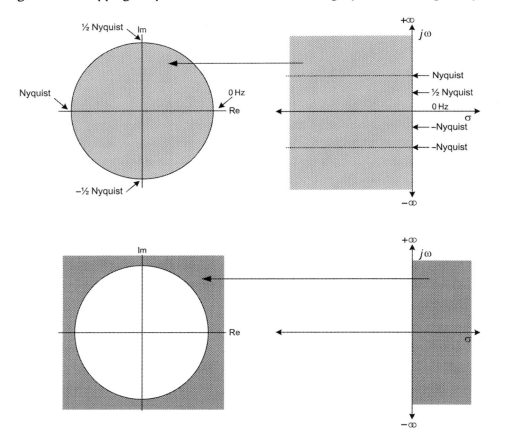

Figure 6.12: Mapping the infinitely large left-hand plane into the finite space inside the unit circle and the right-hand plane into the exterior of the unit circle.

We wish to convert an analog transfer function $H(s)$ into a sampled equivalent $H(z)$. Effectively, this means we need a way to sample the continuous s-plane to produce the sampled z-plane version. In other words, we need to create a sampled analog transfer function $H_s(s)$ where the subscript s stands for "sampled." We seek to find a function $g(z)$ such that Equation 6.9 holds true.

$$H(s) \xrightarrow{s=g(z)} H(z) \tag{6.9}$$

Since the sample interval is T, then the term $e^{j\omega T}$ would correspond to one sample in time. So, if we evaluate $H(z)$ by letting $z = e^{j\omega T}$ then we will arrive at the sampled analog transfer function $H_s(s)$ in Equation 6.10:

$$H(s) \xrightarrow{s=g(z)} H(z) \xrightarrow{z=e^{j\omega T}} H_s(s) \tag{6.10}$$

To solve, we note that $s = j\omega$ (the analog definition of s) and $z = e^{j\omega T}$ to get Equation 6.11:

$$z = e^{sT}$$
$$\ln(z) = \ln(e^{sT})$$
$$\ln(z) = sT \tag{6.11}$$

or

$$sT = \ln(z)$$

We now have the relationship between s and z, but we need to deal with taking the natural log of z, which doesn't have a closed form solution. If we use the Taylor series expansion for $\ln()$, we get Equation 6.12 for $\mathrm{Re}(z) > 0$:

$$sT = 2\left[\frac{z-1}{z+1} + \frac{1}{3}\left(\frac{z-1}{z+1}\right)^3 - \frac{1}{5}\left(\frac{z-1}{z+1}\right)^5 + \frac{1}{7}\left(\frac{z-1}{z+1}\right)^7 - \ldots\right] \tag{6.12}$$

Taking the first term only

$$s = \frac{2}{T}\frac{z-1}{z+1}$$

This first term *approximation* of the general solution is the bilinear transform. The bilinear transform we use is Equation 6.13:

$$s = \frac{z-1}{z+1} \tag{6.13}$$

The $2/T$ term washes out mathematically, so we can neglect it. This equation does the mapping by taking values at the infinite frequencies and mapping them to Nyquist. So, a pole or zero at $-\infty$ or $+\infty$ is mapped to Nyquist. The other frequencies between Nyquist and $-\infty$ or $+\infty$ Hz are squeezed into the little region right around Nyquist, just inside the unit circle, as shown in Figure 6.13.

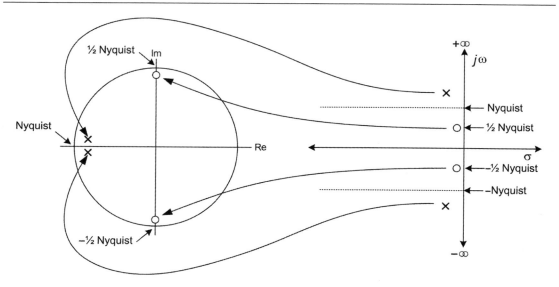

Figure 6.13: The bilinear transform maps the area outsize the Nyquist zone on the left-hand plane to an area just inside the circle near Nyquist.

The bilinear transform maps analog frequencies to their corresponding digital frequencies nonlinearly via the tan() function (Equation 6.14):

$$\omega_a = \tan\left[\frac{\omega_d/f_s}{2}\right]$$

where

$\omega_a =$ the analog frequency
$\omega_d =$ the mapped digital frequency
$f_s =$ the sample rate

(6.14)

The tan() function is linear at low values of ω but becomes more nonlinear as the frequency increases. At low frequencies, the analog and digital frequencies map closely. At high frequencies, the digital frequencies become warped and do not map properly to the analog counterparts. This means that a given analog design with a cutoff frequency f_c might have the wrong digital cutoff frequency after the conversion.

The solution is to pre-warp the *analog* filter so that its cutoff frequency is in the wrong location in the analog domain, but will wind up in the correct location in the digital domain. To pre-warp the analog filter, you just use the same equation (Equation 6.14) applied to the cutoff frequency of the analog filter. When you combine all the operations you get the bilinear Z transform, or BZT. The step-by-step method of conversion is as follows (Ifeachor 1993):

1. Start with an analog filter with a normalized transfer function $H(s)$—"normalized" means the cutoff frequency is set to $\omega = 1$ rad/sec; this is the typical way analog transfer functions are described and specified.
2. Choose the cutoff frequency of the digital filter ω_d; get the pre-warped analog corner frequency using Equation 6.15:

$$\omega_a = \tan\left[\frac{\omega_d/f_s}{2}\right] \quad (6.15)$$

3. Scale the normalized filter's cutoff frequency out to the new analog cutoff ω_a by replacing s with s/ω_a in the analog transfer function.
4. Apply the bilinear transform by replacing s with Equation 6.13:

$$s = \frac{z-1}{z+1} \quad (6.13)$$

5. Manipulate the transfer function $H(z)$ algebraically to get it into the familiar form so you can identify the coefficients (format the numerator and denominator to match the transfer functions you studied in the last chapter); this is often the most difficult part.

NOTE: This only works for LPF designs but, fortunately, there are easy conversions. The difference is in the step where you scale the analog filter's cutoff frequency just before applying the bilinear transform. How it works is described next.

For LPF and HPF:

- Specify ω_d, the desired digital cutoff frequency.
- Calculate the pre-warped analog cutoff with Equation 6.14:

$$\omega_a = \tan\left[\frac{\omega_d/f_s}{2}\right]$$

where (6.14)

$\omega_a = $ the analog frequency
$\omega_d = $ the mapped digital frequency
$f_s = $ the sample rate

For BPF and BSF:

- Specify ω_{dL} and ω_{dH}, the lower and upper corner frequencies of the digital filter.
- Calculate the two analog corner frequencies with Equation 6.16:

$$\omega_{aL} = \tan\left[\frac{\omega_{dL}/f_s}{2}\right]$$

$$\omega_{aH} = \tan\left[\frac{\omega_{dH}/f_s}{2}\right] \quad (6.16)$$

$$\omega_0^2 = \omega_{aL}\omega_{aH}$$

$$W = \omega_{aH} - \omega_{aL}$$

Next, scale the filter with Equation 6.17:

$$\begin{array}{ll} \text{Filter Type} & \text{Scaling Factor} \\ \text{LPF} & s = \dfrac{s}{\omega_a} \\[1em] \text{HPF} & s = \dfrac{\omega_a}{s} \\[1em] \text{BPF} & s = \dfrac{s^2 + \omega_0^2}{Ws} \\[1em] \text{BSF} & s = \dfrac{Ws}{s^2 + \omega_0^2} \end{array} \qquad (6.17)$$

Example:

Convert the basic resistor-capacitor (RC) analog LPF in Figure 6.14 into a digital LPF. The sample rate is 44.1 kHz and the desired cutoff frequency is 1 kHz.

Step 1: Get the normalized $H(s)$:

$$H(j\omega) = \frac{1}{j\omega RC + 1}$$
$$\text{Let } s = j\omega$$
$$\text{Normalize by setting RC} = 1 \qquad (6.18)$$
$$H(s) = \frac{1}{s + 1}$$

Step 2: Calculate the pre-warped analog cutoff:

$$f_c = 1 \text{ kHz}$$
$$\omega_d = 2\pi f_c = 6283.2 \text{ rad/sec} \qquad (6.19)$$

$$\omega_a = \tan\left(\frac{\omega_d T}{2}\right) = \tan\left[\frac{(6283.2)(1/44100)}{2}\right] = 0.07136 \qquad (6.20)$$

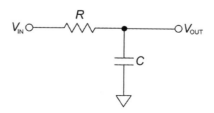

Figure 6.14: A simple analog RC low-pass filter.

Step 3: De-normalize the analog transfer function $H(s)$ with the appropriate factor:

$$H(s) = \frac{1}{s+1}$$

$$s = \frac{s}{\omega_a} \text{ because this is an LPF} \quad (6.21)$$

$$H'(s) = \frac{1}{s/\omega_a + 1} = \frac{1}{s/0.07136 + 1}$$

$$= \frac{0.07136}{s + 0.07136}$$

Step 4: Apply the BZT:

$$H(z) = H'(s)|_{s=(z-1)/(z+1)}$$

$$= \frac{0.07136}{\dfrac{z-1}{z+1} + 0.07136}$$

$$= \frac{0.07136(z+1)}{z - 1 + 0.07136(z+1)} \quad (6.22)$$

$$= \frac{0.07136z + 0.07136}{z - 1 + 0.07136z + 0.07136}$$

$$= \frac{0.07136z + 0.07136}{z + 0.07136z - 0.9286}$$

Get into standard $H(z)$ format:

$$= \frac{0.07136 + 0.07136z^{-1}}{1 + 0.07136 - 0.9286z^{-1}}$$

$$H(z) = \frac{0.0667 + 0.0667z^{-1}}{1 - 0.8667z^{-1}} = \frac{a_0 + a_1 z^{-1}}{1 + b_1 z^{-1}}$$

Equation 6.22 is in the format that we need with the numerator and denominator properly formed to observe the coefficients. From the transfer function, you can see that:

- $a_0 = 0.0667$
- $a_1 = 0.0667$
- $b_1 = -0.8667$

The difference equation is Equation 6.23:

$$y(n) = 0.0667x(n) + 0.0667x(n-1) + 0.8667y(n-1) \quad (6.23)$$

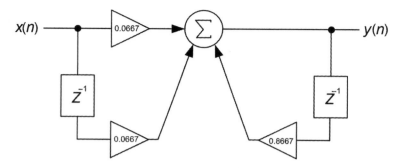

Figure 6.15: The digital equivalent of the analog RC low-pass filter.

The block diagram of the completed design reveals a pole-zero filter. This makes sense—the original analog filter had a pole at $s = -1$ and a zero at infinity. The digital equivalent has a pole at $z = 0.8667$ and a zero at $z = -1$ (Nyquist) both on the real axis (Figure 6.15).

6.4.1 Challenge

The analog transfer function of a second-order LPF is Equation 6.24:

$$H(s) = \frac{1}{s^2 + (1/Q)s + 1} \qquad (6.24)$$

The analog LPF has the following specifications: $Q = 1, f_c = 1$ kHz, $f_s = 44.1$ kHz. Apply the bilinear transform and some algebra to find the coefficients. (Answer: $a_0 = 0.0047$, $a_1 = 0.0095$, $a_2 = 0.0047$, $b_1 = -1.8485$, $b_2 = 0.8673$.)

6.5 Effect of Poles or Zeros at Infinity

In the analog transfer function (Equation 6.19) of the previous example, you can see that there is an analog pole at $s = -1$ since that would make the transfer function go to infinity, and there is a zero at $s = \infty$ because that will cause $H(s)$ to become 0.0. There is also a zero at $s = -\infty$. Interestingly, these two infinity values are in the same location because the reality is that the σ and $j\omega$ axes actually wrap around an infinitely large sphere and touch each other at $\pm\infty$. So, in this first-order case engineers only show the single zero at infinity and they choose to use the one at $-\infty$ so this transfer function's pole and zero would be plotted like Figure 6.16 in the s-plane. For low-pass zeros at infinity, the bilinear transform maps the zero at infinity to $z = -1$ (Nyquist) (Figure 6.17).

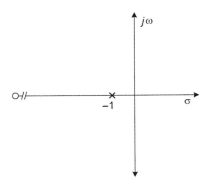

Figure 6.16: The pole at $s = -1$ and the zero at $s = \pm\infty$ plotted in the s-plane.

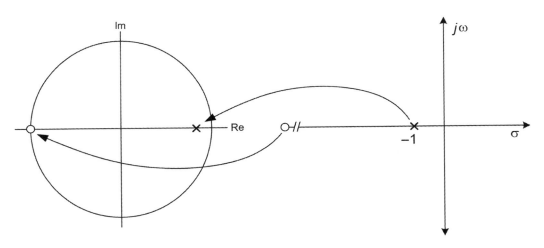

Figure 6.17: The bilinear transform maps real zeros at infinity to the Nyquist frequency in the z-plane.

Next consider the second-order analog low-pass filter transfer function:

$$H(s) = \frac{1}{s^2 + (1/Q)s + 1}$$

which factors as

(6.25)

$$H(s) = \frac{1}{(s - P)(s - P^*)}$$

This transfer function has a pair of conjugate poles at locations P and P^* or $(a + bj\omega)$ and $(a - bj\omega)$ as well as a pair of zeros at $\pm\infty$. The bilinear transform maps the poles on the left side of the *s*-plane to locations inside the unit circle. Once again, it maps the zeros at $\pm\infty$ to $z = -1$ or the Nyquist frequency (Figures 6.18 and 6.19).

There are several methods of overcoming this problem (Orfanidis 1997; Massberg 2011). The Massberg approach is discussed further in Section 6.9. Even with the error in the BZT, it is still an excellent tool for converting existing designs. You will notice that in the design methods, we start with the desired cutoff frequency (and Q, where applicable) and calculate the coefficients.

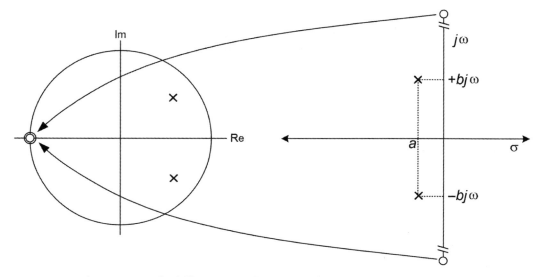

Figure 6.18: The bilinear transform maps imaginary zeros at infinity to Nyquist in the *z*-plane.

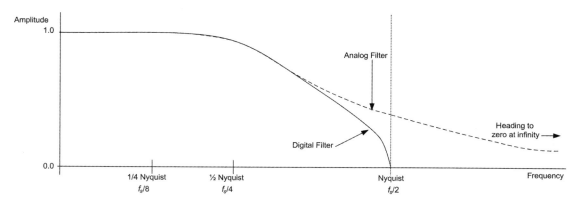

Figure 6.19: The zero at infinity causes an error in the upper part of the frequency. The error is worst at the Nyquist frequency; the analog error is exaggerated for the purposes of illustration.

6.6 Generic Bi-Quad Designs

The following classical analog filters are converted to digital and implemented as bi-quad topologies:

- LPF (low-pass filter)
- HPF (high-pass filter)
- BPF (band-pass filter)
- BSF (band-stop filter)
- Second-order Butterworth LPF and HPF
- Second-order Linkwitz–Riley LPF and HPF (good for crossovers)
- First- and second-order all-pass filters (APF)

Low-pass and high-pass filters are characterized by their corner frequency f_c and (for second-order and higher) their Q or resonant peaking value. A Q of 0.707 is the highest value Q can assume before peaking occurs. It is called the Butterworth or maximally flat response. With a Q of 0.707 the -3 dB point of the filter will be exactly at f_c. For these designs, once the Q rises above 0.707, it will correspond to the peak frequency and not the -3 dB frequency. Equations 6.26 through 6.29 relate the Q, peak frequency, -3 dB frequency, and the peak magnitude values.

$$f_{-3\mathrm{dB}} = f_c \sqrt{\left(\frac{1}{2Q^2} - 1\right) + \sqrt{\left(\frac{1}{2Q^2} - 1\right)^2 + 1}} \qquad (6.26)$$

$$f_c = \frac{f_{-3\mathrm{dB}}}{\sqrt{\left(\frac{1}{2Q^2} - 1\right) + \sqrt{\left(\frac{1}{2Q^2} - 1\right)^2 + 1}}} \qquad (6.27)$$

$$\text{Peak gain} = \frac{Q^2}{\sqrt{Q^2 - 0.25}} \, (Q > 0.707 \text{ only}) \qquad (6.28)$$

$$\text{Peak}_{\mathrm{dB}} = 20\log(\text{peak gain}) \qquad (6.29)$$

Band-pass, band-stop, graphic, and parametric EQs are specified by their center frequency f_c and Q or bandwidth (Equation 6.30). Depending on the filter, the arithmetic (Equation 6.31) or geometric (Equation 6.32) mean is used to relate f_c and the band edges f_U and f_L.

$$Q = \frac{f_c}{BW} \qquad (6.30)$$

$$\text{Arithmetic mean} = f_c = \frac{f_U - f_L}{2} \qquad (6.31)$$

$$\text{Geometric mean} = f_c = \sqrt{f_U f_L} \qquad (6.32)$$

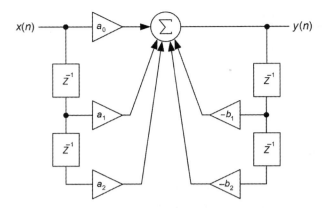

Figure 6.20: Generic bi-quad structure.

The block diagram is shown in Figure 6.20.

The difference equation is as follows:

$$y(n) = a_0 x(n) + a_1 x(n-1) + a_2 x(n-2) - b_1 y(n-1) - b_2 y(n-2) \qquad (6.33)$$

6.6.1 First-Order LPF and HPF

Specify:

- f_c, corner frequency; see Figure 6.21 for examples

The design equations are as follows:

$$
\begin{array}{ll}
\textbf{LPF} & \textbf{HPF} \\
\theta_c = 2\pi f_c/f_s & \theta_c = 2\pi f_c/f_s \\
\gamma = \dfrac{\cos \theta_c}{1 + \sin \theta_c} & \gamma = \dfrac{\cos \theta_c}{1 + \sin \theta_c} \\
a_0 = \dfrac{1-\gamma}{2} & a_0 = \dfrac{1+\gamma}{2} \\
a_1 = \dfrac{1-\gamma}{2} & a_1 = -\left(\dfrac{1+\gamma}{2}\right) \\
a_2 = 0.0 & a_2 = 0.0 \\
b_1 = -\gamma & b_1 = -\gamma \\
b_2 = 0.0 & b_2 = 0.0
\end{array}
\qquad (6.34)
$$

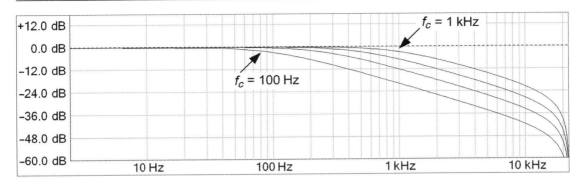

Figure 6.21: First-order LPF with f_c = 100 Hz, 250 Hz, 500 Hz, and 1 kHz.

6.6.2 Second-Order LPF and HPF

Specify:

- f_c, corner frequency
- Q, quality factor controlling resonant peaking; Figure 6.22 shows the effect of Q and peaking

The design equations are as follows:

$$
\begin{array}{ll}
\text{LPF} & \text{HPF} \\
\theta_c = 2\pi f_c / f_s & \theta_c = 2\pi f_c / f_s \\
d = \dfrac{1}{Q} & d = \dfrac{1}{Q} \\
\beta = 0.5 \dfrac{1 - \dfrac{d}{2}\sin\theta_c}{1 + \dfrac{d}{2}\sin\theta_c} & \beta = 0.5 \dfrac{1 - \dfrac{d}{2}\sin\theta_c}{1 + \dfrac{d}{2}\sin\theta_c} \\
\gamma = (0.5 + \beta)\cos\theta_c & \gamma = (0.5 + \beta)\cos\theta_c \\
a_0 = \dfrac{0.5 + \beta - \gamma}{2.0} & a_0 = \dfrac{0.5 + \beta + \gamma}{2.0} \\
a_1 = 0.5 + \beta - \gamma & a_1 = -(0.5 + \beta + \gamma) \\
a_2 = \dfrac{0.5 + \beta - \gamma}{2.0} & a_2 = \dfrac{0.5 + \beta + \gamma}{2.0} \\
b_1 = -2\gamma & b_1 = -2\gamma \\
b_2 = 2\beta & b_2 = 2\beta
\end{array}
\quad (6.35)
$$

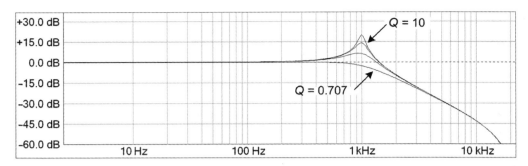

Figure 6.22: Second-order LPF responses: f_c = 1 kHz, Q = 0.707, 2, 5, 10. Notice that as the Q rises above 0.707, f_c becomes the peak frequency.

6.6.3 Second-Order BPF and BSF

Specify:

- f_c, corner frequency
- Q, quality factor controlling width of peak or notch = $1/BW$; Figure 6.23 shows the BSF version

Note: These filter coefficients contain the tan() function, which is undefined at $\pi/2$ and then flips negative between $\pi/2$ and π. The argument is $\theta_c/2Q$, so care must be taken with these filters to ensure that $\theta_c/2Q$ does not fall on the range of $\pi/2$ to π. One solution is to clamp the value of $\theta_c/2Q$ so that it never equals or exceeds $\pi/2$.

The design equations are as follows:

$$
\begin{array}{ll}
\textbf{BPF} & \textbf{BSF} \\
\theta_c = 2\pi f_c/f_s & \theta_c = 2\pi f_c/f_s \\
\beta = 0.5 \dfrac{1 - \tan(\theta_c/2Q)}{1 + \tan(\theta_c/2Q)} & \beta = 0.5 \dfrac{1 - \tan(\theta_c/2Q)}{1 + \tan(\theta_c/2Q)} \\
\gamma = (0.5 + \beta)\cos\theta_c & \gamma = (0.5 + \beta)\cos\theta_c \\
a_0 = 0.5 - \beta & a_0 = 0.5 + \beta \\
a_1 = 0.0 & a_1 = -2\gamma \\
a_2 = -(0.5 - \beta) & a_2 = 0.5 + \beta \\
b_1 = -2\gamma & b_1 = -2\gamma \\
b_2 = 2\beta & b_2 = 2\beta
\end{array}
\qquad (6.36)
$$

6.6.4 Second-Order Butterworth LPF and HPF

Specify:

- f_c, corner frequency

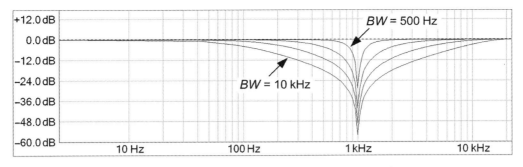

Figure 6.23: Second-order BSF responses: f_c = 1 kHz, bandwidth = 500 Hz (narrowest), 2 kHz, 5 kHz, and 10 kHz (widest).

Butterworth low-pass and high-pass filters are specialized versions of the ordinary second-order low-pass filter. Their Q values are fixed at 0.707, which is the largest value it can assume before peaking in the frequency response is observed.

The design equations are as follows:

$$\begin{array}{ll} \text{LPF} & \text{HPF} \\[6pt] C = \dfrac{1}{\tan(\pi f_c / f_s)} & C = \tan(\pi f_c / f_s) \\[10pt] a_0 = \dfrac{1}{1 + \sqrt{2}C + C^2} & a_0 = \dfrac{1}{1 + \sqrt{2}C + C^2} \\[10pt] a_1 = 2a_0 & a_1 = -2a_0 \\ a_2 = a_0 & a_2 = a_0 \\ b_1 = 2a_0(1 - C^2) & b_1 = 2a_0(C^2 - 1) \\ b_2 = a_0(1 - \sqrt{2}C + C^2) & b_2 = a_0(1 - \sqrt{2}C + C^2) \end{array} \qquad (6.37)$$

6.6.5 Second-Order Butterworth BPF and BSF

Specify:

- f_c, corner frequency
- BW, bandwidth of peak/notch $= f_c/Q$

Butterworth BPF and BSF are made by cascading (BPF) or paralleling (BSF) a Butterworth LPF and Butterworth HPF.

Note: These filter coefficients contain the tan() function, which is undefined at $\pi/2$ and then flips negative between $\pi/2$ and π. The argument is $\pi f_c BW/f_s$, so care must be taken with these filters to ensure that $\pi f_c BW/f_s$ does not fall on the range of $\pi/2$ to π. One solution is to clamp the value of $\pi f_c BW/f_s$ so that it never equals or exceeds $\pi/2$.

The design equations are as follows:

$$
\begin{array}{ll}
\text{BPF} & \text{BSF} \\
C = \dfrac{1}{\tan(\pi f_c BW/f_s)} & C = \tan(\pi f_c BW/f_s) \\
D = 2\cos(2\pi f_c/f_s) & D = 2\cos(2\pi f_c/f_s) \\
a_0 = \dfrac{1}{1+C} & a_0 = \dfrac{1}{1+C} \\
a_1 = 0.0 & a_1 = -a_0 D \\
a_2 = -a_0 & a_2 = a_0 \\
b_1 = -a_0(CD) & b_1 = -a_0 D \\
b_2 = a_0(C-1) & b_2 = a_0(1-C)
\end{array}
\qquad (6.38)
$$

6.6.6 Second-Order Linkwitz–Riley LPF and HPF

Specify:

- f_c, corner frequency (-6 dB)

Second-order Linkwitz–Riley LPFs and HPFs are designed to have an attenuation of -6 dB at the corner frequency rather than the standard -3 dB, shown in Figure 6.24. When these filters are placed in parallel with the same cutoff frequency, their outputs sum exactly and the resulting response is flat. They are often used in crossovers. We use them for the spectral dynamics processor later on.

The design equations are as follows:

$$
\begin{array}{ll}
\text{LPF} & \text{HPF} \\
\theta_c = \pi f_c/f_s & \theta_c = \pi f_c/f_s \\
\Omega_c = \pi f_c & \Omega_c = \pi f_c \\
\kappa = \dfrac{\Omega_c}{\tan(\theta_c)} & \kappa = \dfrac{\Omega_c}{\tan(\theta_c)} \\
\delta = \kappa^2 + \Omega_c^2 + 2\kappa\Omega_c & \delta = \kappa^2 + \Omega_c^2 + 2\kappa\Omega_c \\
a_0 = \dfrac{\Omega_c^2}{\delta} & a_0 = \dfrac{\kappa^2}{\delta} \\
a_1 = 2\dfrac{\Omega_c^2}{\delta} & a_1 = \dfrac{-2\kappa^2}{\delta} \\
a_2 = \dfrac{\Omega_c^2}{\delta} & a_2 = \dfrac{\kappa^2}{\delta} \\
b_1 = \dfrac{-2\kappa^2 + 2\Omega_c^2}{\delta} & b_1 = \dfrac{-2\kappa^2 + 2\Omega_c^2}{\delta} \\
b_2 = \dfrac{-2\kappa\Omega_c + \kappa^2 + \Omega_c^2}{\delta} & b_2 = \dfrac{-2\kappa\Omega_c + \kappa^2 + \Omega_c^2}{\delta}
\end{array}
\qquad (6.39)
$$

All pass filters have interesting designs that yield equally interesting results. Their frequency responses are perfectly flat from DC to Nyquist. However, the phase responses are the same as those in first- and second-order LPFs. You get all of the phase shift but none of the frequency response change. These filters can be found in crossovers and reverb algorithms. They are also used for the phaser effect. APFs are designed with pole-zero pairs whose pole/zero radii are reciprocals of one another. For a first-order APF, the pole lies somewhere on the real axis inside the unit circle at radius R. The zero lies outside the unit circle at radius $1/R$ (Figure 6.25).

If you think about the geometric interpretation and analysis of the transfer function, as you move around the unit circle, your vector distances from the pole and zero will always be reciprocals, or 1/each other. The amplitude response is then flat but the phase response *does* change because it is not based on the distance to the point in question but rather on the angle of incidence of the ray line drawn from the analysis point to the pole or zero. The second-order APF has complementary poles and zeros also at reciprocal radii (Figure 6.26).

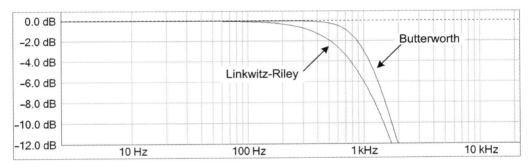

Figure 6.24: A comparison between Butterworth and Linkwitz–Riley filters each with f_c = 1 kHz; the Linkwitz–Riley filter is down −6 dB at f_c.

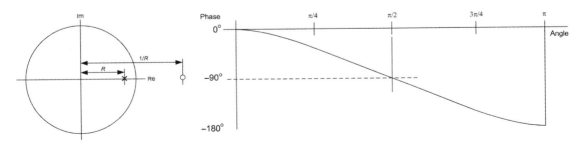

Figure 6.25: The first-order APF has a flat frequency response but shifts the phase of the signal by −90 degrees at f_c ($\pi/2$ in this example).

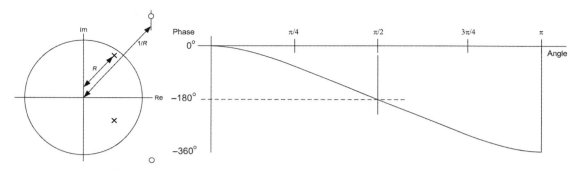

Figure 6.26: The second-order APF adds another −90 degrees of phase shift at f_c ($\pi/2$ here).

6.6.7 First- and Second-Order APF

Specify:

- f_c, corner frequency
- Q, steepness of phase shift at f_c (second-order only)

The design equations are as follows:

$$\text{First-Order APF} \qquad \text{Second-Order APF}$$

$$\alpha = \frac{\tan(\pi f_c/f_s) - 1}{\tan(\pi f_c/f_s) + 1} \qquad \alpha = \frac{\tan(\pi Q/f_s) - 1}{\tan(\pi Q/f_s) + 1}$$

$$\beta = -\cos\theta_c$$

$$\begin{aligned} a_0 &= \alpha \\ a_1 &= 1.0 \\ a_2 &= 0.0 \\ b_1 &= \alpha \\ b_2 &= 0.0 \end{aligned} \qquad \begin{aligned} a_0 &= -\alpha \\ a_1 &= \beta(1-\alpha) \\ a_2 &= 1.0 \\ b_1 &= \beta(1-\alpha) \\ b_2 &= -\alpha \end{aligned} \qquad (6.40)$$

6.7 Audio Specific Filters

The basic classical filters provide many functions in audio and can be very musical (e.g., resonant low-pass filters in synthesizers) but you also need filters that are very audio specific. These filters are found all throughout plug-ins and have been used in mixing consoles and other audio gear for decades. These designs are often not found in DSP textbooks because of their specific audio-only functions. These filters include:

- Shelving filters
- Parametric EQ
- Graphic EQ

These all require a slightly modified bi-quad structure. The reason is that these filters require mixing the original, unfiltered input directly with the output in a mix ratio. The ratio is controlled by two more coefficients, c_0 and d_0.

6.7.1 Modified Bi-Quad

You can see that the filter in Figure 6.27 is a bi-quad with two extra coefficients, c_0 and d_0, which make the wet-dry mix ratio needed for these filter types.

6.7.2 First-Order Shelving Filters

Low shelving

Specify:

- f_c, low shelf frequency
- Low-frequency gain/attenuation in dB

High shelving

Specify:

- f_c, high shelf frequency
- High-frequency gain/attenuation in dB

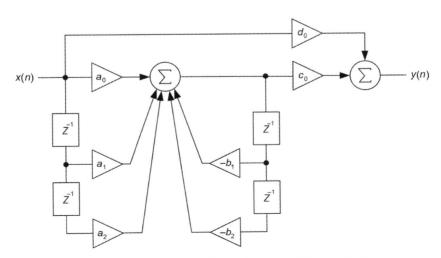

Figure 6.27: The modified bi-quad required for audio filters.

The design equations are as follows:

$$
\begin{array}{ll}
\textbf{Low Shelving} & \textbf{High Shelving} \\
\theta_c = 2\pi f_c / f_s & \theta_c = 2\pi f_c / f_s \\
\mu = 10^{Gain(\text{dB})/20} & \mu = 10^{Gain(\text{dB})/20} \\
\beta = \dfrac{4}{1+\mu} & \beta = \dfrac{1+\mu}{4} \\
\delta = \beta \tan(\theta_c/2) & \delta = \beta \tan(\theta_c/2) \\
\gamma = \dfrac{1-\delta}{1+\delta} & \gamma = \dfrac{1-\delta}{1+\delta} \\
a_0 = \dfrac{1-\gamma}{2} & a_0 = \dfrac{1+\gamma}{2} \\
a_1 = \dfrac{1-\gamma}{2} & a_1 = -\left(\dfrac{1+\gamma}{2}\right) \\
a_2 = 0.0 & a_2 = 0.0 \\
b_1 = -\gamma & b_1 = -\gamma \\
b_2 = 0.0 & b_2 = 0.0 \\
c_0 = \mu - 1.0 & c_0 = \mu - 1.0 \\
d_0 = 1.0 & d_0 = 1.0
\end{array}
\quad (6.41)
$$

Shelving filters are used in many tone controls, especially when there are only two, bass and treble, which are almost always implemented as shelf types. The filters have a corner frequency and gain or attenuation value. Figure 6.28 shows a family of shelving filter response curves.

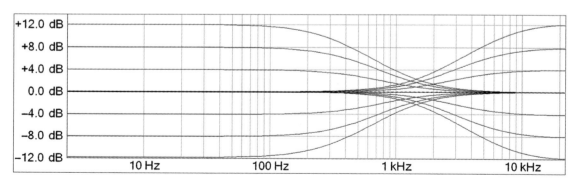

Figure 6.28: Low and high shelving filter responses. The low shelf frequency = 400 Hz, high shelf frequency = 5 kHz, with a variety of boost/cut settings.

6.7.3 Second-Order Parametric/Peaking Filter: Non-Constant-Q

Specify:

- f_c, center frequency
- Q quality factor
- Gain/attenuation in dB

Parametric EQs allow you to adjust the center frequency, Q and boost or cut creating any arbitrary bumps or notches in the frequency response (Figure 6.30). The parametric EQ is a variation on the ordinary band-pass and band-stop filters that generates symmetrical boost/cut curves and mixes in the dry signal to create the final response. A true digital parametric EQ not only has independent controls, but each control only varies one coefficient in the filter. The parametric EQs in this section afford the same frequency response but adjustments in any parameter require a recalculation of all the coefficients. These filters are also called peaking filters.

This parametric EQ is not constant-Q, which means the bandwidth varies depending on the boost/cut value. Some analog filters have the same issue, although there is occasional debate over whether or not this is desirable in an EQ design. Figure 6.29 shows this EQ with three different boost curves with a center frequency of 1 kHz and $Q = 1.0$; therefore, the bandwidth should also be 1000 Hz.

Note: These filter coefficients contain the tan() function, which is undefined at $\pi/2$ and then flips negative between $\pi/2$ and π. The argument is $\theta_c/2Q$, so care must be taken with these filters to ensure that $\theta_c/2Q$ does not fall on the range of $\pi/2$ to π. One solution is to clamp the value of $\theta_c/2Q$ so that it never equals or exceeds $\pi/2$.

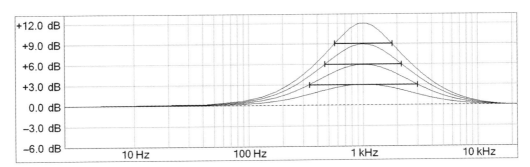

Figure 6.29: The non-constant-Q peaking filter has different bandwidths for different gain values; the bandwidth widens as the gain is reduced.
f_c = 1 kHz, Q = 0.707.

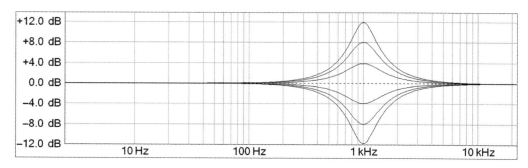

Figure 6.30: A set of responses for the non-constant-Q parametric/peaking filter with a variety of boost/cut settings. f_c = 1 kHz, Q = 2.

The design equations are as follows:

$$\begin{aligned}
\theta_c &= 2\pi f_c/f_s \\
\mu &= 10^{Gain(dB)/20} \\
\zeta &= \frac{4}{1+\mu} \\
\beta &= 0.5 \frac{1 - \zeta \tan(\theta_c/2Q)}{1 + \zeta \tan(\theta_c/2Q)} \\
\gamma &= (0.5 + \beta)\cos\theta_c \\
a_0 &= 0.5 - \beta \\
a_1 &= 0.0 \\
a_2 &= -(0.5 - \beta) \\
b_1 &= -2\gamma \\
b_2 &= 2\beta \\
c_0 &= \mu - 1.0 \\
d_0 &= 1.0
\end{aligned} \quad (6.42)$$

6.7.4 Second-Order Parametric/Peaking Filter: Constant-Q

Specify:

- f_c, center frequency
- Q, quality factor
- Gain/attenuation in dB

This design creates an almost perfect constant-Q filter with only a small amount of error for low-boost (or cut) values (Figure 6.31). The effect of the constant-Q design is clearly evidenced in the frequency response plot (Figure 6.32).

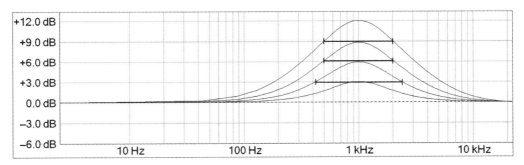

Figure 6.31: The constant-Q peaking filter preserves the bandwidth over most of the range of boost/cut values. f_c = 1kHz, Q = 0.707.

The design equations are as follows:

$$K = \tan(\pi f_c / f_s)$$
$$V_0 = 10^{boost/cut(\text{dB})/20}$$
$$d_0 = 1 + \frac{1}{Q}K + K^2 \quad (6.43)$$
$$e_0 = 1 + \frac{1}{V_0 Q}K + K^2$$

	Boost	Cut
$\alpha = 1 + \frac{V_0}{Q}K + K^2$	$a_0 = \frac{\alpha}{d_0}$	$a_0 = \frac{d_0}{e_0}$
$\beta = 2(K^2 - 1)$	$a_1 = \frac{\beta}{d_0}$	$a_1 = \frac{\beta}{e_0}$
$\gamma = 1 - \frac{V_0}{Q}K + K^2$	$a_2 = \frac{\gamma}{d_0}$	$a_2 = \frac{\delta}{e_0}$
$\delta = 1 - \frac{1}{Q}K + K^2$	$b_1 = \frac{\beta}{d_0}$	$b_1 = \frac{\beta}{e_0}$
$\eta = 1 - \frac{1}{V_0 Q}K + K^2$	$b_2 = \frac{\delta}{d_0}$	$b_2 = \frac{\eta}{e_0}$
	$c_0 = 1.0$	$c_0 = 1.0$
	$d_0 = 0.0$	$d_0 = 0.0$

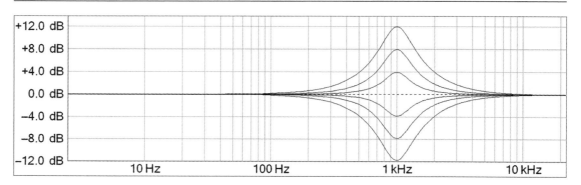

Figure 6.32: A set of responses for the constant-Q parametric/peaking filter with a variety of boost/cut settings. f_c = 1 kHz, Q = 2.

6.7.5 Cascaded Graphic EQ: Non-Constant-Q

Specify:

- f_c, center frequency
- Gain/attenuation in dB

The graphic EQ is a fixed Q variation on the parametric EQ. In a graphic EQ, the user only has control over the boost and cut of a set of filters tuned to fixed center frequencies. The Q that is used will depend on the boost or cut settings as well as the number of bands of EQ. For non-constant-Q graphic EQs, the method of finding the proper Q can be difficult because the bandwidth will change for various settings of the boost/cut. The value for Q is sometimes chosen so that when all the controls are set to the maximum boost or cut values, the frequency response is as flat as possible.

The Q for constant-Q filters is easy to calculate knowing the number of bands split across the 10 octaves of audio. It is a good starting point for developing a non-constant-Q EQ plug-in; experimentation may be needed to find the ideal Q for your particular design. You might also allow the user to adjust the value within reason. The equation relating Q to the number of EQ bands is Equation 6.44.

$$N = \frac{\text{Number of modules}}{10}$$
$$Q = \frac{\sqrt{2^N}}{2^N - 1} \tag{6.44}$$

For a 10-band graphic EQ, $Q = 1.414$, while for a 31-band (aka 1/3 octave) EQ, $Q = 4.32$. The center frequencies of the bands are usually found with the following International Organization for Standardization (ISO) standard equation:

$$q = \frac{\text{Number of bands}}{10}$$
$$f_n = 1000 * 2^{n/q} \quad (6.45)$$
$$n = 0, \pm 1, \pm 2, \pm 3, \text{etc} ...$$

For a 10-band EQ, this works out (with rounding) to 32 Hz, 64 Hz, 125 Hz, 250 Hz, 500 Hz, 1 kHz, 2 kHz, 4 kHz, 8 kHz, and 16 kHz.

The cascaded graphic EQ topology is shown in Figure 6.33. It consists of a simple series cascade of each of the filter modules. The non-constant-Q graphic uses the same non-constant-Q peaking filter algorithm, but with the Q fixed according to Equation 6.42. You can get a good feel for how the Q affects the overall response by looking at a plot with all controls set to the maximum boost (or cut). For a 10-band non-constant-Q graphic, we observe the combined response shown in Figure 6.34.

6.7.6 Cascaded Graphic EQ: Constant-Q

Specify:

- f_c, center frequency
- Gain/attenuation in dB

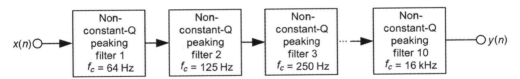

Figure 6.33: The cascaded graphic EQ topology; this is for a 10-band design using the ISO center frequencies. The design equations for each module use the non-constant-Q algorithm above.

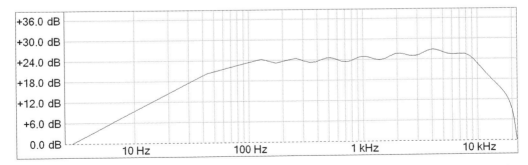

Figure 6.34: At the prescribed constant-Q value of $Q = 1.414$ we observe rippling and an increased high-frequency response with all controls at maximum boost.

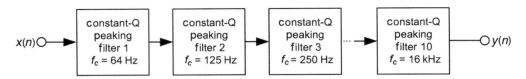

Figure 6.35: The cascaded graphic EQ topology; this is for a 10-band design using the exact unrounded ISO center frequencies. The design equations for each module use the constant-Q algorithm above.

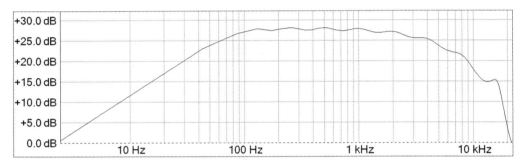

Figure 6.36: The 10-band constant-Q graphic EQ with the prescribed $Q = 1.414$ produces less low-frequency rippling and a decreased high-frequency response with all controls at maximum boost.

The constant-Q graphic EQ follows the same design pattern (Figure 6.35) except that you use the constant-Q peaking filter in each of the modules. The equations for finding the center frequencies and Q are the same as above. Bohn (1986) recommends not rounding the ISO center frequencies but rather use the exact values. Figure 6.36 shows the composite response with all controls set to full boost.

6.8 Design a Resonant LPF Plug-In

We'll continue with a resonant LPF design. The filter will use the second-order bilinear transform design and the bi-quad structure for implementation. After that you will redesign the filter using the Massberg technique for analog emulation and examine the differences. Note: This filter is found in RackAFX's module menu item under "HP/LP Filter" for you to experiment with and use to check your own implementations.

For this project, we'll be using our first built-in object. I've already prepared a bi-quad object for you to use in this (or any other) project. It is not a RackAFX plug-in but rather a simple C++ object that accomplishes only one task: the bi-quad structure realization. The bi-quad can be used to make any first- or second-order feed-forward or feed-back type filter. Let's take a moment to check out the CBiquad object. You can find the interface in the pluginconstants.h

Table 6.2: The CBiquad object interface.

CBiquad	
Member Variables	**Purpose**

protected:
    float m_f_Xz_1
    float m_f_Xz_2
    float m_f_Yz_1
    float m_f_Yz_2
	Implements the four delay elements needed for the bi-quad: $x(n-1)$, $x(n-2)$, $y(n-1)$ and $y(n-2)$ in these protected variables

public:
    float m_f_a0
    float m_f_a1
    float m_f_a2
    float m_f_b1
    float m_f_b2
	The filter coefficients

Member Functions	
void flushDelays()	Initialize all delay elements with 0.0s
float doBiQuad(float f_xn)	Do one cycle of filter operation; return value is $y(n)$
Parameters	
• float f_xn	Input: the current input sample, $x(n)$

file and the implementation in the pluginobjects.cpp files, respectively. The CBiquad object has the members and variables in Table 6.2.

You can see that this object is simple; it only has two functions, one to reset and the other to do the bi-quad calculation. The coefficients must be calculated externally—this object has no knowledge of the filter type that it is implementing, the sample rate, and so on. You will see that this simple object will also save you a lot of coding for filters in later chapters. We will need to set up the user interface (UI) and then link the variables to the bi-quad objects in userInterfaceChange().

6.8.1 Project: ResonantLPF

Create the project and name it "ResonantLPF," then add the sliders for the graphical user interface (GUI).

6.8.2 ResonantLPF GUI

Figure 6.37 shows what your final GUI will look like. You will need the controls shown in Table 6.3.

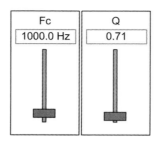

Figure 6.37: The resonant LPF GUI.

Table 6.3: The GUI controls for the resonant LPF.

Slider Property	Value	Slider Property	Value
Control Name	Fc	Control Name	Q
Units	Hz	Units	
Variable Type	float	Variable Type	float
Variable Name	m_f_fc_Hz	Variable Name	m_f_Q
Low Limit	100	Low Limit	0.5
High Limit	5000	High Limit	20
Initial Value	1000	Initial Value	0.707

6.8.3 ResonantLPF.h File

RackAFX provides you with the built-in bi-quad object named CBiquad and you don't need to do anything at all to add it to your project. You can just declare your bi-quad objects directly in the .h file like you would any other object:

CBiquad m_LeftLPF; // or whatever you like to name it

Here's my .h file with the left and right LPF objects declared:

```
// Add your code here: ------------------------------------------------ //
CBiquad m_LeftLPF;
CBiquad m_RightLPF;

// END OF USER CODE --------------------------------------------------- //
```

We'll also need a function to calculate the bi-quad's coefficients (a_0, a_1, a_2, b_1, and b_2) and we can share this between the left and right objects (i.e., both LPFs will have the same cutoff and Q, so they will also have the same coefficients). Declare this now:

```
// Add your code here: ------------------------------------------------ //
CBiquad m_LeftLPF;
CBiquad m_RightLPF;
```

```
void calculateLPFCoeffs(float fCutoffFreq, float fQ);

// END OF USER CODE ------------------------------------------------------ //
```

6.8.4 ResonantLPF.cpp File

Write the calculateLPFCoeffs() function in the .cpp file by using Equation 6.35. I have used the same intermediate coefficient names here too.

```
void CResonantLPF::calculateLPFCoeffs(float fCutoffFreq, float fQ)
{
    // use same terms as in book:
    float theta_c = 2.0*pi*fCutoffFreq/(float)m_nSampleRate;
    float d = 1.0/fQ;

    // intermediate values
    float fBetaNumerator = 1.0 - ((d/2.0)*(sin(theta_c)));
    float fBetaDenominator = 1.0 + ((d/2.0)*(sin(theta_c)));

    // beta
    float fBeta = 0.5*(fBetaNumerator/fBetaDenominator);

    // gamma
    float fGamma = (0.5 + fBeta)*(cos(theta_c));

    // alpha
    float fAlpha = (0.5 + fBeta - fGamma)/2.0;

    // left channel
    m_LeftLPF.m_f_a0 = fAlpha;
    m_LeftLPF.m_f_a1 = 2.0*fAlpha;
    m_LeftLPF.m_f_a2 = fAlpha;
    m_LeftLPF.m_f_b1 = -2.0*fGamma;
    m_LeftLPF.m_f_b2 = 2.0*fBeta;

    // right channel
    m_RightLPF.m_f_a0 = fAlpha;
    m_RightLPF.m_f_a1 = 2.0*fAlpha;
    m_RightLPF.m_f_a2 = fAlpha;
    m_RightLPF.m_f_b1 = -2.0*fGamma;
    m_RightLPF.m_f_b2 = 2.0*fBeta;
}
```

prepareForPlay()

- Flush the bi-quad buffers.
- Cook the raw variables.

```
bool __stdcall CResonantLPF::prepareForPlay()
{
```

```
            // Add your code here:
            m_LeftLPF.flushDelays();
            m_RightLPF.flushDelays();

            // calculate the initial values
            calculateLPFCoeffs(m_f_fc_Hz, m_f_Q);

            return true;
    }
```

userInterfaceChange()

- If the user moves either slider then we have to recalculate all the coefficients because they are interdependent on both f_c and Q.
- There is no need to check the slider nControlIndex value here.

```
    bool __stdcall CResonantLPF::userInterfaceChange(int nControlIndex)
    {
            // update coeffs for next time
            calculateLPFCoeffs(m_f_fc_Hz, m_f_Q);

            return true;
    }
```

processAudioFrame()

- Call the methods on the filter objects to do the processing.
- They will automatically update their own delay lines and maintain themselves so the code is simple.

```
    bool __stdcall CResonantLPF::processAudioFrame(float* pInputBuffer, float* pOutputBuffer,
                                    UINT uNumInputChannels, UINT uNumOutputChannels)
    {
            // Do LEFT (MONO) Channel; there is always at least one input/one output
            // (INSERT Effect)
            pOutputBuffer[0] = m_LeftLPF.doBiQuad(pInputBuffer[0]);

            // Mono-In, Stereo-Out (AUX Effect)
            if(uNumInputChannels == 1 && uNumOutputChannels == 2)
                    pOutputBuffer[1] = pOutputBuffer[0]; // Just copying

            // Stereo-In, Stereo-Out (INSERT Effect)
            if(uNumInputChannels == 2 && uNumOutputChannels == 2)
                    pOutputBuffer[1] = m_RightLPF.doBiQuad(pInputBuffer[1]);

            return true;
    }
```

Build and test your resonant LPF. Use the analyzer's frequency and phase response features to verify that all is working as you would expect. You now have a working resonant LPF, but the Nyquist response is always clamped to 0.0, so let's investigate a technique to make the response more closely match the original analog filter.

6.9 The Massberg Analog-Matched Low-Pass Filter

At the 131st Audio Engineering Society Convention in 2011, Michael Massberg presented a solution to the clamping problem, producing first- and second-order low-pass filters that very closely match their analog counterparts near and at the Nyquist frequency. It takes advantage of the fact that a shelving filter with a very high upper shelf frequency has a roll-off portion that very closely resembles the roll-off of a low-pass filter. In a nutshell, the idea is to design a shelving filter whose upper shelf frequency is above Nyquist. In the range from DC to Nyquist, it will resemble a low-pass filter (Figure 6.38). A unique matching scheme is applied to force the two responses to match their gains exactly at a point halfway between the gain at DC and the gain at Nyquist in decibels, with a slight error at points above that frequency. The result is a filter that has an excellent approximation to the analog response it is trying to mimic.

Massberg used first- and second-order shelving filters to produce analog-matched low-pass filters applying the bilinear transform to produce the final result. Figure 6.39 shows a comparison of the standard digital LPF and the Massberg LPF.

6.9.1 First-Order Massberg LPF

Specify:

- f_c, corner frequency

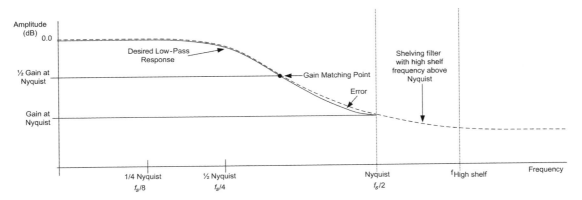

Figure 6.38: The Massberg filter uses a shelving filter as a prototype. In the range from DC to Nyquist, it very closely resembles the desired low-pass filter. The two curves are displaced slightly.

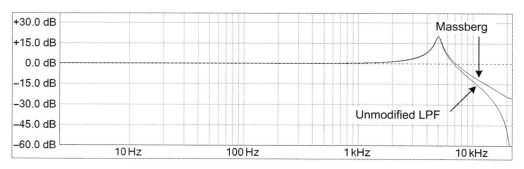

Figure 6.39: The Massberg and unmodified LPF responses with f_c = 5 kHz and Q = 10. The difference in the high-frequency response is evident.

The design equations are as follows:

$$g_1 = \frac{2}{\sqrt{4 + \left(\frac{f_s}{f_c}\right)^2}} ac$$

$$g_m = \max(\sqrt{0.5}, \sqrt{g_1})$$

$$\omega_m = \frac{2\pi f_c \sqrt{1 - g_m^2}}{g_m}$$

$$\Omega_m = \tan\left(\frac{\omega_m}{2 f_s}\right)$$

$$\Omega_s = \Omega_m \frac{\sqrt{(g_m^2 - g_1^2)(1 - g_m^2)}}{1 - g_m^2}$$

$$\gamma_0 = \Omega_s + 1$$

$$\alpha_0 = \Omega_s + g_1$$

$$\alpha_1 = \Omega_s - g_1$$

$$\beta = \Omega_s - 1$$

$$a_0 = \frac{\alpha_0}{\gamma_0}$$

$$a_1 = \frac{\alpha_1}{\gamma_0}$$

$$a_2 = 0$$

$$b_1 = \frac{\beta_1}{\gamma_0}$$

$$b_2 = 0$$

(6.46)

6.9.2 Second-Order Massberg LPF

Specify:

- f_c, corner frequency
- Q, quality factor controlling resonant peaking

The design equations are as follows:

$$\theta_c = 2\pi f_c / f_s$$

$$g_1 = \frac{2}{\sqrt{\left(2 - \left(\frac{\sqrt{2\pi}}{\theta_c}\right)^2\right) + \left(\frac{2\pi}{Q\theta_c}\right)^2}}$$

Calculate Ω_s depending on the value of Q:

$Q > \sqrt{5}$?

$$g_r = \frac{2Q^2}{\sqrt{4Q^2 - 1}}$$

$$\omega_r = \theta_c \sqrt{1 - \frac{1}{2Q^2}}$$

$$\Omega_r = \tan\left(\frac{\omega_r}{2}\right)$$

$$\Omega_s = \Omega_r \left(\frac{g_r^2 - g_1^2}{g_r^2 - 1}\right)^{1/4}$$

$Q \leq \sqrt{5}$?

$$\omega_m = \theta_c \sqrt{\frac{2 - \frac{1}{2Q^2} + \sqrt{\frac{1 - 4Q^2}{Q^4} + \frac{4}{g_1}}}{2}}$$

$$\Omega_m = \tan\left(\frac{\omega_m}{2}\right)$$

$$\Omega_s = \frac{\theta_c (1 - g_1^2)^{1/4}}{2} \qquad (6.47)$$

$$\Omega_s = \min(\Omega_s, \Omega_m)$$

Calculate the pole and zero frequencies (ω), gains (g) and Qs

$$\omega_p = 2 \arctan(\Omega_s) \qquad \omega_z = 2 \arctan\left(\frac{\Omega_s}{\sqrt{g_1}}\right)$$

$$g_p = \frac{1}{\sqrt{\left(1 - \left(\frac{\omega_p}{\theta_c}\right)^2\right) + \left(\frac{\omega_p}{Q\theta_c}\right)^2}} \qquad g_z = \frac{1}{\sqrt{\left(1 - \left(\frac{\omega_z}{\theta_c}\right)^2\right) + \left(\frac{\omega_z}{Q\theta_c}\right)^2}}$$

$$Q_p = \sqrt{\frac{g_1(g_p^2 - g_z^2)}{(g_1 + g_z^2)(g_1 - 1)^2}} \qquad Q_p = \sqrt{\frac{g_1^2(g_p^2 - g_z^2)}{g_z^2(g_1 + g_p^2)(g_1 - 1)^2}}$$

$$\gamma_0 = \Omega_s^2 + \frac{1}{Q_p}\Omega_s + 1$$

$$\alpha_0 = \Omega_s^2 + \frac{\sqrt{g_1}}{Q_z}\Omega_s + g_1$$

$$\alpha_1 = 2(\Omega_s^2 - g_1) \qquad \beta_1 = 2(\Omega_s^2 - 1)$$

$$\alpha_2 = \Omega_s^2 - \frac{\sqrt{g_1}}{Q_z}\Omega_s + g_1 \qquad \beta_2 = \Omega_s^2 - \frac{1}{Q_p}\Omega_s + 1 \qquad (6.48)$$

$$a_0 = \frac{\alpha_0}{\gamma_0}$$

$$a_1 = \frac{\alpha_1}{\gamma_0} \qquad b_1 = \frac{\beta_1}{\gamma_0}$$

$$a_2 = \frac{\alpha_2}{\gamma_0} \qquad b_2 = \frac{\beta_2}{\gamma_0}$$

Challenge: Modify your digital resonant LPF plug-in to add the Massberg filter option, then experiment with high-fidelity music and listen for the differences in sound.

Bibliography

Allred, R. 2003. Second-order IIR filters will support cascade implementations: 5 part digital audio application tutorial. *EE Times Design Guide*. http://www.eetimes.com/design/audio-design/4009473/Second-order-IIR-Filters-will-support-cascade-implementations—5-Part-digital-audio-application-tutorial. Accessed August 15, 2012.

Berners, D. P. and Abel, J. S. 2003. Discrete-time shelf filter design for analog modeling. *Journal of the Audio Engineering Society*, preprint 5939.

Bohn, D. 1986. Constant-Q graphic equalizers. *Journal of the Audio Engineering Society* 34(9): 611–25.

Bohn, D. 2008. Bandwidth in octaves versus Q in band-pass filters. Application Note 170. Mukilteo, WA: Rane Corp.

Dodge, C. and Jerse, T. 1997. *Computer Music Synthesis, Composition and Performance*, Chapter 6. New York: Schirmer.

Giles, M., ed. 1980. *The Audio/Radio Handbook*. Santa Clara, CA: National Semiconductor Corp.

Kwakernaak, H. and Sivan, R. 1991. *Modern Signals and Systems*, Chapter 9. Englewood Cliffs, NJ: Prentice-Hall.

Lane, J. et al. 2001. *DSP Filters*, Chapters 4–10 and 20. Carmel, IN: Howard W. Sams & Co.

Lane, J. and Hillman, G. 1991. *Implementing IIR/FIR Filters with Motorola's DSP56000/DSP56001*. APR7/D Rev1. Schomberg, ON: Motorola, Inc.

Leach, M. 1999. *Introduction to Electroacoustics and Audio Amplifier Design*, Chapter 6. Dubuque, IA: Kendall-Hunt.

Lindquist, C. 1977. *Active Network Design*, Chapter 2. Miami: Steward & Sons.

Lindquist, C. 1999. *Adaptive and Digital Signal Processing*, Chapter 5. Miami: Steward & Sons.

Massberg, M. 2011. Digital low-pass filter design with analog-matched magnitude response. *Journal of the Audio Engineering Society*, preprint 8551 (Massberg 2011).

Moore, R. 1990. *Elements of Computer Music*, Chapter 2. Englewood Cliffs, NJ: Prentice-Hall.

Motorola, Inc. 1991. *Digital Stereo 10-Band Graphic Equalizer Using the DSP56001*. APR2/D. Schomberg, ON: Motorola, Inc.

Oppenheim, A. V. and Schafer, R. W. 1999. *Discrete-Time Signal Processing*, 2nd ed., Chapter 7. Englewood Cliffs, NJ: Prentice-Hall.

Smith, J. O. and Angell, J. B. 1982. A constant gain digital resonator tuned by a single coefficient. *Computer Music Journal* 4(4): 36–40.

Zöler, U. 2011. *Digital Audio Effects*, 2nd ed., Chapter 2. West Sussex, U.K.: John Wiley & Sons.

References

Ifeachor, Emmanuel C. and Jervis, Barrie W. 1993. *Digital Signal Processing: A Practical Approach*. Menlo Park: Addison-Wesley. pp. 398–400.

Orfanidis, Sophacles. 1997. Digital Parametric Equalizer Design with Prescribed Nyquist-Frequency Gain. *Journal of the Audio Engineering Society* 45(6): 444–55.

CHAPTER 7

Delay Effects and Circular Buffers

Before we can start looking at some finite impulse response (FIR) algorithms, we need to deal with the concept of long delay lines or circular buffers. Not only are they used for the delay effects, but also they are needed to make long FIR filters. In this chapter we'll take a break from the DSP filter algorithms and develop some digital delays. If you think back to the infinite impulse response (IIR) filters you've worked on so far you will remember that after implementing the difference equation, you need to shuffle the z^{-1} delay element values. You do this by overwriting the delays backwards, like this:

   m_f_z2 = m_f_z1;
   m_f_z1 = xn;   // xn is the input sample

Suppose you had a filter that was higher than a second-order one and you had to implement z^{-4} using discrete variables for each delay element. You might wind up writing something like this to do the shuffling:

   m_f_z4 = m_f_z3;
   m_f_z3 = m_f_z2;
   m_f_z2 = m_f_z1;
   m_f_z1 = xn;   // xn is the input sample

But what happens when the delay line gets really long, like z^{-1024} or, for a 1-second digital delay, z^{-44100}? It's going to be difficult to implement the delay shuffling this way. Not only would it be tedious to code, it would also be very inefficient to have to implement all those read/write operations each sample period. The answer to the problem of long delay lines is called circular buffering.

Digital signal processors (DSPs) have an interesting feature in the address generation unit (AGU), which allows you to declare a buffer of data to be addressed circularly. This kind of addressing is not built into the C++ programming language's array access operation. We will have to code the mechanism ourselves. When you declare a buffer, the addressing is linear.

This means that the pointer will add or subtract the offset you provide and move linearly to the next location. Suppose you declare a buffer of floats like this:

```
float Buffer[1024];
```

and a pointer to the buffer:

```
float* pBuffer = &Buffer[0];
```

and then you enter a loop which accesses the data, incrementing by five addresses each time through the loop:

```
for(int n=0; n<someNumber; n+=5)
{
    float data = pBuffer[n];
}
```

What happens if the pointer is accessed outside the buffer, as shown in Figure 7.1? Usually a crash or debug halt. You always have to be careful when setting up a loop like that one to make sure the pointer never accesses outside the buffer. But what if you are stuck with a certain sized buffer and a different sized loop so that you are never sure if the pointer is going to go outside the buffer on the next iteration? In a DSP chip, when you declare a buffer to be circular, you are setting up a pointer access mechanism. Reads and writes are made with a pointer, which is incremented or decremented by some amount to generate the next access location. If the amount of offset causes the pointer to go outside the bounds of the buffer, it automatically wraps to the other side, including the amount of offset, as depicted in Figure 7.2.

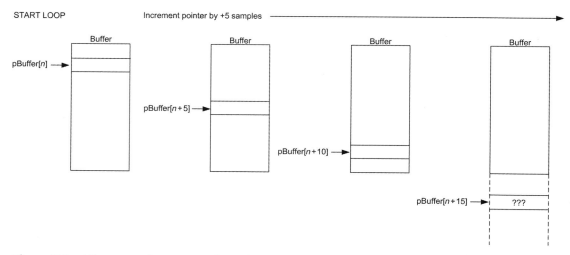

Figure 7.1: After several accesses, the pointer goes outside the buffer into an unknown location, usually resulting in a crash.

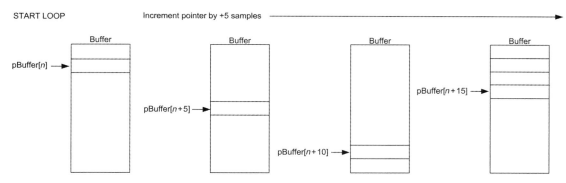

Figure 7.2: In a circular buffer, the pointer is automatically wrapped back to the top and offset by the proper amount to continue the access-by-five-samples loop.

Circular buffers are useful in audio signal processing. You can create circular buffers of audio samples or buffers of coefficients and loop through and access them automatically. Let's start with the most obvious use and make a digital delay effect.

7.1 The Basic Digital Delay

The digital delay effect or DDL (digital delay line) consists of a long buffer for storing audio samples. The samples enter one end of the buffer and come out the other end after D samples of delay, which corresponds to D sample periods. A feedback path allows for delay regeneration, or repeated echoes, as shown in Figure 7.3. With the feedback control *fb* set to 0, there is only a single delayed sample. With any other value, repeated echoes will form at the rate corresponding to the length of the delay line. The block diagram is shown in Figure 7.3.

The difference equation is as follows:

$$y(n) = x(n-D) + fb*y(n-D) \tag{7.1}$$

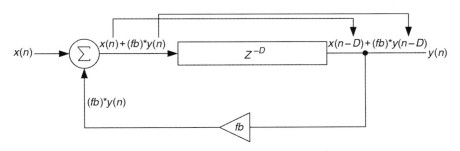

Figure 7.3: Basic DDL.

From the difference equation in Equation 7.1, you can see that the output consists of an input sample delayed by D samples plus a scaled version of the output at that time, $fb*y(n-D)$. The sequence of accessing the delay line during the processAudioFrame() function is as follows:

1. Read the output value, y(n), from the DDL and write it to the output buffer.
2. Form the product $fb*y(n)$.
3. Write the input value, $x(n) + fb*y(n)$, into the delay line.

You might notice that something is missing here: the shuffling of the samples through the delay. If we use a circular buffer as the delay line, then we don't have to shuffle data around, but we do have to keep track of the read and write access locations in the buffer and wrap the pointers as needed. In order to understand how the buffer operates to make a delay, consider a circular buffer that we've been writing samples into each sample period and automatically wrapping the pointer (or index) back to the top of the buffer as needed. The code would look something like this:

```
// buffer size, 1 second of audio data
int nBufferLength = 44100;

float Buffer[nBufferLength];
.
.
.
. // somewhere else in code:
int nIndex = 0;
.
.

      // inside a loop of some kind:
    pBuffer[nIndex] = audioSample;

    nIndex++;

    // if we go outside the buffer
    if(nIndex >= nBufferLength) // if we hit nBufferLength, we are one sample outside
         nIndex = 0;    // wrap the pointer back to the top for the next iteration
.
.
```

Suppose this has been going on for some time, and we are left with the buffer looking like Figure 7.4 *after* the last write access and just before we increment the pointer index. If pBuffer is pointing to the current sample value $x(n)$

- Where is the $x(n-1)$ sample (the youngest delayed value)?
- Where is the oldest sample in the delay?

In Figure 7.5 the youngest sample, $x(n-1)$, is in the location just before pBuffer[i], that is pBuffer[i−1]. The oldest sample is found by working backwards to the top of the buffer,

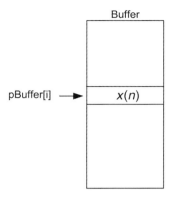

Figure 7.4: After many loops through the buffer, pBuffer[i] points to x(n).

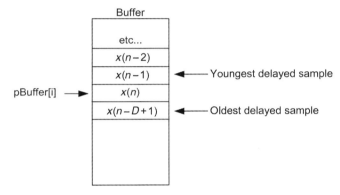

Figure 7.5: The youngest and oldest samples in the delay line.

wrapping back to the bottom, and locating the oldest sample written; it is at pBuffer[i+1]. If the pointer is accessing the samples sequentially from top to bottom, then the youngest sample is just above $x(n)$ and the oldest is just below it. It is easy to understand that the youngest sample is $x(n-1)$ but why is the oldest sample labeled $x(n-D+1)$ rather than $x(n-D)$?

The answer to the question is that we overwrote the actual oldest sample, $x(n-D)$, when we wrote in $x(n)$. This is one of the reasons for our rule about always performing reads before writes in our algorithms. This means that to get the oldest sample in the delay, you must first read out pBuffer[i] before writing into it. In other words, *before* the write access, the buffer looks like Figure 7.6; you can see the oldest value $x(n-D)$ is actually the location of the current write operation.

The steps for creating the delay line in your plug-in are as follows:

1. Decide on the maximum amount of delay you need to provide.
2. Declare read and write index values.

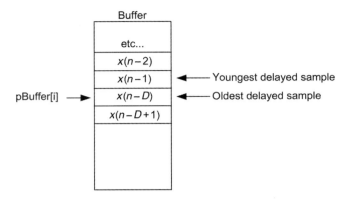

Figure 7.6: The location of the oldest audio sample $x(n - D)$.

3. Declare a float buffer for each channel, right and left: for very long delay lines this is traditionally done with the *new* operator in the constructor of the plug-in.
4. Initialize the buffers with 0.0 using the memset() function.

During the processAudioFrame() function, you will need to

- Use the read and write index values to operate the delay.
- Increment or decrement the indices according to the algorithm you are using.
- Check to see if you need to wrap the index values for the next sample period.

In order to use the delay line, the user needs to provide a value for the delay amount in samples (in Section 7.3 we will allow them to enter the delay in milliseconds instead, but we will convert this to a sample count inside the delay). In your code, there are two basic ways to do this:

1. Subtract the delay amount (in samples) from your write index, wrapping the pointer backwards if needed, to locate the delayed sample.
2. Store a read index and offset the two indices when the user changes the delay amount.

The second option is better because you only have to offset the index values and perform the subtraction/wrap when the user changes the delay value, instead of each time through the processAudioFrame() loop. Suppose the user selects 100 samples of delay time. You have declared two indices, m_nRead and m_nWrite, to use for buffer. During the processAudioFrame() function you will need to do the following five steps.

Step 1: Read out the delayed audio data, $d(n-D-100)$, which is 100 samples behind the current $x(n)$ sample time; this value is $y(n)$, the current output value (Figure 7.7).

```
float yn = pBuffer[m_nRead];
```

Step 2: Form the input combination input + feedback * output:

```
float xn = pInputBuffer[0] + m_fFeedBack*yn;
```

m_fFeedBack is declared in your .h file; this example code is for processing the left channel, pInputBuffer[0].

Step 3: Write the input data into the delay line at the m_nWrite location (Figure 7.8).

Step 4: Increment the read and write pointers by one to set up for the next time through the function; check for wrapping and do that if necessary. Can you think of other ways to do this?

```
// inc/wrap
m_nWrite++;
if(m_nWrite >= m_ nBufferLength)
        m_nWrite = 0;

m_nRead++;
if(m_nRead >= m_ nBufferLength)
        m_nRead = 0;
```

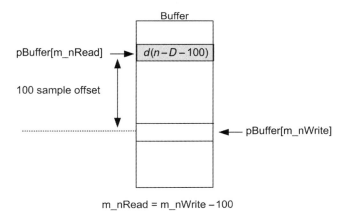

Figure 7.7: The delayed sample is read at location $d(n-D-100)$, 100 samples before the current write location.

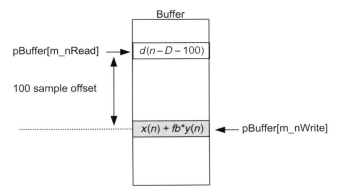

Figure 7.8: The delayed sample plus feedback is written into the current write location.

Notice that we wrap if the incremented index hits nBufferLength because this references the location just one sample outside the delay line.

Step 5: In the event that the user changes the delay time, you need to recalculate the m_nRead index to accommodate it. Note that once set up, the m_nWrite index is never changed except to increment it through the line. In this example, the user has selected nSamples of delay in the plug-in.

```
// user selects nSample delay
// first subtract the index
m_nRead = m_nWrite - nSamples;

// the check and wrap BACKWARDS if the index is negative
if (m_nRead < 0)
      m_nRead += nBufferLength;   // amount of wrap is Read + Length
```

7.2 Digital Delay with Wet/Dry Mix

Although there may be some instances where you don't need a wet/dry mix, in general when you make a delay plug-in, you still want to hear your unaffected signal. This can be done by using the plug-in as an Auxiliary (AUX) Send effect rather than an insert effect. But, to provide both capabilities, you need to provide the user with a wet/dry ratio control. The block diagram is slightly modified as shown in Figure 7.9.

The difference equation is as follows:

$$y(n) = dry*x(n) + wet*[x(n-D) + fb*y(n-D)] \tag{7.2}$$

7.2.1 Frequency and Impulse Responses

Consider the basic delay with *no* feedback applied and with the wet/dry ratio at 1:1. The block diagram and difference equation would then reduce down to Figure 7.10.

The difference equation is as follows:

$$y(n) = x(n) + x(n-D) \tag{7.3}$$

To find the frequency response, first take the *z* transform of the difference equation and form the transfer function:

$$\begin{aligned} y(n) &= x(n) + x(n-D) \\ Y(z) &= X(z) + X(z)z^{-D} \\ &= X(z)(1+z^{-D}) \\ H(z) &= \frac{Y(z)}{X(z)} = 1+z^{-D} \end{aligned} \tag{7.4}$$

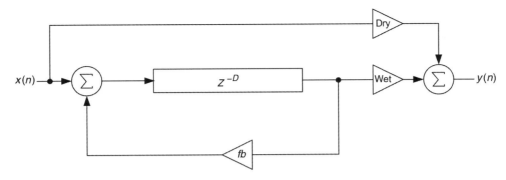

Figure 7.9: A more useful digital delay incorporates wet and dry controls.

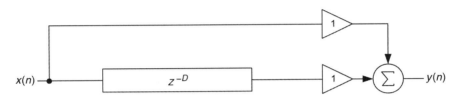

Figure 7.10: The simplified DDL of *D* samples delay.

Next, calculate the poles and zeros of the transfer function. We can see that this is a pure feed-forward filter in its current state so there are only zeros. We need to find the roots of the equation for $H(z)$:

$$0 = 1 + z^{-D}$$
$$= z^D + 1$$
$$z^D = -1$$

$$\text{Let } z = e^{j\omega} \quad (7.5)$$
$$e^{jD\omega} = -1$$

Apply Euler
$$\cos(D\omega) + j\sin(D\omega) = -1$$

The roots (zeros) of the function lie wherever $D\omega$ causes the equation to evaluate to $-1 + j0$, and we know this will happen when Equation 7.6 holds true.

$$\cos(\Theta) + j\sin(\Theta) = -1 + j0$$
$$\text{if} \quad (7.6)$$
$$\Theta = \pm\pi, \pm 3\pi, \pm 5\pi, \text{etc...}$$

Notice that both $\pm\pi$ produce the desired result of $-1 + j0$ as well as all the odd multiples of π: $\pm 3\pi$, $\pm 5\pi$, and so on. So the actual solution to find the roots becomes Equation 7.7:

$$\cos(D\omega) + j\sin(D\omega) = -1 + j0$$
if
$$D\omega = \pm\pi, \pm 3\pi, \pm 5\pi, ..., \pm N\pi$$
until
$$N > D-1 \qquad (7.7)$$

or

$$\text{zeros at } \omega = \pm\frac{k\pi}{D} \quad k = 1, 3, 5, ..., D$$

After $N > D-1$, the whole mathematical sequence repeats again, cycling through odd multiples of π. This means that there are D zeros spread equally around the unit circle. This makes sense—the fundamental theorem of algebra predicts D roots for a polynomial of order D. Now consider the simple case of $D = 2$ samples; we get Equation 7.8:

$$\cos(2\omega) + j\sin(2\omega) = -1$$
if
$$\omega = \pm\frac{k\pi}{D} \quad k = 1, 3, 5, ..., D \qquad (7.8)$$
$$\omega = \pm\frac{\pi}{2}$$

There are two zeros, one at $+\pi/2$ and the other at $-\pi/2$. Plot those on the unit circle in the z-plane and you can see what the frequency response will be, shown in Figure 7.11. You can see from Figure 7.11 that the zeros produce a notch (zero of transmission) at $\pm\pi/2$. In fact, when the delay time is very small, your ears hear the effect as a frequency response change; your ears cannot discriminate two samples that are only 23 uS apart as separate echoes. Now consider what would happen if we increase the delay amount to four samples, as in Figure 7.12. Finally, what would happen if the delay is an odd value, like $D = 5$ (Figure 7.13)?

$$\cos(D\omega) + j\sin(D\omega) = -1 + j0$$
$$\cos(4\omega) + j\sin(4\omega) = -1 + j0$$
$$\omega = \pm\frac{\pi k}{D} \quad k = 1, 3, 5, ..., D \qquad (7.9)$$
$$\omega = \pm\frac{\pi}{4}, \pm\frac{3\pi}{4}$$

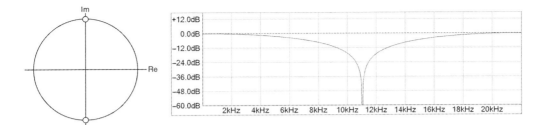

Figure 7.11: The z-plane pole/zero plot and resulting frequency response.

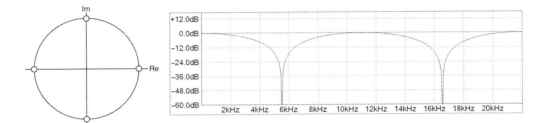

Figure 7.12: The z-plane pole/zero plot and resulting frequency response for $D = 4$ samples.

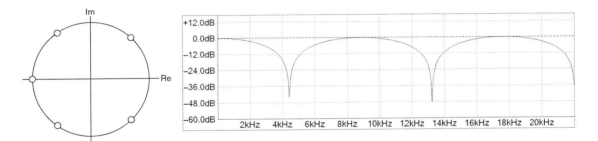

Figure 7.13: The z-plane pole/zero plot and resulting frequency response for $D = 5$ samples.

$$\cos(D\omega) + j\sin(D\omega) = -1 + j0$$
$$\cos(5\omega) + j\sin(5\omega) = -1 + j0$$
$$\omega = \pm\frac{k\pi}{5} \quad k = 1, 3, 5, ..., D-1$$
$$\omega = \pm\frac{\pi}{5}, \pm\frac{3\pi}{5}, \pm\frac{5\pi}{5}$$
$$\omega = \pm\frac{\pi}{5}, \pm\frac{3\pi}{5}, \pm\pi$$

(7.10)

This kind of frequency response in Figure 7.13 is called *inverse comb filtering*. As we add more and more samples of delay, we add more and more notches to the response. You can use the built in module in RackAFX to experiment. Figures 7.14 and 7.15 show the frequency response for 32 samples of delay—it's an inverse comb filter with 16 zeros in the positive frequency domain.

7.2.2 The Effect of Feedback

When you add feedback to the delay, two things happen: first, for long delays your ear will hear discrete repeating echoes that decay away after the signal is removed. As the delay time gets shorter and shorter, the echoes merge, begin to ping, and then the actual filtering (which has been going on all the time) now dominates what you hear.

In Figure 7.16, you can see the effect of feedback on the impulse response. The initial impulse is the one through the dry path, and the next identical-sized one is the first output of the delay line. The echoes decay, being reduced 90% on each trip through the filter. To understand the effect on frequency response, write the transfer function and take the z transform. To make it easier to find the pole frequencies, let the feedback value become 100%

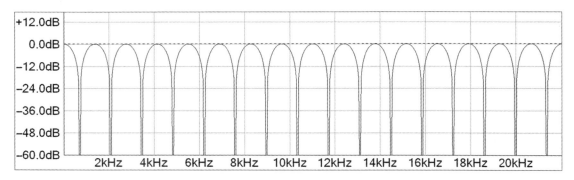

Figure 7.14: Frequency response (linear) with D = 32 samples.

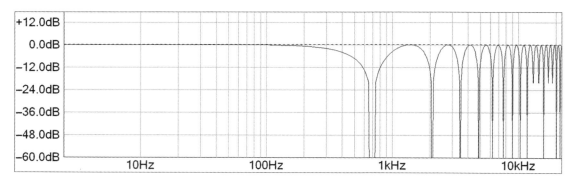

Figure 7.15: Frequency response (log) with D = 32 samples.

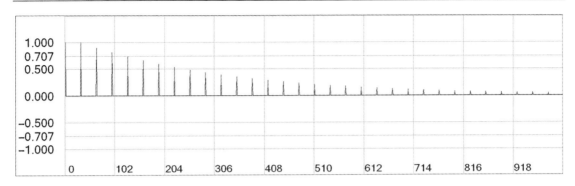

Figure 7.16: Impulse response with 90% feedback, 32 samples of delay.

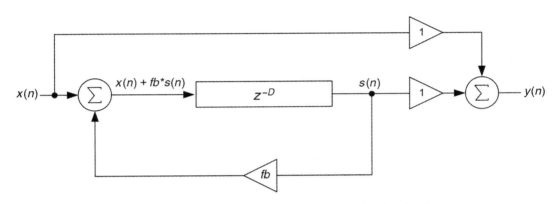

Figure 7.17: Block diagram of the DDL with feedback.

or 1.0, as shown in the block diagram in Figure 7.17; even though we know this would result in oscillation, it will make calculating the frequencies easier.

The difference equation is as follows:

$$y(n) = x(n) + x(n-D) - fb*x(n-d) + fb*y(n-D) \qquad (7.11)$$

To derive the difference equation, label the output of the delay line $s(n)$ (Equation 7.12):

The input to the delay line $= x(n) + fb*s(n)$

therefore

$$s(n) = x(n-D) + fb*s(n-D) \qquad (7.12)$$

and

$$y(n) = x(n) + s(n)$$
$$= x(n) + x(n-D) + fb*s(n-D)$$

rearranging:
$$s(n) = y(n) - x(n)$$
$$s(n-D) = y(n-D) - x(n-D) \tag{7.13}$$

Substituting Equation 7.13 into Equation 7.12 gives you the following difference equation:

$$\begin{aligned} y(n) &= x(n) + x(n-D) + fb*[y(n-D) - x(n-D)] \\ &= x(n) + x(n-D) + fb*y(n-D) - fb*x(n-D) \end{aligned} \tag{7.14}$$

To find the transfer function, separate variables and take the z transform:

$$\begin{aligned} y(n) - fb*y(n-D) &= x(n) + x(n-D) - fb*x(n-D) \\ Y(z) - fbY(z)z^{-D} &= X(z) + X(z)z^{-D}[1 - fb] \\ Y(z)[1 - fbz^{-D}] &= X(z)[1 + z^{-D} - fbz^{-D}] \\ H(z) = \frac{Y(z)}{X(z)} &= \frac{1 + z^{-D} - fbz^{-D}}{1 - fbz^{-D}} \\ H(z) &= \frac{1 + (1-fb)z^{-D}}{1 - fbz^{-D}} \end{aligned} \tag{7.15}$$

The new transfer function has both zeros (which we already calculated) and poles. The poles are caused by the feedback and will occur whenever the denominator becomes zero. If we let $fb = 1.0$, then equation Equation 7.15 reduces to Equation 7.16.

$$H(z) = \frac{1}{1 - z^{-D}}$$
$$0 = 1 - z^{-D}$$
$$0 = z^D - 1$$
$$z^D = 1$$

Let $z = e^{j\omega}$
$$e^{jD\omega} = 1 \tag{7.16}$$

Apply Euler
$$\cos(D\omega) + j\sin(D\omega) = 1$$

$$\cos(\Theta) + j\sin(\Theta) = 1 + j0$$
if
$$\Theta = 0, \pm 2\pi, \pm 4\pi, \pm 6\pi, \text{etc...}$$

$$\cos(D\omega) + j\sin(D\omega) = 1 + j0$$

if

$$D\omega = 0, \pm\frac{2\pi}{D}, \pm\frac{4\pi}{D}, \pm\frac{6\pi}{D}, \dots, \pm\frac{N\pi}{D}$$

until

$$N > D-1$$

(7.17)

or

$$\text{poles at } \omega = \pm\frac{k\pi}{D} \quad k = 0, 2, 4, 6, \dots, D$$

Equation 7.17 shows that the poles will occur at even multiples of π and DC (0 Hz) when Euler's equation becomes $1 + j0$; the analysis is nearly identical to the zero case. Consider the case of four samples:

$$\cos(D\omega) + j\sin(D\omega) = 1 + j0$$
$$\cos(4\omega) + j\sin(4\omega) = 1 + j0$$
$$\omega = \pm\frac{\pi k}{D} \quad k = 0, 2, 4, 6, \dots, D$$
$$\omega = 0, \pm\frac{\pi}{2}, \pm\frac{4\pi}{4}$$
$$\omega = 0, \pm\frac{\pi}{2}, \pm\pi$$

(7.18)

Figure 7.18 shows the effect of 100% feedback – the response is technically infinite at the pole frequencies. It produces a comb filter (with the teeth of the comb pointing up) rather than the inverse comb filter you saw when examining the zeros. The amount of feedback will affect the radius of the poles and the zeros, but not the pole or zero frequencies, which are

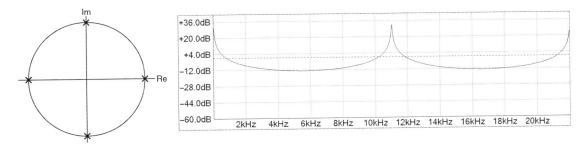

Figure 7.18: The z-plane pole/zero plot and resulting frequency response for $D = 4$ samples with $fb = 1.0$. The peaks theoretically go to infinity.

only dependent on the amount of delay. Consider the transfer function with a feedback value of 0.75:

$$H(z) = \frac{1 + (1-fb)z^{-D}}{1 - fbz^{-D}}$$

$$H(z)|_{fb=0.75} = \frac{1+0.25z^{-D}}{1-0.75z^{-D}} \qquad (7.19)$$

The poles will have a radius of 0.75, while the zeros will have a radius of 0.25. This will result in the z-plane plot and frequency response in Figure 7.19. You can see that the lowered radius results in less gain at the pole frequencies. The peaks are now softer and the overall gain is reduced down to about +8 dB from infinity. If you continue to drop the feedback to 50% (0.5) then the poles and zeros will be distributed at equal radii, as shown in Figure 7.20.

As the feedback gets smaller and smaller, the response will turn into the pure inverse comb filtering when the poles disappear (by converging on $z = 0$) and the response goes to 0.0 at the zero frequencies. What would happen if we inverted the feedback? This would mean that the feedback value is a negative percentage. Consider −50% feedback and look at the transfer function in Equation 7.20:

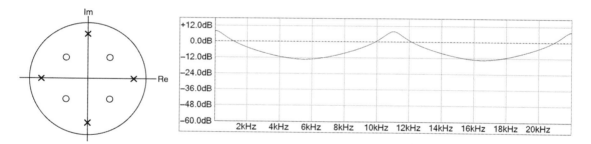

Figure 7.19: The z-plane pole/zero plot and resulting frequency response for D = 4 samples, 75% feedback.

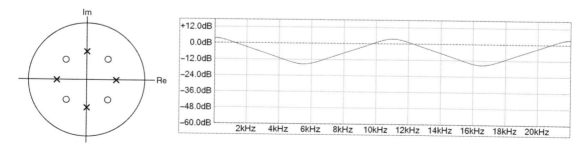

Figure 7.20: The z-plane pole/zero plot and resulting frequency response for D = 4 samples, 50% feedback.

$$H(z) = \frac{1 + (1-fb)z^{-D}}{1-fbz^{-D}}$$

$$H(z)|_{fb=-1} = \frac{1+1.5z^{-D}}{1+0.5z^{-D}}$$
(7.20)

If you look at Equation 7.20, you can figure out that the pole frequencies are going to lie at the zero frequencies (notice the signs of the coefficients). The zeros will be at a radius of 1.5, while the poles will be at 0.5. A frequency lying on the unit circle will be under the influence of all four poles and zeros.

For the four-sample delay, a feedback value of –62% will make the frequency response perfectly flat, but with –3 dB of attenuation seen in Figure 7.21. The poles will have radii of 0.62 with the zeros at radii of 1.38. This means you can create a delay that has no comb/inverse comb filtering, but only at this particular value. Other negative feedback values will give varying degrees of cancellation. In practice, the poles will dominate and small peaks can appear at high inverted feedback values. In the time domain, the echoes will alternate between positive and negative values each time they are flipped in the feedback path, shown in Figure 7.22.

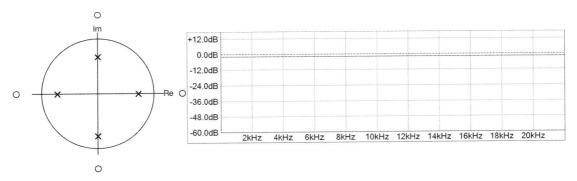

Figure 7.21: At –62% feedback with 4 samples of delay, the frequency response becomes flat but slightly attenuated.

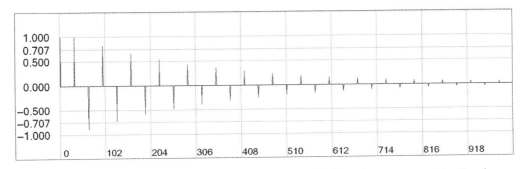

Figure 7.22: The effect of inverted feedback on the impulse response; feedback is –90% here.

7.3 Design a DDL Module Plug-In

In the previous projects, it was easy enough to simply declare left and right delay elements and coefficients for our simple filters. However, as the filters become more complex, this becomes more tedious and is also bad coding practice since we have replicated some code. For educational purposes, it is better to split the code out at first, but now it's time to think modularly. More complicated delay effects like the stereo cross feedback or LCR (left center right) delay will be easier to implement if we have a flexible delay module to work with. In this project, you will create a DDL module that you can use in many other projects. It is going to be a slight modification on the above DDL + wet/dry control. Here are the specifications:

- Implements an *n*-sample delay line, user controllable, up to 2 seconds of delay.
- Delay is given in milliseconds.
- Has feedback capable of –100% to +100% operation.
- Has wet/dry mix ratio control; 50/50 is an equal mix while 100/0 is full wet (delay only).
- Allows the feedback sample to be taken from the delay line, or supplied "outside" the module; the reason for this will become evident later.
- Allows access to the current feedback output sample. This allows you to insert other effects into the feedback path by using the switch to choose feedback_in; it also allows for cross coupling the feedback paths of stereo modules.

The block diagram is given in Figure 7.23.

For this design, let's implement a mono-only version. We can then make stereo versions by making two of our member variables the module object. That's right—you can use one plug-in inside another by making it a member object. First, let's get started on the DDL module. Note: For the initial DDL modules, don't worry about the feedback path switching. We need to get the delay debugged and running first.

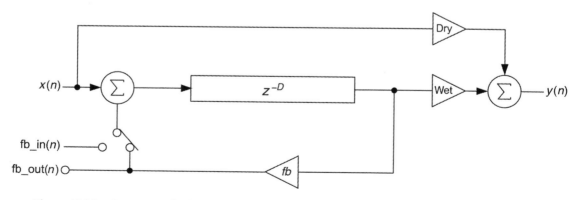

Figure 7.23: Our more flexible DDL module; a switch allows the user to choose where the feedback sample is taken. Here it is using the normal output feedback path. The feedback output is available for use at the fb_out pin.

7.3.1 Project: DDLModule

By now, you should be getting good at RackAFX programming, so we can move more quickly through the process. Create the project and add the sliders.

7.3.2 DDLModule GUI

Here is the final graphical user interface (GUI) for the DDL module in Figure 7.24. You can use my variable names or make up your own. You will need the controls shown in Table 7.1. We do not need a switch for the feedback option on the UI; it will only be needed by the super plug-in that includes this module as a member object.

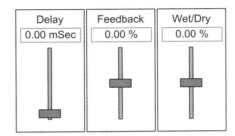

Figure 7.24: The DDL Module GUI.

Table 7.1: GUI controls for the DDL module

Slider Property	Value
Control Name	Delay
Units	mSec
Variable Type	float
Variable Name	m_fDelay_ms
Low Limit	0
High Limit	2000
Initial Value	0

Slider Property	Value
Control Name	Feedback
Units	%
Variable Type	float
Variable Name	m_f_Feedback_pct
Low Limit	−100
High Limit	100
Initial Value	0

(continued)

Table 7.1: GUI controls for the DDL module (Continued)

Slider Property	Value
Control Name	Wet/dry
Units	%
Variable Type	float
Variable Name	m_f_WetLevel_pct
Low Limit	0
High Limit	100
Initial Value	50

7.3.3 DDLModule.h File

In the .h file, add the cooked variables, m_fDelayInSamples, m_fFeedback, and m_fWetLevel:

```
// Add your code here: ------------------------------------------------- //
float m_fDelayInSamples;
float m_fFeedback;
float m_fWetLevel;
// END OF USER CODE --------------------------------------------------- //


// ADDED BY RACKAFX -- DO NOT EDIT THIS CODE!!! ----------------------- //
//    **--0x07FD--**

float m_fDelay_ms;
float m_fFeedback_pct;
float m_fWetLevel_pct;
//    **--0x1A7F--**
// ------------------------------------------------------------------- //
```

Note: I named the RackAFX controls with _ms and _pct appended to remind me that these values need to be cooked to be used.

Note: The delay time cooked variable is a float, m_fDelayInSamples rather than an integer number of samples. This is because we will allow *fractional delay* in a future version of the module. For now, though, we will treat it as an integer.

7.3.4 DDLModule .cpp File

Constructor

- Initialize variables.

```
CDDLModule::CDDLModule()
{
  <SNIP SNIP SNIP>
  // Finish initializations here
  m_fDelayInSamples = 0;
  m_fFeedback = 0;
  m_fDelay_ms = 0;
  m_fFeedback_pct = 0;
  m_fWetLevel = 0;

}
```

The formula for figuring out the delay time in samples from the delay time in milliseconds is Equation 7.21.

$$\text{Samples delay} = (D \text{ mSec}) \left[\frac{\text{sample rate}}{1000} \right] \quad (7.21)$$

Cooking the feedback value is easy—just divide by 100 to convert the percent to a raw multiplier. The same thing is true for the wet level value. In this project, we'll introduce the concept of a cooking function to handle the work. Because we are planning on using this plug-in as a module for future development, it will be a good idea. From this point on, you should get in the habit of making a cooking function. In this case, we will keep it simple and have the cooking function recalculate all the plug-in's variables regardless of which ones actually change. This is an effort to educate first. You can always go back and streamline your functions and code after the plug-in has been tested and is functioning. First, declare the cooking function in the .h file:

```
// Add your code here: ----------------------------------------- //
float m_fDelayInSamples;
float m_fFeedback;
float m_fWetLevel;

void cookVariables();
// END OF USER CODE -------------------------------------------- //
```

Write the function:

```
// function to cook the variables
void CDDLModule::cookVariables()
{
  m_fFeedback = m_fFeedback_pct/100.0;
  m_fWetLevel = m_fWetLevel_pct/100.0;
  m_fDelayInSamples = m_fDelay_ms*((float)m_nSampleRate/1000.0);
}
```

Then, we can add the cooking function to

- The end of the constructor()
- prepareForPlay()
- userInterfaceChange()

Constructor

```
CDDLModule::CDDLModule()
{
        <SNIP SNIP SNIP>

        // Finish initializations here
        <SNIP SNIP SNIP>
        m_fFeedback_pct = 0;
        m_fWetLevel = 0;
        cookVariables();
}
```

prepareForPlay()

```
bool __stdcall CDDLModule::prepareForPlay()
{
        // cook
        cookVariables();

        return true;
}
```

Notice the big change made here—rather than a switch/case statement, we just go ahead and cook all the data.

userInterfaceChange()

```
bool __stdcall CDDLModule::userInterfaceChange(int nControlIndex)
{
        // cook
        cookVariables();

        return true;
}
```

Now that the UI details are taken care of, we can get to the business of declaring the delay line, initializing it, and finally implementing the effect.

7.3.5 Declare and Initialize the Delay Line Components

For a delay line, you will need the following variables:

- A float* which points to a buffer of samples
- An integer read index
- An integer write index
- An integer that is the size of the buffer in samples

Add them to your .h file:

```
// Add your code here: ------------------------------------------------------- //
float m_fDelayInSamples;
float m_fFeedback;
float m_fWetLevel;

float* m_pBuffer;
int m_nReadIndex;
int m_nWriteIndex;
int m_nBufferSize;
// END OF USER CODE --------------------------------------------------------- //
```

The delay line will be created dynamically. It will be destroyed in the destructor. The problem is that we don't yet know what the sample rate will be; we won't know that until the user loads a new file and begins playing it. Just before RackAFX calls your prepareForPlay() function, it sets the sample rate on your plug-in's m_nSampleRate variable. Therefore, we will have to dynamically create and flush out the buffer each time prepareForPlay() is called. In the constructor, we set the m_pBuffer to NULL as a flag to know that it is uninitialized, as well as zero the buffer size and read and write index values.

Constructor

```
CDDLModule::CDDLModule()
{
<SNIP SNIP SNIP>
    m_fFeedback_pct = 0;
    m_fWetLevel = 0;

    // reset
    m_nReadIndex = 0;
    m_nWriteIndex = 0;

    // no buffer yet because we don't have a sample rate yet
    m_pBuffer = NULL;
    m_nBufferSize = 0;

    // cook
    cookVariables();
}
```

prepareForPlay()

- Create the buffer now that we know the sample rate.

```
bool __stdcall CDDLModule::prepareForPlay()
{
    // setup our delay line
    m_nBufferSize = 2*m_nSampleRate;     // 2 seconds delay @ fs
```

```
        // delete it if it exists
        if(m_pBuffer)
                delete [] m_pBuffer;

        // create the new buffer
        m_pBuffer = new float[m_nBufferSize];

        return true;
}
```

Destructor

- Delete the buffer.

```
CDDLModule::~CDDLModule(void)
{
        // delete buffer if it exists
        if(m_pBuffer)
                delete [] m_pBuffer;
}
```

To initialize the buffer with 0.0, use the the memset function. That memset() flushes the buffer of data, and we need to do this each time prepareForPlay() is called so we don't play out old data at the onset. We are going to be flushing and resetting the buffer in several places in code, so it is also a good thing to make into a function.

7.3.6 DDLModule.h File

```
// Add your code here: ------------------------------------------------- //
float m_fDelayInSamples;
float m_fFeedback;
float m_fWetLevel;

void cookVariables();
void resetDelay();
// END OF USER CODE --------------------------------------------------- //
```

7.3.7 DDLModule.cpp File

```
// function to flush buffer and set Write pointer back to top
// read pointer will be calculated based on write pointer location
void CDDLModule::resetDelay()
{
        // flush buffer
        if(m_pBuffer)
                memset(m_pBuffer, 0, m_nBufferSize*sizeof(float));

        // init read/write indices
        m_nWriteIndex = 0; // reset the Write index to top
        m_nReadIndex = 0;  // reset the Read index to top
}
```

You can also modify the cooking function to add the code for updating the read index; set the read and write indices depending on the amount of delay time, so you can use a modified version of the code from Section 7.1. Note: The delay in samples is cast to an integer using the casting method (int).

```
// function to cook the variables
void CDDLModule::cookVariables()
{
        m_fFeedback = m_fFeedback_pct/100.0;
        m_fWetLevel = m_fWetLevel_pct/100.0;
        m_fDelayInSamples = m_fDelay_ms*((float)m_nSampleRate /1000.0);

        // subtract to make read index
        m_nReadIndex = m_nWriteIndex - (int)m_fDelayInSamples; // cast as int!

        //check and wrap BACKWARDS if the index is negative
        if (m_nReadIndex < 0)
                m_nReadIndex += m_nBufferSize;      // amount of wrap is Read + Length
}
```

prepareForPlay()

- Reset the delay in prepareForPlay() after creating the buffer and before cookVariables().

```
bool __stdcall CDDLModule::prepareForPlay()
{
        // setup our delay line
        m_nBufferSize = 2*m_nSampleRate;                 // 2 seconds delay @ fs

        // delete it if it exists
        if(m_pBuffer)
                delete [] m_pBuffer;

        // create the new buffer
        m_pBuffer = new float[m_nBufferSize];

        // reset
        resetDelay();

        // then cook
        cookVariables();

        return true;
}
```

processAudioFrame()

- The flowchart for the processAudioFrame() function is shown in Figure 7.25.

Note: We have one minor detail to deal with, and this is going to happen when we use the delay line in a read-then-write fashion. If the user has chosen 0.00 mSec of delay, then

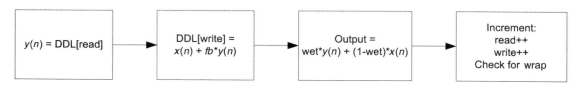

Figure 7.25: Flowchart for processAudioFrame().

the write pointer and read pointer will be the same. This also occurs if the user selects the *maximum* delay value since we want to read the oldest sample before writing it. So, we need to make a check to see if there is no delay at all and deal with it accordingly.

```
bool __stdcall CDDLModule::processAudioFrame(float* pInputBuffer, float* pOutputBuffer,
UINT uNumInputChannels, UINT uNumOutputChannels)
{
        // SYNC CODE: DO NOT REMOVE - DO NOT PLACE CODE BEFORE IT
        WaitForUIChangeDone();
        setProcessAudioDone(false);
        // END SYNC CODE

        // Do LEFT (MONO) Channel
        // Read the Input
        float xn = pInputBuffer[0];

        // Read the output of the delay at m_nReadIndex
        float yn = m_pBuffer[m_nReadIndex];

        // if zero delay, just pass the input to output
        if(m_fDelayInSamples == 0)
                yn = xn;

        // write the input to the delay
        m_pBuffer[m_nWriteIndex] = xn + m_fFeedback*yn;

        // create the wet/dry mix and write to the output buffer
        // dry = 1 - wet
        pOutputBuffer[0] = m_fWetLevel*yn + (1.0 - m_fWetLevel)*xn;

        // incremnent the pointers and wrap if necessary
        m_nWriteIndex++;
        if(m_nWriteIndex >= m_nBufferSize)
                m_nWriteIndex = 0;

        m_nReadIndex++;
        if(m_nReadIndex >= m_nBufferSize)
                m_nReadIndex = 0;

        // Mono-In, Stereo-Out (AUX Effect)
        if(uNumInputChannels == 1 && uNumOutputChannels == 2)
                pOutputBuffer[1] = pOutputBuffer[0]; // copy MONO!
```

```
        // DDL Module is MONO - just do a copy here too
        // Stereo-In, Stereo-Out (INSERT Effect)
        if(uNumInputChannels == 2 && uNumOutputChannels == 2)
                pOutputBuffer[1] = pOutputBuffer[0]; // copy MONO!

        // SYNC CODE: DO NOT REMOVE
        setProcessAudioDone();

        return true;
}
```

Build and test the module in RackAFX to make sure it works. You should get a generic delay effect with up to two seconds of delay. We only have two items to take care of to complete this first version of the module. The first is easy: we need to provide the user of the module with a way to choose external feedback and have access to the feedback variable. The second is more difficult: we need to be able to handle fractional delay.

In order for someone to use this module inside another plug-in and get access to the feedback path, we need to provide some functions that will only be called by a plug-in parent. We also need to modify the processAudioFrame() function to use an externally supplied feedback sample.

7.4 Modifying the Module to Be Used by a Parent Plug-In

It's actually pretty easy to modify this module to work as an internal module for a plug-in. We need to provide a variable for the feedback sample and allow the user to set this value. We also need to provide a way to allow the user to get the current feedback value. Finally, we need a switch to allow the user to select the external feedback mode. That switch will be in the form of a boolean flag. After we get the variables set up, we can modify the processAudioFrame() function to use the external feedback sample.

7.4.1 DDLModule.h File

Declare the following new variables:

```
bool m_bUseExternalFeedback;  // flag for enabling/disabling
float m_fFeedbackIn;          // the user supplied feedback sample value
```

Also, declare and implement the get/set functions. They are so short you can just put them in the .h file rather than implementing them in the .cpp file.

```
// Add your code here: ----------------------------------------------- //
float m_fDelayInSamples;
float m_fFeedback;
float m_fWetLevel;

void cookVariables();
void resetDelayLine();
```

```
float* m_pBuffer;
int m_nReadIndex;
int m_nWriteIndex;
int m_nBufferSize;

bool m_bUseExternalFeedback; // flag for enabling/disabling
float m_fFeedbackIn;         // the user supplied feedback sample value

// current FB is fb*output
float getCurrentFeedbackOutput(){return m_fFeedback*m_pBuffer[m_nReadIndex];}

// set the feedback sample
void setCurrentFeedbackInput(float f){m_fFeedbackIn = f;}

// enable/disable external FB source
void setUsesExternalFeedback(bool b){m_bUseExternalFeedback = false;}
// END OF USER CODE ----------------------------------------------------- //
```

The current feedback sample output is found at m_pBuffer[m_nReadIndex] and multiplied by the feedback coefficient as per the block diagram. By allowing the user to get and set the feedback, we allow them to break into the loop and insert other effects, or use the feedback output for some other purpose altogether.

7.4.2 DDLModule.cpp File

processAudioFrame()

- Modify the function to allow the use of externally supplied feedback samples:

```
bool __stdcall CDDLModule::processAudioFrame(float* pInputBuffer, float* pOutputBuffer,
UINT uNumInputChannels, UINT uNumOutputChannels)
{
        // Do LEFT (MONO) Channel
        // Read the Input
        float xn = pInputBuffer[0];

        // Read the output of the delay at m_nReadIndex
        float yn = m_pBuffer[m_nReadIndex];

        // if zero delay, just pass the input to output
        if(m_fDelayInSamples == 0)
             yn = xn;

        // write the input to the delay
        if(!m_bUseExternalFeedback)
             m_pBuffer[m_nWriteIndex] = xn + m_fFeedback*yn; // normal
        else
             m_pBuffer[m_nWriteIndex] = xn + m_fFeedbackIn; // external feedback
```

```
            // create the wet/dry mix and write to the output buffer
            // dry = 1 - wet
            pOutputBuffer[0] = m_fWetLevel*yn + (1.0 - m_fWetLevel)*xn;

        etc…
```

You can see that the change required is very minor—just a switch to change the feedback sample value. Of course if the user enables this option, then they are responsible for placing meaningful data in the fFeedbackIn variable.

Rebuild and test the code to make sure it still works properly. In the next section, we will use the module to make two different plug-ins:

1. Stereo digital delay
2. Stereo crossed-feedback delay

7.5 Modifying the Module to Implement Fractional Delay

Before we work on the bigger projects, we need to take care of the problem of fractional delay. We would like the user to be able to get any amount of delay they want. By implementing only sample-based delay, we are limiting the user to choosing delays that are multiples of the sample period, about 23 uSec. You might think that is enough accuracy; however, there are several instances where this won't be enough. The first is the case of a delay that is beats-per-minute (BPM) synchronized (e.g., delay is set to a multiple of the song's BPM to create synchronized echoes or percussion effects). The second case is that of modulated delays like chorus and flanging, which require smooth modulations from one delay time to another. Linear interpolation will provide acceptable results.

Suppose our delay is in a state where we have calculated our delay position to be at sample location 23.7183 samples. We need to find the value of the data at the location 0.7183 between sample 23 and sample 24. In Figure 7.26 you can see a graphic representation of the interpolation method. Since it's linear interpolation, a line is drawn between the adjacent samples and the interpolated value is found on the line at 0.7183 the distance between the two samples.

In polynomial interpolation such as LaGrange interpolation, a curve is drawn between the points (or a series of points), and then the interpolated value is found on that curve. There are several ways to implement the linear interpolation but the easiest method is to treat it like a DSP filter. Another way of thinking about interpolation is as a weighted sum. For example, if the interpolation point is 0.5 between samples 1 and 2, then the interpolated value is made up of 50% of sample 1 plus 50% of sample 2. In the above case, our interpolated distance is 0.7183, so we can view the output as

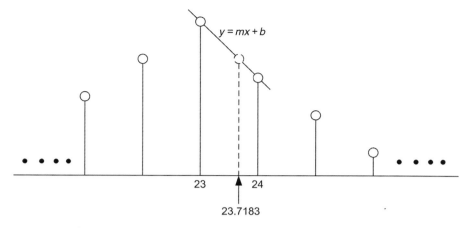

Figure 7.26: Linear interpolation of sample values.

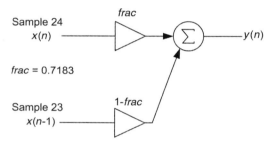

Figure 7.27: Linear interpolation as a kind of feed-forward filter. The z^{-1} element is removed since we do not know if the interpolated samples will always be exactly one sample apart. For example, what if the next fractional delay sample is at location 56.2394?

$$\text{interp\_output} = (0.7183)(\text{Sample 2}) + (0.2817)(\text{Sample 1})$$

Here is a linear interpolation function you can use; it is already declared in your pluginconstants.h file:

```
float dLinTerp (float x1, float x2, float y1, float y2, float x);
```

You give it a pair of data points ($x1,y1$) and ($x2,y2$), plus a distance between them on the x-axis (x), and it returns the interpolated value using the weighted sum method. The first part of the code checks for a potential divide by zero fault that technically should not happen. You should also be aware that linear interpolation is a form of feed-forward filtering. The block diagram of the above function would look like Figure 7.27.

Thus, in the processAudioFrame(), we need to find the two sample values that our actual delay time needs, then do the interpolation. It is pretty straightforward except the situation where we are interpolating across the wrap boundary (from the last sample in the buffer to the first one). Suppose the user enters a delay time that corresponds to 2.4 samples of delay. In the cookVariables() function, we locate the read index to be two samples before the write pointer because we cast the value to an integer, stripping out the fractional part. The actual delay we really want is 0.4 samples between the current sample and the one before it. In other words, $x(n-2.4)$ rather than $x(n-2)$. You can see that we need a sample that is between $x(n-2)$ and $x(n-3)$; in Figure 7.28 it happens to be a distance of 0.4 between them.

We need to modify our code as follows:

- We are already calculating and updating the first sample that we need to interpolate; it is located with the index value at m_nReadIndex, so there is nothing to change in the cooking function.
- In the processAudioFrame() function, we need to interpolate between our current read location m_nReadIndex and the location just *before* it in the buffer, m_nReadIndex-1.
- Since we are only focused on a single pair of samples at any time and we know they are one sample apart, we can use the values 0 and 1 for the interpolation function as $x1 = 0$, $x2 = 1$; then we interpolate the fractional distance between them. You can use m_nReadIndex but you will get in trouble when you are interpolating across the wrap boundary of the buffer.
- We will need to check for a wrap condition backward if m_nReadIndex−1 takes us outside the top of the buffer.

We can get the fractional value from our m_fDelayInSamples in several ways; here is one of them:

float fFracDelay = m_fDelayInSamples − (int)m_fDelayInSamples

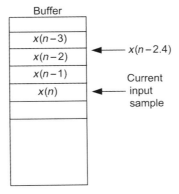

Figure 7.28: Fractional interpolation.

It really only comes down to locating the sample 1 behind our current read index, then using the linear interpolation code to get the fractional value. There are two extreme conditions to consider:

- At the maximum delay time, the read and write indices will be equal and the fractional part of the delay in samples will be 0 so no interpolation will occur—this is OK.
- If the delay in samples is less than 1, the read and write indices will also be equal, but this will be a problem. In this case, we need to interpolate between the current $x(n)$ and the sample at $x(n-1)$, one sample behind the read location. Branching will be necessary to catch this event and handle it.

7.5.1 DDLModule.cpp File

processAudioFrame()

- Modify the code to do the interpolation.

```
bool __stdcall CDDLModule::processAudioFrame(float* pInputBuffer, float* pOutputBuffer,
UINT uNumInputChannels, UINT uNumOutputChannels)
{
        // Do LEFT (MONO) Channel
        // Read the Input
        float xn = pInputBuffer[0];

        // Read the output of the delay at m_nReadIndex
        float yn = m_pBuffer[m_nReadIndex];

        // if delay < 1 sample, interpolate between input x(n) and x(n-1)
        if(m_nReadIndex == m_nWriteIndex && m_fDelayInSamples < 1.00)
        {
                // interpolate x(n) with x(n-1), set yn = xn
                yn = xn;
        }
        // Read the location ONE BEHIND yn at y(n-1)
        int nReadIndex_1 = m_nReadIndex - 1;
        if(nReadIndex_1 < 0)
                nReadIndex_1 = m_nBufferSize-1; // m_nBufferSize-1 is last location

        // get y(n-1)
        float yn_1 = m_pBuffer[nReadIndex_1];

        // interpolate: (0, yn) and (1, yn_1) by the amount fracDelay
        float fFracDelay = m_fDelayInSamples - (int)m_fDelayInSamples;

        // linerp: x1, x2, y1, y2, x
        float fInterp = dLinTerp(0, 1, yn, yn_1, fFracDelay); // interp frac between them

        // if zero delay, just pass the input to output
        if(m_fDelayInSamples == 0)
                yn = xn;
```

```
        else
                yn = fInterp;

        // write the intput to the delay
        if(!m_bUseExternalFeedback)
                m_pBuffer[m_nWriteIndex] = xn + m_fFeedback*yn;  // normfInterpal
        else
                m_pBuffer[m_nWriteIndex] = xn + m_fFeedbackIn;  // external feedback
                    sample
}
```

Now, build and test the module. Try a variety of delay settings. If you hear a repetitive click in your output that is exactly the same rate as the delay time, then there is something wrong during the interpolation across the wrap boundary of buffer[0] to buffer[size–1]. Only advance to the next section when you have this project fully debugged and functional. We will be including it in the next project, a stereo digital delay.

7.6 Design a Stereo Digital Delay Plug-In

In this project, we use two DDL modules in one parent plug-in. RackAFX makes it easy to do this by allowing you to add other plug-in components (.h and .cpp files) into a new project. It will automatically #include the components too. However, if you use external modules or other files you might need to manually #include these. In Figure 7.29 you can see that we now have two DDL modules declared as member objects of the new plug-in. The plug-in implements its own interface of three sliders, which we use to control our modules.

7.6.1 Project: StereoDelay

Create a project named "StereoDelay." When you create the project, you have the option of including other modules in your code, seen in Figure 7.30. RackAFX finds all of the existing RackAFX projects in the default directory you supply and lists them here. You use the Add button to move them into your project. If you have a project located in another directory that is not the default, you will need to move the files on your own (copy them to the new project directory and #include them in the new <project>.h file and add them into the compiler). RackAFX will automatically copy them and #include whichever modules you choose. In this case, choose the DDL module.

> When you use a plug-in as a module for another parent plug-in you must create and implement a new UI. The child plug-in objects will *not* expose their sliders to RackAFX, but you can manipulate the UI variables. All other aspects of the child objects work as expected. In this plug-in, we will implement another UI to control the modules. See Appendix A.2 for advanced control of the UI variables.

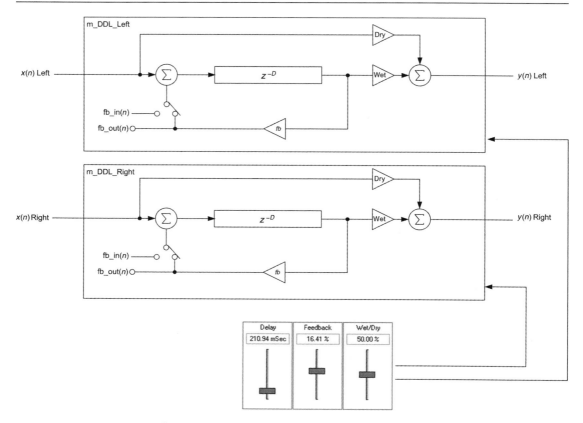

Figure 7.29: Block diagram of our stereo delay.

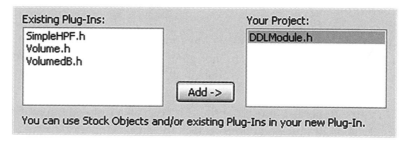

Figure 7.30: Adding existing modules can be done programmatically through RackAFX.

You can see in Figure 7.30 that I added "DDLModule.h" to the project. After completing the new project dialog, check your new <plugin>.h file:

```
#pragma once

#include "pluginconstants.h"
```

```
#include "DDLModule.h"
#include "plugin.h"

// abstract base class for DSP filters
class CStereoDelay : public CPlugIn
{
public: // Plug-In API Functions
        //
        // 1. One Time Initialization
        CStereoDelay();

etc ...
```

7.6.2 StereoDelay GUI

Your GUI will look like that in Figure 7.29 and you use the same setup from the DDLModule in Table 7.1.

7.6.3 StereoDelay.h File

In the .h file, declare two member objects of type CDDLModule. Also, add a function called setDelayVariables() to transfer our global delay variables to the member objects and optionally have the member objects cook the data:

```
// Add your code here: ----------------------------------------------- //
CDDLModule m_DDL_Left;
CDDLModule m_DDL_Right;

// function to pass our variables to member delays
void setDelayVariables(bool bCook);
// END OF USER CODE -------------------------------------------------- //
```

Our DDLModules are just C++ objects, so you can treat them as such and call their member functions and set their member variables. Before implementing the function above, go ahead and add your UI sliders, exactly the same as before with the same variable names. You will have something like this:

```
// ADDED BY RACKAFX -- DO NOT EDIT THIS CODE!!! ---------------------- //
// **--0x07FD--**

    float m_fDelay_ms;
    float m_fFeedback_pct;
    float m_fWetLevel_pct;

// **--0x1A7F--**
// ------------------------------------------------------------------- //
```

7.6.4 StereoDelay.cpp File

Write the setDelayVariables() function. The Boolean flag allows you to optionally call the cooking functions on the members:

```
void CStereoDelay::setDelayVariables(bool bCook)
{
        // forward our variables over to the member objects
        m_DDL_Left.m_fDelay_ms = m_fDelay_ms;
        m_DDL_Right.m_fDelay_ms = m_fDelay_ms;

        m_DDL_Left.m_fFeedback_pct = m_fFeedback_pct;
        m_DDL_Right.m_fFeedback_pct = m_fFeedback_pct;

        m_DDL_Left.m_fWetLevel_pct = m_fWetLevel_pct;
        m_DDL_Right.m_fWetLevel_pct = m_fWetLevel_pct;

        // cook, if desired
        if(bCook)
        {
                m_DDL_Left.cookVariables();
                m_DDL_Right.cookVariables();
        }
}
```

Constructor

- Initialize the delay variables and cook them; this version does not use the external feedback option, so set the bool accordingly.

```
CStereoDelay::CStereoDelay()
{
        <SNIP SNIP SNIP>

        // Finish initializations here
        m_DDL_Left.m_bUseExternalFeedback = false;
        m_DDL_Right.m_bUseExternalFeedback = false;

        // set and (true) cook the delays
        setDelayVariables(true);

}
```

prepareForPlay()

- Set the delay variables.
- Forward the calls to prepareForPlay() on the member objects, which will cook them.

The DDL module will handle flushing the buffers and so on.

```
bool __stdcall CStereoDelay::prepareForPlay()
{
        setDelayVariables(false);

        m_DDL_Left.prepareForPlay();
        m_DDL_Right.prepareForPlay();

        return true;
}
```

userInterfaceChange()

- Set the delay variables and cook them whenever one of our controls changes. This could be streamlined for better efficiency.

```
bool __stdcall CStereoDelay::userInterfaceChange(int nControlIndex)
{
        // set and cook the variables
        setDelayVariables(true);

        return true;
}
```

processAudioFrame()

- Forward the processAudioFrame() call to the member objects.
- Remember that we need to send it only one channel at a time and make sure it is the correct one. Note the "address of" operator (&) to point to the single memory location we pass it.

```
bool __stdcall CStereoDelay::processAudioFrame(float* pInputBuffer, float*
pOutputBuffer, UINT uNumInputChannels, UINT uNumOutputChannels)
{
        // Do LEFT (MONO) Channel; there is always at least one input/one output
        // forward call to sub-object: pInput, pOutput, 1 input ch, 1 output ch
        m_DDL_Left.processAudioFrame(&pInputBuffer[0], &pOutputBuffer[0], 1, 1);

        // Mono-In, Stereo-Out (AUX Effect)
        if(uNumInputChannels == 1 && uNumOutputChannels == 2)
                pOutputBuffer[1] = pOutputBuffer[0]; // just copy

        // Stereo-In, Stereo-Out (INSERT Effect)
        if(uNumInputChannels == 2 && uNumOutputChannels == 2)
                // forward call to sub-object pInput, pOutput, 1 input ch, 1 output ch
                m_DDL_Right.processAudioFrame(&pInputBuffer[1], &pOutputBuffer[1], 1, 1);
        return true;
}
```

Rebuild and test the project and you now have a stereo version of the previous project. Hopefully, you have a better idea of how powerful it can be to create modules that are combined, though it does take a bit of extra work on the module since you have to think ahead and implement functions or variables that are not required for standalone (simple) plug-in operation. Next, we'll exercise the module by converting this plug-in to a crossed feedback delay. We're going to do this in a specific way so we can later use an enumerated variable to switch between normal and crossed-feedback operation.

7.7 Design a Stereo Crossed-Feedback Delay Plug-In

A crossed-feedback delay is implemented by crossing the feedback paths of the two delay lines. We will add this functionality to the existing StereoDelay project. This is where we will use the external feedback option that we built into the module. In Figure 7.31 you can trace the feedback paths to see that they are crossed into the opposite delay line. You will be surprised at how easy it is to convert the delay into a crossed-feedback delay:

- Enable the external feedback function with the boolean flag.
- Use the getCurrentFeedbackOutput() and setCurrentFeedbackInput() functions to "cross the beams" of the feedback paths.

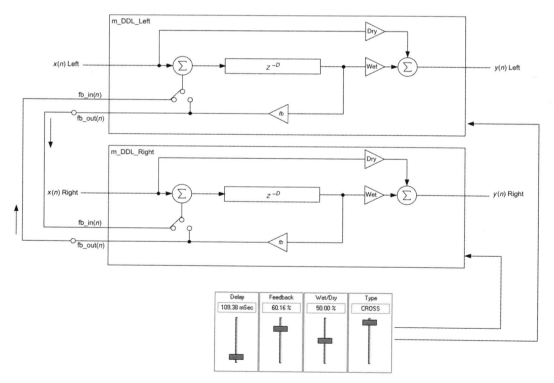

Figure 7.31: Block diagram of the crossed-feedback delay.

7.8 Enumerated Slider Variables

You can see from Figure 7.31 that there is a new slider control for the GUI to select between normal and crossed-feedback operation. You can take advantage of RackAFX's enumerated UINT variable to make a slider that selects between multiple enumerated values, like a multi-position switch. In our case, we only have two positions right now: normal and cross.

Right-click on a slider to bring up the properties dialog and add a new slider/variable combination, as shown in Figure 7.32. Choose "enum" for the data type and create a variable called m_uDelayType—the variable will be an unsigned integer type (UINT). The low, high, and initial value cells will no longer be editable; RackAFX will figure them out depending on what you type in the enumerated variable box. You type in the values separated by commas. The first value will be the default value.

In the box at the bottom, type in the strings which will be shown in the slider control (keep them short); you can have as many enumerated variables as you want, but in this

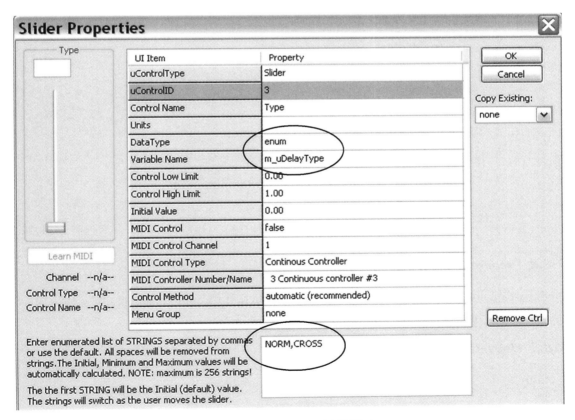

Figure 7.32: The enumerated UINT data type.

case, we only need NORM and CROSS. Go to your plug-in's .h file to see the new variables:

```
// ADDED BY RACKAFX -- DO NOT EDIT THIS CODE!!! ------------------------------- //
// **--0x07FD--**

float m_fDelay_ms;
float m_fFeedback_pct;
float m_fWetLevel_pct;
UINT m_uDelayType;
enum{NORM,CROSS};
// **--0x1A7F--**
// --------------------------------------------------------------------------- //
```

For enumerated variables, the first one in the list will be the default, with the slider at the bottom. The other strings will appear in sequence when you move the slider up. The UINT will automatically be updated. We can use this as a directly controlled variable so there's nothing to add in userInterfaceChange(). Make the edits discussed next.

7.8.1 Constructor

- Initialize the delay type to NORM.

```
CStereoDelay::CStereoDelay()
{
        <SNIP SNIP SNIP>

        // Finish initializations here
        m_DDL_Left.m_bUseExternalFeedback = false;
        m_DDL_Right.m_bUseExternalFeedback = false;

        // set and (true) cook the delays
        setDelayVariables(true);

        // init the delay type
        m_uDelayType = NORM;
}
```

7.8.2 PrepareForPlay()

Nothing to do; we don't care what mode we're in and don't want to reset the mode each time the user plays audio.

7.8.3 UserInterfaceChange()

Nothing to do; we are using this as a direct control variable.

7.8.4 ProcessAudioFrame()

- Use the enumerated variable in a switch/case statement to modify the feedback as required. For CROSS operation:
  - Set the external feedback flag to true.
  - Get the left feedback output and write it to the right feedback input.
  - Get the right feedback output and write it to the left feedback input.
- Call the processAudioFrame() functions on the DDL modules.

```
bool __stdcall CStereoDelay::processAudioFrame(float* pInputBuffer, float*
pOutputBuffer, UINT uNumInputChannels, UINT uNumOutputChannels)
{
        switch(m_uDelayType)
        {
        case CROSS:
                // CROSS FB DELAY --- NOTE: MUST HAVE STEREO FILE! --------------- //
                m_DDL_Left.m_bUseExternalFeedback = true;
                m_DDL_Right.m_bUseExternalFeedback = true;

                // cross the Feedbacks
                m_DDL_Left.setCurrentFeedbackInput(m_DDL_Right.getCurrent
                FeedbackOutput());
                m_DDL_Right.setCurrentFeedbackInput(m_DDL_Left.getCurrent
                FeedbackOutput());
                break;

        case NORM: // clear the flags
                m_DDL_Left.m_bUseExternalFeedback = false;
                m_DDL_Right.m_bUseExternalFeedback = false;
                break;

        default: // deault is NORM operation; clear the flags
                m_DDL_Left.m_bUseExternalFeedback = false;
                m_DDL_Right.m_bUseExternalFeedback = false;
                break;
        }

        // SHARED CODE ----------------------------------------------------- //
        // Do LEFT (MONO) Channel; there is always at least one input/one output
        // forward call to sub-object: pInput, pOutput, 1 input ch, 1 output ch
        m_DDL_Left.processAudioFrame(&pInputBuffer[0], &pOutputBuffer[0], 1, 1);

        // Mono-In, Stereo-Out (AUX Effect)
        if(uNumInputChannels == 1 && uNumOutputChannels == 2)
                pOutputBuffer[1] = pOutputBuffer[0]; // just copy
```

```
        // Stereo-In, Stereo-Out (INSERT Effect)
        if(uNumInputChannels == 2 && uNumOutputChannels == 2)
            // forward call to sub-object pInput, pOutput, 1 input ch, 1 output ch
            m_DDL_Right.processAudioFrame(&pInputBuffer[1], &pOutputBuffer[1], 1, 1);

    return true;
}
```

Rebuild and test the plug-in; exercise the new delay type control as well.

Recapping some of the things you learned:

- Declaring, flushing, and manipulating delay lines as circular buffers.
- Implementing fractional delay times using interpolation.
- Using a plug-in as a child module for a parent plug-in.
- How to add an enumerated UINT variable for type selection or using a slider as a multi-pole switch.

7.9 More Delay Algorithms

Here are some ideas for a more complex DDL module and other delay plug-ins.

7.9.1 Advanced DDL Module

The more flexible module in Figure 7.33 allows the parent to have access to the input and pre-wet output of the delay line and the input and output of the feedback path. Another module (low-pass filter, for example) could be inserted here or it could be used for more crossed-delay modes.

7.9.2 Delay with LPF in Feedback Loop

Analog delays suffer a high frequency loss on each path through the delay line/feedback path. This can be modeled with a LPF in the feedback path, as shown in Figure 7.34. You can

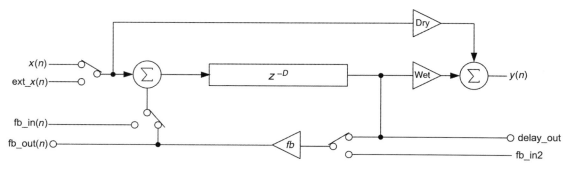

Figure 7.33: A DDL module with more input and feedback path options.

try first- or second-order filters with or without resonance for a variety of interesting delay effects. For other ideas, try changing the type (LPF, high-pass filter, band-pass filter, band-stop filter) or location (before or after the *fb* coefficient). Be careful with adding resonant or high-gain filters in the feedback loop. You could also inject some filtered white noise into the feedback path to simulate the noisy characteristics of typical analog delays.

7.9.3 Multi-Tap Delay

The multi-tap delay line shown in Figure 7.35 produces four separate delays from one unit. Only the longest delay is fed back in this version. You create the multiple taps by simply declaring more read index values (m_nReadIndexTap_1, m_nReadIndexTap_2, etc..); the multiple taps are used and incremented exactly the same as the single tap case.

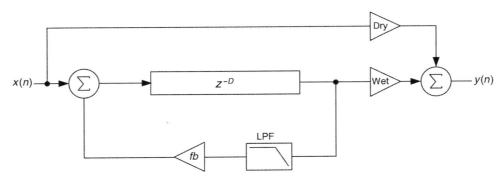

Figure 7.34: An analog delay modeled with an LPF in the feedback loop.

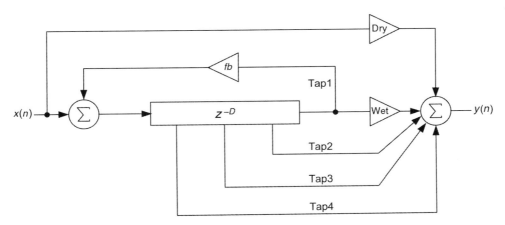

Figure 7.35: A four-tap multi-tap delay line.

You can also experiment with multiple feedback paths, filters, or setting your tap times using BPM and musical rhythms (e.g., delays on eighth and quarter notes).

7.9.4 Ping-Pong Delay

Figure 7.36 shows how the ping-pong delay cycles the delay back and forth between the left and right channels.

7.9.5 LCR Delay

The LCR delay in Figure 7.37 based on the Korg Triton® has both LPF and HPF in the feedback path (both are switchable or bypass-able) for multiple feedback tone shaping options.

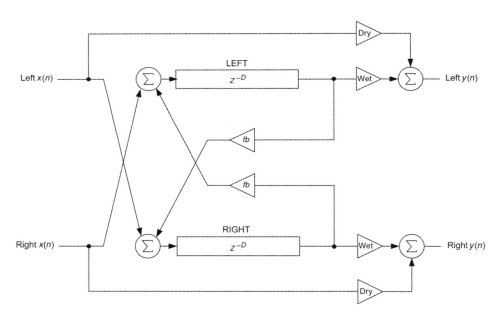

Figure 7.36: The ping-pong delay builds on the cross-feedback delay by crossing the inputs as well as the feedback paths to produce the back and forth ping-pong effect. You will probably want to design the advanced DDL module first and use its input, output, and feedback ports.

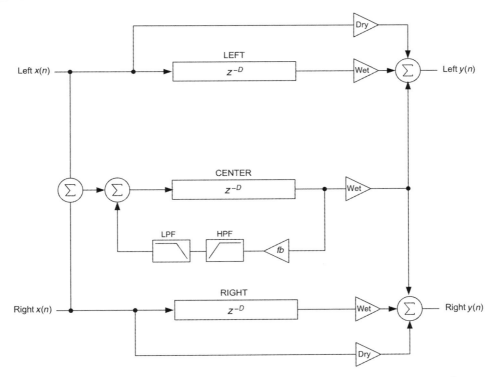

Figure 7.37: The LCR delay, based on the LCR delay in the Korg Triton®.

Bibliography

Coulter, D. 2000. *Digital Audio Processing*, Chapter 11. Lawrence, KS: R&D Books.
DSP56KFAM/AD. Schomberg, ON: Motorola, Inc.
Korg, Inc. 2000. *Triton-Rack Parameter Guide*, Tokyo: Korg, Inc.
Motorola, Inc. 1992. *DSP56000 Digital Signal Processor Family Manual*.
Roads, C. 1996. *The Computer Music Tutorial*, Chapter 3. Cambridge, MA: The MIT Press.

CHAPTER 8
Audio Filter Designs: FIR Filters

Infinite impulse response (IIR) filters have several attractive properties:

- They only require a few delay elements and math operations.
- You can design them directly in the z-plane.
- You can use existing analog designs and convert them to digital with the Bilinear z-Transform (BZT); the fact that IIR topologies somewhat resemble the signal flow in analog filters emphasizes their relationship.
- You can get extreme resonance/gain by placing poles very near the unit circle.
- You can make filters, EQs, and so on with controls that link to the coefficients directly or indirectly for real-time manipulation of the plug-in.

Their main drawback is that they can become unstable and blow up, or oscillate. Their impulse responses (IRs) can be infinite. Finite impulse response (FIR) filters have all zeros and a finite IR. They are unconditionally stable so their designs can never blow up. However, they can put out all zeros, a constant value, a series of clicks or pulses, or other erroneous output, but they don't actually go unstable.

You can also make a linear phase filter with an FIR, just like the simple feed-forward (FF) topology you analyzed in Chapters 5 and 6; a linear phase filter is impossible to make with an IIR topology, although you can approximate it by adding phase compensation filters. However, the one thing that separates FIR filters from all other kinds, including analog counterpart filters, is that their coefficients $a_0, a_1, \ldots, a_N$ are the IR of the filter. You proved that when you manually pushed an impulse through the simple FF filter in Chapter 4, and then again when taking the z transform of the IR of the same filter in Chapter 5.

8.1 The IR Revisited: Convolution

In Chapter 1 you saw how the digitized signal was reconstructed into its analog version by filtering through an ideal low-pass filter (LPF). When the series of impulses is filtered, the resulting set of $\sin(x)/x$ pulses overlap with each other and their responses all add up linearly. The addition of all the smaller curves and damped oscillations reconstructs the inter-sample curves and damped fluctuations (Figure 8.1).

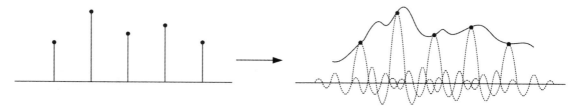

Figure 8.1: The sin(x)/x outputs of the LPF are summed together to reconstruct the original band-limited input waveform.

In the time domain, you can see how the IR of each sample is overlaid on the others and that the summing together of the peaks and valleys of the sin(x)/x shape ultimately creates the portions in between the samples which appeared to have been lost during the sampling process.

The process of overlaying the IR on top of the input stream of impulses *x(n)* and adding up the results to get the final time domain output *y(n)* is called convolution. Convolution is a mathematical operation used in many fields of science; digital audio processing is just one of them. The mathematical symbol for convolution is * which can be confusing because this is used to represent the multiplication operation in C/C++. In the above example, you convolved the input signal *x(n)* with the IR *h(n)* by overlaying the *h(n)* signal on top of each input impulse, then summing up signals to create the final output *y(n)*. Mathematically, you would write this as Equation 8.1:

$$y(n) = x(n) * h(n) \tag{8.1}$$

Visually, it's easy to understand the concept of overlaying the signals and adding them to get the final result, but how do you describe that mathematically? The answer is that this kind of operation is a special circumstance of a more generalized operation of convolution. Mathematically, convolution for discrete signals is described in Equation 8.2:

$$c(n) = f(n) * g(n) = \sum_{m=-\infty}^{+\infty} f(n)g(n-m) \tag{8.2}$$

In this case, *f* and *g* are two generalized signals and neither of them has to be an IR. Convolution is commutative, so that $f * g = g * f$, or Equation 8.3:

$$\begin{aligned} c(n) &= f(n) * g(n) = \sum_{m=-\infty}^{+\infty} f(n)g(n-m) \\ c(n) &= g(n) * f(n) = \sum_{m=-\infty}^{+\infty} g(n)f(n-m) \end{aligned} \tag{8.3}$$

> If you know how a system affects one single impulse, you can exactly predict how it will affect a stream of impulses (i.e., a signal) by doing the time domain overlay. If you have the IR of a system, you have the *algorithm* for the system coded in a single function.

The operation this equation is describing is not simple. On the right-hand side of Equation 8.3 the function $f(n)$ is one signal while the function $g(n - m)$ represents the second signal reversed in time. The multiplication/summation of the two across $-\infty$ to $+\infty$ describes the process of sliding the two signals over each other to create overlapping areas. On each iteration, the area under the curve of overlap between $g(n - m)$ and $f(n)$ is computed. This results in a third (output) signal $c(n)$. This signal $c(n)$ is made up of the overlapping area of the two input signals. This operation is shown graphically in Figures 8.2 and 8.3.

Thus, the convolution of two arbitrary signals is quite involved mathematically. If the two signals have any complexity at all, the resulting convolution signal is generally not distinguishable as a linear combination of the two. If you know the IR of a system $h(n)$, you can convolve it with the input signal $x(n)$ to produce the output signal $y(n)$. This is the equivalent of multiplying the transfer function $H(z)$ with the input signal $X(z)$ to produce

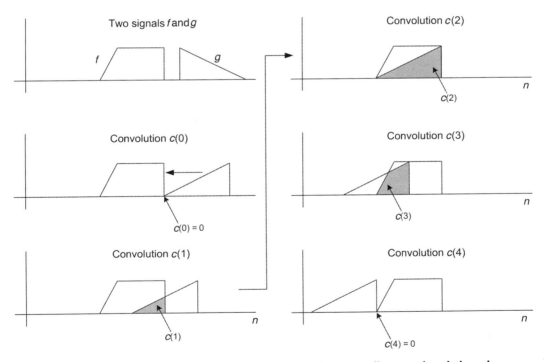

Figure 8.2: Two signals f and g are convolved. These are discrete signals but the sample symbols have been removed to make it easier to see; instead they are shown as continuous. In the first step, $c(0)$, one of the signals is reversed and the two are pushed up next to each other. As the convolution progresses through each step $c(1)$ to $c(4)$, the overlapping areas are calculated and stored.

the output $Y(z)$. Thus convolution in the time domain is multiplication in the frequency (z) domain (Equation 8.4).

$$y(n) = x(n) * h(n) \leftrightarrow Y(z) = X(z)H(z) \tag{8.4}$$

To understand how a FF filter implements convolution first rearrange the block diagram. Let's consider a long FIR filter with $N + 1$ coefficients in Figure 8.4.

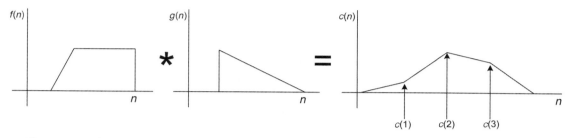

Figure 8.3: The convolution of f and g results in $c(n)$, consisting of five samples $c(0)$ through $c(4)$, which represent the overlap areas.

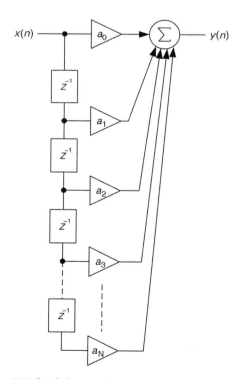

Figure 8.4: The familiar FIR feed-forward structure expanded out to N delay taps with $N + 1$ coefficients. It is important to see that there is one less delay element than coefficients since a_0 is multiplied against the original undelayed signal.

Next, mentally rotate the structure so it looks like Figure 8.5. In Figure 8.5 you can see that at any given time, a portion of the input signal $x(n)$ is trapped in the delay line. On each sample period, the input signal slides to the right and a summation is formed with the product of the coefficients and the samples $x(n - d)$. The words "sliding, summation and product" are key here—they're the same words used to describe convolution.

In Figure 8.6, the input signal $x(n)$ moves through the delay line being convolved with the IR on each sample period. Since each sample in the delay line is an impulse, and each impulse is symmetrical when reversed, this is the same as conceptually overlapping the IR on top of each sample and scaling by the sample value. The result is the final summation of all the peaks and valleys of the IR with the delayed signal $x(n)$. Thus, an FIR filter exactly implements discrete convolution in the time domain. This ultimately gives us a whole new way to filter a signal—by convolving against an impulse.

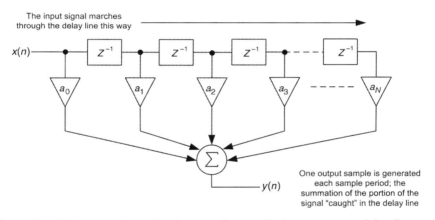

Figure 8.5: This rotated version is sometimes called a transverse delay line or a transverse structure.

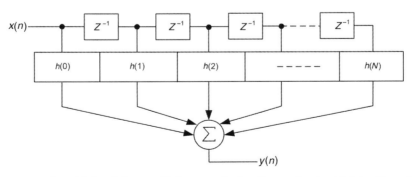

Figure 8.6: You can also think of the coefficients as being frozen in the $h(n)$ buffer while the input signal marches one sample to the right on each iteration.

So, if an ideal LPF has an IR in the shape $\sin(x)/x$ and we sample the IR, we get a discrete IR $h(n)$. The more samples we take, the more accurate our version of the IR becomes. This introduces the first way to design a FIR filter: find an IR you like, sample it, and convolve with it to produce the filtered output.

8.2 Using RackAFX's Impulse Convolver

RackAFX has a built-in module to do impulse convolution and a directory of IR files that you can experiment with. The impulses are stored in a directory called IR1024 and they are all 1024-point IRs. Some of them came from RackAFX itself—you can save IRs of any plug-in you make, then load them into the convolver module. You will also learn to write your own convolution plug-in and tell RackAFX that your software would like to receive IRs any time a user loads or creates one using the built-in tools.

First, let's look at the Impulse Convolver tool. Open the modules menu item in RackAFX. There are two built-in FIR modules: the Impulse Convolver 1024 and the FIR Designer located at the bottom of the list. Choose the Impulse Convolver 1024 module. The analyzer will then appear with the IR directory populated and additional IR buttons enabled. These will allow you to load an IR file into the convolver.

8.2.1 Loading IR Files

On the right side of the analyzer (Figure 8.7), you will see a box full of the IRs in your IR1024 directory. You might not have the exact same list as this one but you will have the file "optimal.64.sir" in the list. All the IR files are named with the ".sir" suffix and must be created in RackAFX or loaded using the RackAFX IR file format (see the website for details). RackAFX automatically saves your IRs as .wav files so you can use them in other applications. You can find these .wav files in your IR1024 directory. If you want to convolve with your own .wav files, see the website for code examples.

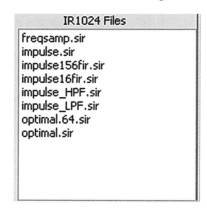

Figure 8.7: The IR files are visible in the analyzer window.

At the bottom right, you will see the buttons for loading and saving IR files. The first two, Save *h(n)* and Load *h(n)*, will save and load the .sir files from the IR1024 directory. The lower two buttons save and load the IR to the clipboard. The IR is actually C++ code, and you can use the clipboard to paste this code directly into your own source code. You might do this to hard-code a particular IR or for testing purposes, or to set a default IR.

To get started, click on the file named "optimal.64.sir" and then click the Load h(n) File button (you can also double-click on the file name to load it). The IR for the file will load into the analyzer as well as the Impulse Convolver module itself. You will automatically be switched into the impulse view to see the imported file data (Figure 8.8). Next, click on the Frequency Response button to see the filter's frequency response (Figure 8.9). This LPF was designed to have a cutoff frequency of 500 Hz and a very steep roll-off of about 50 dB/oct. Load a wave file and play the file through the convolver. It will implement this 500-Hz LPF exactly. Try loading some other IR files and playing audio through them.

8.2.2 Creating IR Files

You can capture the IR of any RackAFX plug-in, including your own. As an example let's capture the IR of a built-in module first. Open the module named "HP/LP Filter" from the module menu. Then, open the analyzer, click on the Frequency button and adjust the slider

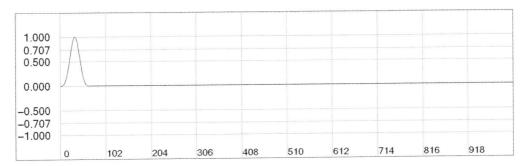

Figure 8.8: The IR for the optimal.64.sir file.

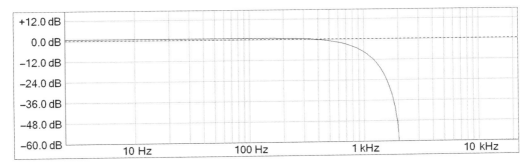

Figure 8.9: The frequency response for the optimal.64.sir file.

controls to give you a unique shape. For example, I will make a highly resonant LPF by setting the Q to 12 (Figure 8.10).

Click on the Impulse button in the analyzer. The IR of the filter is shown in Figure 8.11. This IR completely captures the filter at these particular settings (f_c = 1 kHz, Q = 12). If we store the IR of the IIR filter, we can load it into the convolver and turn it into a FIR filter instead.

Click on the Save h(n) File button and name it. It will then appear in the list of available IRs. Play a wave file through the module and remember what it sounds like. Now, go back and open the Impulse Convolver 1024 module from the module menu. You will see your freshly created IR in the list. Double-click on it to load it and you will see the original IR. Click on the Frequency button and you will see the original frequency response. Next, play a wave file through the convolver and you should hear something that is remarkably similar to the original IIR filter.

But, how similar is it? The IR convolver module can convolve up to 1024-point IRs. If the IR of the original filter is longer than 1024 points, then the resulting FIR filter will not be exactly perfect. However, if it is shorter than 1024 points (meaning that the IR becomes 0.0 and remains that way at some point in the IR duration) then the resulting FIR will be a dead-on accurate version of the original. If you look at the IR for the original filter, you can see that it

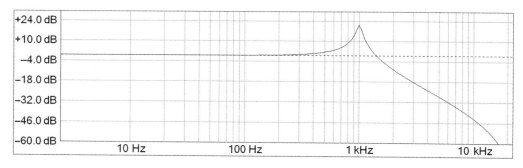

Figure 8.10: The frequency response for the resonant LPF test filter.

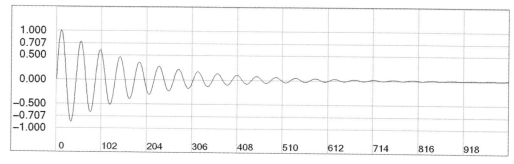

Figure 8.11: The ringing IR of the resonant LPF.

is still just barely ringing right there at the end, so we are not getting an exact duplicate, but upon listening you should hear that they are nearly identical.

8.2.3 The IR File Format

The IR file actually contains C++ code and you can quickly understand how it works by using the clipboard functions. In the analyzer window that you still have pulled up, click on the button $h(n) \rightarrow$ Clipboard and after the success message, open a text editor or a C++ compiler. Then, use the operating system (OS) paste function to paste the clipboard data into your editor. You will see something like this at the top:

```
// h(n) Impulse Response
// Length = 1024

m_nIRLength = 1024;
m_h_Left[0] = 0.00000000;
m_h_Left[1] = 0.00503618;
m_h_Left[2] = 0.01998402;
m_h_Left[3] = 0.03938961;
m_h_Left[4] = 0.05777339;
m_h_Left[5] = 0.07477701;
m_h_Left[6] = 0.09007415;
m_h_Left[7] = 0.10337673;
```

etc...

This is the IR data starting with the length (1024) and then the left and right channels respectively. If you scroll to the bottom, you will see the last few samples of the IR:

```
m_h_Right[1016] = -0.00000850;
m_h_Right[1017] = 0.00004104;
m_h_Right[1018] = 0.00008918;
m_h_Right[1019] = 0.00013495;
m_h_Right[1020] = 0.00017745;
m_h_Right[1021] = 0.00021589;
m_h_Right[1022] = 0.00024953;
m_h_Right[1023] = 0.00027775;
```

This was my resonant LPF and you can see that it is still ringing, even after 1024 samples. If you listen to a wave file through a filter like this, you can hear pinging sounds at the peak resonant frequency. These pinging noises are the ringing of the filter. If you look at the C++ code you can tell that the IR data appears to be some kind of member variable information for a C++ object because of the "m_" Hungarian notation used to describe the data.

> All RackAFX plug-ins already have two default IR arrays declared as m_h_Left[1024] and m_h_Right[1024] and another variable m_nIRLength that defines how much of the 1024 point IR buffer is being used. The FIR designer will let you create IRs with variable sizes—in many cases, you don't need all 1024 points to describe the IR of the system. You tell RackAFX that you want it to populate your IR arrays by setting a flag m_bWantIRs in your plug-in constructor. When a user loads or creates an IR in the analyzer, it is automatically delivered to your plug-in.

8.3 Using RackAFX's FIR Designer

RackAFX has a powerful built-in module called FIR Designer that lets you use two popular methods to design FIR filters: the optimal method and the frequency sampling method. The optimal method is sometimes called the "Parks–McClellan algorithm." When the module first opens, you will see the new controls at the right side, as shown in Figure 8.12.

This module creates IRs. You can save the IRs to a file or the clipboard using the same buttons as before. If your plug-in has the m_bWantIRs flag set, any time the user hits the Calculate button to make a new IR, it will automatically be delivered and copied into the plug-in's default IR arrays. Even though the FIR Designer defaults to the optimal method for design, let's begin with the frequency sampling method since the optimal method relies on it.

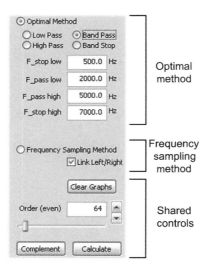

Figure 8.12: The FIR designer controls consist of three parts. The order slider and edit box and the Calculate and Complement buttons are shared between both methods. You enable the method of choice using the radio button controls.

8.4 The Frequency Sampling Method

The frequency sampling method is really interesting because it lets you design any kind of frequency response you want—it can be any arbitrary shape and it does not have to be a classical filter type (LPF, high-pass filter [HPF], band-pass filter [BPF], band-stop filter [BSF]), but you can make these types if you want to. The frequency sampling method involves these steps:

1. Decide on a desired frequency response and plot it in the frequency domain.
2. Sample the frequency response at evenly spaced intervals determined by the filter order you choose.
3. Take the inverse discrete cosine-transform (DCT) of the sampled frequency response to get the sampled IR (the DCT is simply the real part of the fast Fourier transform [FFT]).
4. Load the sampled IR into a convolver and go.

8.4.1 Linear-Phase FIR Using the Frequency Sampling Method

Choose:

N = number of coefficients

Calculate:

For N = odd:
- $(N + 1)/2$ = number of samples in frequency domain, starting at 0 Hz

For N = even
- $N/2$ = number of samples in frequency domain, starting at 0 Hz

$\Delta f = f_s/N$ = frequency spacing, starting at 0 Hz

Calculate the filter coefficients a_0 to $a_{N/2}$ with Equation 8.5:

For N = odd

$$a_n = \frac{1}{N}\left[H(0) + 2\sum_{i=1}^{(N-1)/2}|H(i)|\cos\left[2\pi i\left[n - \frac{N-1}{2}\right]/N\right]\right]$$

(8.5)

For N = even

$$a_n = \frac{1}{N}\left[H(0) + 2\sum_{i=1}^{N/2-1}|H(i)|\cos\left[2\pi i\left[n - \frac{N-1}{2}\right]/N\right]\right]$$

Note: This produces half the coefficients; the other half are a mirror image, as shown in the example below. Because the IR is guaranteed to be symmetrical about its center, it produces a linear phase FIR filter every time.

Example: Design an LPF with a cutoff of 5.5 kHz, $f_s = 44.1$ kHz, $N = 16$.

Solution:

1. $N = 16$, which produces eight sampled points in the frequency domain with a spacing of 2.756 kHz.
2. Sample the plot, producing the magnitude response, $H(i)$ (Figure 8.13).

For this plot notice that:

- The plot is linear in frequency. To design a frequency sampling method filter by hand, it is often easier to start with a linear frequency axis.
- The point at Nyquist is not used; it is there to pin down the response that would occur after Nyquist.

The sampled frequency response is read directly off the plot: (0 dB = 1.0, −60 dB = 0.001).

$$|H(i)| = \{1.0, 1.0, 1.0, 0.001, 0.001, 0.001, 0.001, 0.001\}$$

1. Use Equation 8.5 to extract IR $h(n)$, which are the coefficients, a_n.

$a_0 = 0.04858366$
$a_1 = 0.00364087$
$a_2 = -0.05199205$
$a_3 = -0.07047625$
$a_4 = -0.02194221$
$a_5 = 0.08695625$
$a_6 = 0.21101949$
$a_7 = 0.29421023$
$a_8 = 0.29421023$
$a_9 = 0.21101949$
$a_{10} = 0.08695625$
$a_{11} = -0.02194221$
$a_{12} = -0.07047625$
$a_{13} = -0.05199205$
$a_{14} = 0.00364087$
$a_{15} = 0.04858366$

Notice the symmetry about a_7 to a_8 boundary.

2. Measure the response (Figure 8.14).

The resulting filter is guaranteed to exactly match the desired frequency response at the sampled points only. In between the sampled points, the frequency response can do anything;

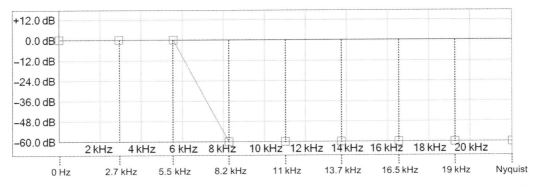

Figure 8.13: Here is our prototype LPF with cut-off at 5.5 kHz; its response becomes −60 dB (0.001) at 8.2 kHz, which is a steep roll-off.

rippling in the pass band and stop band can occur as shown here. You can see that this is a pretty bad rendition of our desired response.

The first lobe in the stop band only produces about −14 dB of attenuation, which is poor considering that we wanted a filter that would have a stop-band attenuation of −60 dB. To improve the response, you have several choices:

- Relax the specifications.
- Add more points to the desired response.
- Increase the filter order.

Relax the specifications and roll off the response less steeply by changing the point at 8.2 kHz to −12 dB (0.25) instead of −60 dB (Figure 8.15). Now, the sampled frequency response is

$$|H(i)| = \{1.0, 1.0, 1.0, 0.25, 0.0, 0.0, 0.0, 0.0\}$$

The resulting magnitude response $|H(f)|$ is shown in Figure 8.16.

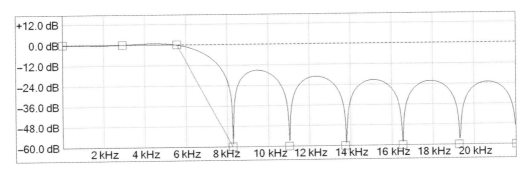

Figure 8.14: The resulting low-pass filter magnitude response.

Figure 8.15: The same design with specifications relaxed; the slope is less steep.

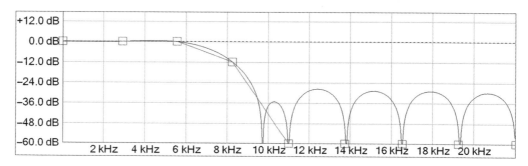

Figure 8.16: The relaxed magnitude response shows improved stop-band attenuation. Now, the first lobe in the stop band has moved to a magnitude of about −30 dB, an improvement of about 15 dB.

8.5 Complementary Filter Design for Linear Phase FIR Filters

This technique results in a complementary filter, rotated about the center of the Nyquist bandwidth, that is, rotated about Nyquist/2. To convert an LPF to HPF or vice versa on a linear phase FIR:

For N = even
- Multiply the even-numbered coefficients by −1.

For N = odd
- Multiply the odd-numbered coefficients by −1.

This will rotate the frequency response around Nyquist/2 such that an LPF will become an HPF. However, they will not share the same cutoff frequency, but will rather be mirror images of each other. Thus, the first filter design above with a cutoff point of 5.5 kHz would produce an HPF with a cutoff frequency 5.5 kHz *below Nyquist*. Table 8.1 shows the coefficients, while Figure 8.17 shows the frequency response.

Table 8.1: The LPF and complementary HPF coefficients for the current design.

Low-Pass Filter	Complementary High-Pass Filter
$a_0 = 0.02598719$	$a_0 = -0.02598719$
$a_1 = 0.01331121$	$a_1 = 0.01331121$
$a_2 = -0.02018474$	$a_2 = 0.02018474$
$a_3 = -0.05761180$	$a_3 = -0.05761180$
$a_4 = -0.04827195$	$a_4 = 0.04827195$
$a_5 = 0.04957144$	$a_5 = 0.04957144$
$a_6 = 0.20692666$	$a_6 = -0.20692666$
$a_7 = 0.33027202$	$a_7 = 0.33027202$
$a_8 = 0.33027202$	$a_8 = -0.33027202$
$a_9 = 0.20692666$	$a_9 = 0.20692666$
$a_{10} = 0.04957144$	$a_{10} = -0.04957144$
$a_{11} = -0.04827195$	$a_{11} = -0.04827195$
$a_{12} = -0.05761180$	$a_{12} = 0.05761180$
$a_{13} = -0.02018474$	$a_{13} = -0.02018474$
$a_{14} = 0.01331121$	$a_{14} = -0.01331121$
$a_{15} = 0.02598719$	$a_{15} = 0.02598719$

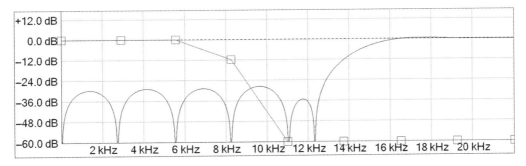

Figure 8.17: The complementary HPF has its cut-off point at 16.5 kHz and you can see the rotation about the ½ Nyquist frequency.

8.6 Using RackAFX's Frequency Sampling Method Tool

RackAFX allows you to implement frequency sampling method filters by drawing responses directly on the analyzer's frequency response plots. You can place as many points as you like as well as choose the order (even only) of the filter. The previous examples were done using the frequency sampling method tool. Let's do an example similar to the last one but in the log frequency domain instead. First, click on the radio button marked frequency sampling method and you will see a red horizontal line appear on the graphs. Make sure the frequency axis is log. In this design, let's try to make a resonant LPF with a peak frequency around 1 kHz.

You will see two boxes on the red line, one at DC, the other at Nyquist. These points cannot be removed. To enter points and move them, use the following rules:

- Right-click on the red line to add a new point.
- Click on the new point and drag it up or down.
- To remove a point, right-click on it and choose "delete point" from the pop-up menu.
- Hit the Calculate button when you are done to view the resulting filter.

Hit the Frequency Sampling button and note the filter order of 64 (if it is not 64, change it with the slider or the edit control). Create a resonant LPF with a peak frequency of 1 kHz. In Figure 8.18 you can see the bandwidth of the peak is 1.5 kHz and the response goes to −60 dB at 5 kHz. Next, hit the Calculate button to get Figure 8.19.

You can use the order slider to slowly increase the filter order until the response matches your sampled points as closely as possible. You can move the slider, or type an order into the edit box and hit the Tab button to inject it. You can also use the small up/down arrow buttons to advance the order up or down in even increments. Typically, the way you would use this is to first get close with the slider, then use the up/down arrows to fine tune the design. Figure 8.20 shows the same filter with the order set to 164.

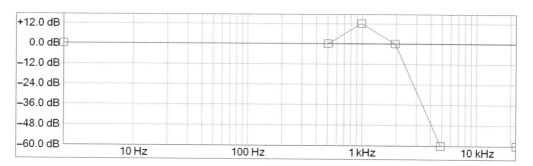

Figure 8.18: The prototype resonant LPF sampled points.

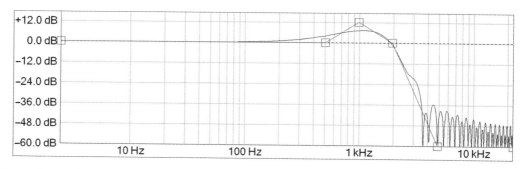

Figure 8.19: The 64-tap FIR filter produces marginal results with poor stop-band attenuation.

Play an audio file through the new filter and listen to the resonant LPF characteristics. Here are some interesting things you can do in RackAFX while the audio file is playing or looping:

- You can move the order control; the filter will be updated in real time and you can hear the results.
- You can add or remove points on the desired response or move them around, then hit Calculate to update the filter in real time, and you can also hear the results.
- You can save the IR as a file, then load it into the Impulse Convolver module just as before.

8.7 Designing a Complementary Filter

You can convert any design into a complementary design by hitting the Complementary button. With the current 164th-order resonant LPF, first switch to the linear scale (Figure 8.21).

Hit the Complement button to create the complementary HPF filter. The original design points are left to show you the complementary nature of the filter. You can clearly see the rotation about ½ Nyquist here (Figure 8.22). You can perform this operation while audio

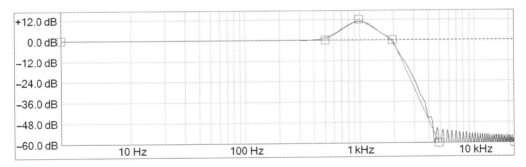

Figure 8.20: The 164th-order FIR filter produces an excellent match to the specifications.

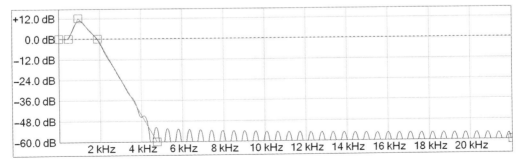

Figure 8.21: The linear domain frequency response of the 1-kHz LPF.

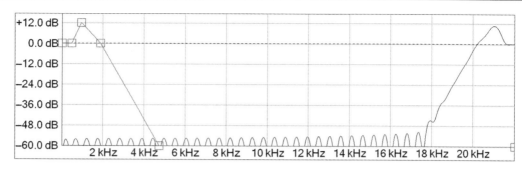

Figure 8.22: The complementary HPF shows the mirror image nature of complementary filter designs; the new resonant peak is 1 kHz below Nyquist.

files are playing in RackAFX as well. If you look at the IRs in the .sir file you can see the operation of negating the odd-numbered coefficients.

8.8 The Optimal (Parks–McClellan) Method

FIR filters produce all zeros in the *z*-plane. You can see the effect of these zeros in the frequency sampling method filters we designed—they create the notches and humps in the frequency response. In general, we seek to minimize the humps in the pass band and set some kind of limit for the largest lobe we see in the stop band. In the frequency sampling filters, we changed the design specifications or the order of the filter to get a response that satisfied our initial desired response to some extent. The optimal method is an algorithm for designing classical filters (LPF, HPF, BPF, BSF). You can create filters with extremely steep roll-off edges that would be difficult to synthesize in the analog domain. The word "optimal" comes from the engineering phrase "optimal in the Chebychev way," which describes a filter design with a certain kind of rippling in the pass band and stop band.

The LPF in Figure 8.23 might be specified like this:

- Pass-band ripple: less than 0.1 dB
- Stop-band attenuation: greater than −24 dB

The pass-band ripple (0.1 dB) and stop-band attenuation (−24 dB) are converted into weighting values using Equation 8.6:

$$\begin{aligned} \text{Pass-band weight} &= (10^{pb\_ripple(dB)/20}) - 1 \\ \text{Stop-band weight} &= (10^{sb\_atten(dB)/20}) \end{aligned} \quad (8.6)$$

The optimal method uses these weighting values in its design calculation. The weightings give acceptable error tolerances in each band. It the weights are equal, then there will be

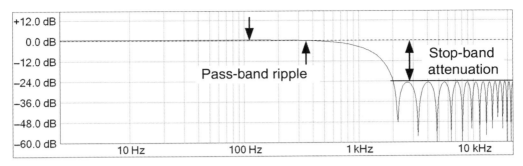

Figure 8.23: An FIR LPF can be specified by the acceptable ripple in the pass band (barely observable here) and the minimum stop-band attenuation.

an equal deviation from the ideal in both bands. This is called an *equiripple* design. The specifications for the optimal method are:

- The pass-band and stop-band frequencies
- The maximum pass-band ripple and minimum stop-band attenuation
- The order of the filter (the number of zeros in the frequency response)

The algorithm works by first distributing the zeros out equally across the frequency axis. This is going to produce a series of humps in the pass band (the ripple) and lobes in the stop band. The ripples and lobes will produce a set of maxima and minima frequencies. The problem is that for a given specification, the locations of the extremal frequencies are not known before hand. The optimal method must use a search algorithm to iteratively move the locations around until a solution is found that is within the tolerance of the specifications. The optimal method uses an algorithm called the Remez exchange method to iteratively move the zeros around until a solution converges. It must be noted that the Remez exchange algorithm is not guaranteed to converge for a given set of specifications.

Once the Remez exchange method finds the locations of the extremal frequencies, the resulting $H(f)$ graph is sampled using the frequency sampling method to produce the IR, which is the set of FIR filter coefficients. To summarize the method:

- Specify the filter parameter and order.
- Equally distribute zeros across the spectrum.
- Set the pass-band and stop-band weights.
- Use the Remez exchange algorithm to iteratively find the location of the zeros.
- Sample the resulting frequency response to produce the final IR.

8.9 Using RackAFX's Optimal Method Tool

RackAFX lets you design any of the four classical filter types using the equiripple weighting for the pass band and stop band. BPF and BSF filters have an additional band edge to specify, but the meaning of pass-band ripple and stop-band attenuation is the same for all filter types. You can start with an LPF design by clicking on the Optimal button. Try the following filter specifications:

- Type: LPF
- F_pass low: 1 kHz (the low edge cut-off frequency)
- F_stop low: 2 kHz (the lowest frequency that must receive the required stop band attenuation)
- Filter order: 16

Now, use the Calculate button to generate the filter. You can see from Figure 8.24 that the filter is not performing exactly to specifications. Although the pass band looks good, the stop band does not. Next, begin increasing the order of the filter using the slider or nudge buttons until you get the stop-band attenuation you desire; for −48 dB of attenuation, you will need a 104th-order filter as shown in Figure 8.25. Finally, check the IR and you will see a truncated version of the $\sin x(x)/x$ function (Figure 8.26).

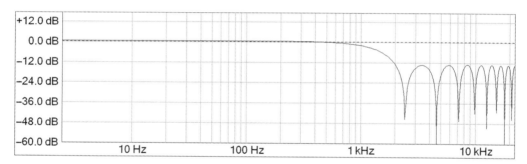

Figure 8.24: The pass-band ripple is well within spec; however, the stop-band attenuation is only getting to about −12 dB.

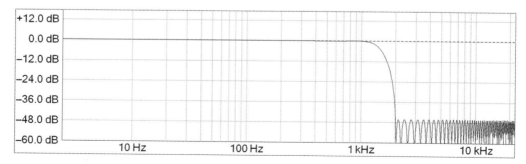

Figure 8.25: The 104th-order optimal method LPF produces the desired stop-band attenuation.

> The Remez exchange algorithm is not guaranteed to converge. You will receive an error message if it does not converge. Even if the algorithm does converge, the resulting IR is not guaranteed to be finite. Increasing the filter order will not necessarily produce a better design.

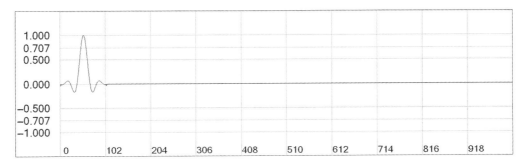

Figure 8.26: The IR for the 104th-order optimal method LPF.

Experiment:

- Load a wave file and audition the filter.
- Try the other filter types (HPF, BPF, BSF).
- Adjust the order control noting when the Remez exchange algorithm fails to converge or the filter blows up.
- Try to find the lowest possible filter order to just match the desired specifications.
- Save IR files with various filters you design.
- Copy and paste the IR code into your own convolution plug-in.

8.10 Design a Convolution Plug-In

In order to implement the convolution (FIR) algorithm you need to use the delay line theory from the last chapter. The filters often need hundreds or thousands of delay elements and you know that a circular buffer works perfectly for storing and updating a sequence of $x(n)$, $x(n-1)$, $x(n-2)$... In addition, the IR will need to be stored in a buffer and accessed sequentially with a pointer like the input buffer. The convolution equation in Equation 8.2 accumulates from $-\infty$ to $+\infty$ which uses both past and future data. We can only use past data and so we only need half of the equation. The generalized FIR convolution equation is Equation 8.7:

$$y(n) = h_0 x(n) + h_1 x(n-1) + h_2 x(n-2) + \ldots + h_M(n-M) \tag{8.7}$$

The number of delay elements required is $M-1$ since the first term $h_0 x(n)$ operates on the current, undelayed input signal. So, a 64-tap FIR requires 63 delay elements. Remember from Chapter 7 that when we access a circular buffer and write the current input sample, we are overwriting the oldest sample, $x(n-M)$ but we can use this to our advantage in this case by using a 64-element circular buffer to implement a 64-tap FIR and by writing in the first sample before doing the convolution operation. This will give us a buffer with $x(n)$ through $x(n - M - 1)$ lined up and ready for access. We will have an identically sized buffer to store the IR, $h(n)$. During the convolution operation we will zip through both buffers at the same time, accumulating the product of each operation. The only tricky thing is that the IR buffer will be reading sequentially from top to bottom exactly once each sample period to create the sequence $h(0), h(1), h(2)$, and so on. The input buffer will be circular and reading *backwards* to create the sequence $x(n), n(n - 1), x(n - 2)$, and so on, shown graphically in Figure 8.27.

If you look at your base class file, PlugIn.h, you will find the declarations of your built-in IR buffers and variables:

```
// impulse response buffers!
float m_h_Left[1024];
float m_h_Right[1024];

// the length of the IR from 0 to 1024
int m_nIRLength;

// flag to set to request impulse responses from the UI
bool m_bWantIRs;
```

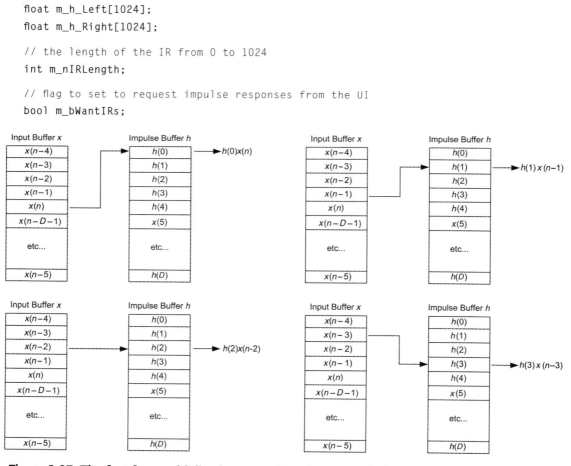

Figure 8.27: The first four multiplication operations for a convolution operation shows how one buffer reads backward while the other reads forward.

Table 8.2: IR variables

Variable	Description
m_h_Left[1024]	The IR buffer for the left channel
m_h_Right[1024]	The IR buffer for the right channel
m_nIRLength	The length of the current convolution
m_bWantIRs	A flag to tell RackAFX to populate your IR buffers automatically whenever the user loads an IR file or creates a filter with the FIR Designer tool

These variables are shown in Table 8.2.

8.10.1 Project: Convolver

Create a new RackAFX project; I named mine "Convolver." It has no GUI elements to set up.

8.10.2 Convolver.h File

Declare the variables you need to implement a stereo convolution and remember that the IR buffers and length variables are already declared for you. We need the following:

- Buffer to hold the x input for the left channel
- Buffer to hold the x input for the right channel
- Read index for the IR delay line (buffers)
- Read index for the input (x) buffers
- Write index for sequentially writing input samples into the delay lines

```
// Add your code here: -------------------------------------------- //
//
// pointers to the left and right input buffers
float* m_pBufferLeft;
float* m_pBufferRight;

// read index for delay lines (input x buffers)
int m_nReadIndexDL;

// read index for impulse response buffers
int m_nReadIndexH;

// write index for input x buffer
int m_nWriteIndex;
```

8.10.3 Convolver.cpp File

Constructor

- Create the buffers.
- Flush the buffers; reset to 0.0.
- Reset all indices.

```
CConvolver::CConvolver()
{
    <SNIP SNIP SNIP>

    // Finish initializations here

    // set our max buffer length for init
    m_nIRLength = 1024;  // 1024max

    // dynamically allocate the input x buffers and save the pointers
    m_pBufferLeft = new float[m_nIRLength];
    m_pBufferRight = new float[m_nIRLength];

    // flush x buffers
    memset(m_pBufferLeft, 0, m_nIRLength*sizeof(float));
    memset(m_pBufferRight, 0, m_nIRLength*sizeof(float));

    // flush IR buffers
    memset(&m_h_Left, 0, m_nIRLength*sizeof(float));
    memset(&m_h_Right, 0, m_nIRLength*sizeof(float));

    // reset all indices
    m_nReadIndexDL = 0;
    m_nReadIndexH = 0;
    m_nWriteIndex = 0;
}
```

You can see that we've allocated the memory for input buffers of maximum size 1024. We've also flushed out all buffers by setting all data to 0.0 with memset() and finally reset the indices to all point to the tops of the buffers.

Destructor

- Since we allocated memory in the constructor, we need to remove it in the destructor:

```
CConvolver::~CConvolver(void)
```

```
{
    // free up our input buffers
    delete [] m_pBufferLeft;
    delete [] m_pBufferRight;
}
```

prepareForPlay()

- We also need to flush the buffers and reset the indices:

```
bool __stdcall CConvolver::prepareForPlay()
{
    // Add your code here:
    // flush buffers
    memset(m_pBufferLeft, 0, m_nIRLength*sizeof(float));
    memset(m_pBufferRight, 0, m_nIRLength*sizeof(float));

    // reset indices
    m_nReadIndexDL = 0;
    m_nReadIndexH = 0;
    m_nWriteIndex = 0;

    return true;
}
```

processAudioFrame()

- Implement the convolution loop.

The loop will need to do the following operations (these are not streamlined in the code for ease of reading; you can always improve on the code by combining steps together). You are implementing the operation in Figure 8.27:

- Read the current sample $x(n)$ and write it into the buffer.
- Reset the delay line read pointer to the current input sample, $x(n)$; the pointer needs to be reset because its value will change and be destroyed later in the loop.
- Reset the IR read index to point to the top of the buffer at location 0.
- Set up the accumulator and create the convolution loop.
- After forming each product $h(i)x(n-i)$, *increment* the IR buffer read pointer and *decrement* the delay line read pointer.
- Check for a wrap after the delay line pointer is decremented.
- Write the output sample.

- Process the second (right) channel the same way.
- Increment the delay line write index and wrap if necessary.

```
bool __stdcall CCConvolver::processAudioFrame(float* pInputBuffer, float* pOutputBuffer,
UINT uNumInputChannels, UINT uNumOutputChannels)
{
    // Do LEFT (MONO) Channel; there is always at least one input/one output
    // Read the Input
    float xn = pInputBuffer[0];

    // write x(n) -- now have x(n) -> x(n-1023)
    m_pBufferLeft[m_nWriteIndex] = xn;

    // reset: read index for Delay Line -> write index
    m_nReadIndexDL = m_nWriteIndex;

    // reset: read index for IR - > top (0)
    m_nReadIndexH = 0;

    // accumulator
    float yn_accum = 0;

    // convolve:
    for(int i=0; i<m_nIRLength; i++)
    {
        // do the sum of products
        yn_accum += m_pBufferLeft[m_nReadIndexDL]*m_h_Left[m_nReadIndexH];

        // advance the IR index
        m_nReadIndexH++;

        // decrement the Delay Line index
        m_nReadIndexDL--;

        // check for wrap of delay line (no need to check IR buffer)
        if(m_nReadIndexDL < 0)
            m_nReadIndexDL = m_ nIRLength -1;
    }
    // write out
    pOutputBuffer[0] = yn_accum;
```

Now you should try to write the code for the second (right) channel. After you're done, check it against the following code:

```
// Mono-In, Stereo-Out (AUX Effect)
if(uNumInputChannels == 1 && uNumOutputChannels == 2)
    pOutputBuffer[1] = pOutputBuffer[0]; // just copy
```

```cpp
        // Stereo-In, Stereo-Out (INSERT Effect)
        if(uNumInputChannels == 2 && uNumOutputChannels == 2)
        {
                // Read the Input
                xn = pInputBuffer[1];

                // write x(n) -- now have x(n) -> x(n-1023)
                m_pBufferRight[m_nWriteIndex] = xn;

                // reset: read index for Delay Line -> write index
                m_nReadIndexDL = m_nWriteIndex;

                // reset: read index for IR - > top (0)
                m_nReadIndexH = 0;

                // accumulator
                yn_accum = 0;

                // convolve:
                for(int i=0; i<m_nIRLength; i++)
                {
                        // do the sum of products
                        yn_accum +=
                                m_pBufferRight[m_nReadIndexDL]*m_h_Right
                                [m_nReadIndexH];

                        // advance the IR index
                        m_nReadIndexH++;

                        // decrement the Delay Line index
                        m_nReadIndexDL--;

                        // check for wrap of delay line (no need to check IR buffer)
                        if(m_nReadIndexDL < 0)
                                m_nReadIndexDL = m_nIRLength-1;
                }
                // write out
                pOutputBuffer[1] = yn_accum;
        }

        // incremnent the pointers and wrap if necessary
        m_nWriteIndex++;
        if(m_nWriteIndex >= m_nIRLength)
                m_nWriteIndex = 0;

        return true;
}
```

Increment the buffer write pointer after the end of the right channel processing. The reason is that it is shared between both left and right (just like the IR buffer read index is shared between the left and right IR buffers). Build and test the code; in order to quickly test the code, you can use a feature of the FIR Designer and copy an IR to the clipboard, then paste it into the constructor. Open the FIR Designer and use the default optimal method LPF design—it will automatically be calculated when you first see the FIR Designer interface. Hit the button h(n) → Clipboard to copy the IR code to your Windows clipboard. Now, go to the constructor and use the paste function Ctrl+V to load in the IR code. Do this at the end of all the initialization stuff you wrote in Step 2. It should look like this:

Constructor

```
CConvolver::CConvolver()
{
        <SNIP SNIP SNIP>

        // reset all indices
        m_nReadIndexDL = 0;
        m_nReadIndexH = 0;
        m_nWriteIndex = 0;

        // h(n) Impulse Response
        // Length = 1024
        m_nIRLength = 65;

        m_h_Left[0] = -0.00057406;
        m_h_Left[1] = -0.00016395;
        m_h_Left[2] = -0.00015001;
        m_h_Left[3] = -0.00007587;
        m_h_Left[4] = 0.00007552;
        m_h_Left[5] = 0.00029672;
        m_h_Left[6] = 0.00063462;
        m_h_Left[7] = 0.00108109;

        etc...

        m_h_Right[1018] = 0.00000000;
        m_h_Right[1019] = 0.00000000;
        m_h_Right[1020] = 0.00000000;
        m_h_Right[1021] = 0.00000000;
        m_h_Right[1022] = 0.00000000;
        m_h_Right[1023] = 0.00000000;
}
```

This IR has 65 samples and will create a 65-tap FIR filter. Compile the dynamic link library (DLL) and load it into RackAFX. Play a wave file through it; it's an LPF at 1 kHz, so this will be easy to verify by ear. Open the analyzer window and look at the frequency and IRs—these will also be identical to the original design. Now that you have verified that your convolution works properly, you can set the IR flag in your constructor to let RackAFX know to deliver your IRs when the user makes, loads, or calculates them. You can also remove or comment-out the default IR code you pasted in to do the initial testing:

```
CConvolver::CConvolver()
{
    <SNIP SNIP SNIP>

    // reset all indices
    m_nReadIndexDL = 0;
    m_nReadIndexH = 0;
    m_nWriteIndex = 0;

    // set the flag for RackAFX to load IRs into our convolver
    m_bWantIRs = true;
}
```

Now, build and load your DLL. Next, open the analyzer window and you will see two new buttons on the right below the IR/FIR design area. Because your plug-in wants IRs, RackAFX reveals these two hidden buttons on the analyzer panel allowing you to use either the FIR Designer or the IR directory to create or load IR files. As soon as this window opens, it delivers the current IR loaded, which is the optimal LPF design by default. If you change the design it will remember it if you close the analyzer.

Load a wave file and hit Play—this will play the optimal LPF design. Now, hit one of the other buttons (HPF, BPF, or BSF) and the IR will automatically be calculated and loaded into your plug-in, even if an audio file is playing. Test the other IR tools, like the frequency sampling method and the IR Directory, and listen to the resulting filters.

8.11 Numerical Method FIR Filters

The last type of FIR filters consists of numerical methods and is not based on IRs or convolution, even though many of them implement a convolution loop. There are books full of numerical methods for performing all kinds of operations, from integration and differentiation to moving averages and interpolation. In this section we will examine a few algorithms and if it interests you, get a book on numerical methods and you will have a plethora of algorithms to try out. Some of these filters can be implemented as FIR topologies while others are better suited for a straight mathematical function call.

8.11.1 Moving Average Interpolator

Choose:

N = number of points to interpolate, minimum = 2

Calculate:

$$a_k = \frac{1}{N}, \text{ where } 0 \leq k \leq N-1 \quad (8.8)$$

The moving average interpolator (or MA filter) in Figure 8.28 implements a sliding window of N samples wide, over which the input samples are averaged. Each sample gets an equal weight of $1/N$.

Example: Design a five-point MA filter.

Solution:

$N = 5$, so $a_k = 1/5 = 0.20$

To code this in RackAFX, you would flush the IR buffers, then set the first five values of each impulse array to 0.2 and set the m_nIRLength variable to 5:

```
// disable RackAFX IRs for the MA filter test
m_bWantIRs = false;

// h(n) Impulse Response
// MA Filter N = 5
```

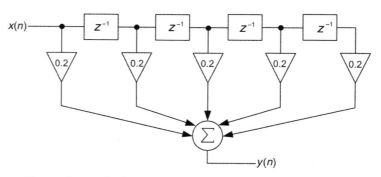

Figure 8.28: The implementation of a five-point MA filter.

```
            m_nIRLength = 5;

            m_h_Left[0] = 0.2;
            m_h_Left[1] = 0.2;
            m_h_Left[2] = 0.2;
            m_h_Left[3] = 0.2;
            m_h_Left[4] = 0.2;

            m_h_Right[0] = 0.2;
            m_h_Right[1] = 0.2;
            m_h_Right[2] = 0.2;
            m_h_Right[3] = 0.2;
            m_h_Right[4] = 0.2;
    }
```

Of course you could also write a function to calculate and populate the IR buffers, but this one is short enough to code by hand if you want.

The frequency response (Figure 8.29) of an MA filter is always an LPF. The more samples that are averaged, the more stop-band attenuation and the more zeros get inserted into the z-plane. For this filter there is a pair of complex conjugate zeros which produce two null points in the response.

You can see this is a pretty poor LPF compared to some of the optimal filter designs, but these filters can be very useful, for example to smooth the response of an audio detector output or to insert in the feedback path of a delay line module to gently roll off high frequencies on each pass through the delay. You can get better stop-band attenuation by increasing the order of the filter. This will also effectively pull in the corner frequency, which looks like about 2.5 kHz here. Experiment with higher-order MA filters, or add a slider to let the user adjust the window size. This exact filter already exists as a RackAFX module called a "Smoothing Filter" for you to experiment with.

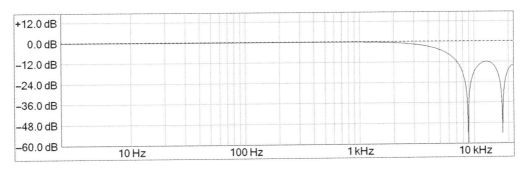

Figure 8.29: The MA filter with a window of 5 samples.

8.11.2 Lagrange Interpolator

Choose:

N = number of points to interpolate, minimum = 2

Calculate:

$$a_j = \prod_{\substack{k \neq j \\ k=0}}^{n} \frac{x - x_k}{x_j - x_k} \tag{8.9}$$

The Lagrange interpolator uses a polynomial of order $j - 1$ to interpolate across the window of points that you give it. The window is of length j in the above equation. This is a complex filter because the coefficients change every sample period and they are based on the window of input values, x_0 to x_j. This filter can be implemented as a pure math function call. To facilitate your coding, a Lagrange interpolation function is implemented in your pluginconstants.h file:

```
/*
Function: lagrpol()     implements n-order Lagrange Interpolation
Inputs:   double* x     Pointer to an array containing the x-coordinates of the
                        input values
          double* y     Pointer to an array containing the y-coordinates of the
                        input values
          int n         The order of the interpolator, this is also the length of
                        the x,y input arrays
          double xbar   The x-coorinates whose y-value we want to interpolate
Returns   The interpolated value y at xbar. xbar ideally is between the middle two
          values in the input array, but can be anywhere within the limits, which
          is needed for interpolating the first few or last few samples in a table
          with a fixed size.
*/
```

8.11.3 Median Filter

Choose:

N = number of points in window = odd

Calculate:

Acquire samples in windows of N values, then sort and choose the median value as the output.

The median filter (Figure 8.30) is a very interesting and somewhat strange algorithm. It has no IR or frequency response. It smoothes an input signal, which is an LPF type of operation, but it preserves transient edges, which is very un-LPF in nature. It has applications in noise reduction without losing high-frequency transients. Its central algorithm uses a sorting mechanism to sort the window of data by amplitude. The median value is chosen from the sort operation as the output. When the next sample arrives, the window is re-sorted and the next median value is obtained. To understand how it smoothes a signal without affecting high-frequency transients, consider the following example.

Example: Design a five-point median filter and test with example.

Consider this input sequence: $x = \{1, 2, 1, 0, 1, 2, 3, 3, 2, 1, 9, 8, 9, 9, 7, 5, 5, 4\}$ in Figure 8.31.

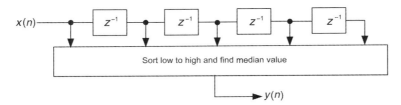

Figure 8.30: The block diagram of a five-point median filter implementation.

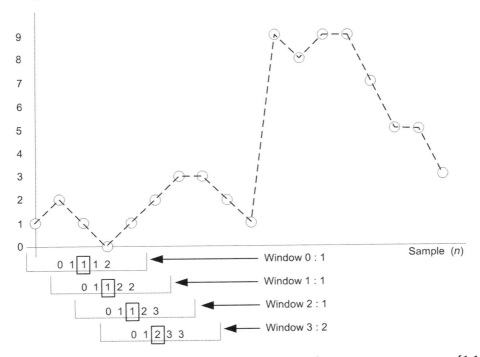

Figure 8.31: The first four windows of the median filter produce an output sequence {1,1,1,2}.

You can see a transient edge where the signal jumps from 1 to 9 and then another transient where it drops from 9 to 7 to 5. The first window operates on the first five samples and sorts them from low to high. Then, the median value is chosen as the output. The median value is shown in a box in the center of each window. You can see the smoothing effect immediately—the first three samples out of the filter are all 1, even though the first three samples vary from 1 to 2. Figures 8.32 and 8.33 show the result of median filtering the signal in Figure 8.31.

FIR filters can be complicated to design and long convolutions in direct form are slow. You can use the FIR design tools when you need to create linear-phase filters with very steep roll offs (optimal method) or filters with arbitrary frequency responses (frequency sampling method). You might also want to investigate other FIR designs such as the windowing method and the recursive frequency sampling method.

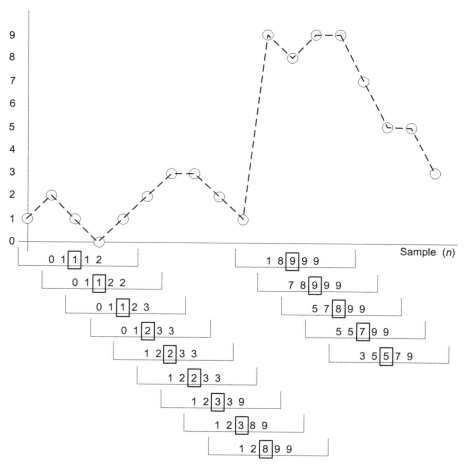

Figure 8.32: The complete set of median filter outputs for the sliding-window-of-five operation.

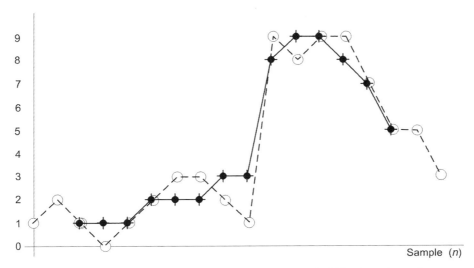

Figure 8.33: Input and output sequences plotted together and shifted to show the smoothing of the steady-state portions and preservation of the transient edges.

Bibliography

Ifeachor, E. C. and Jervis, B. W. 1993. *Digital Signal Processing, A Practical Approach*, Chapters 4 and 6. Menlo Park, CA: Addison-Wesley.

Kwakernaak, H. and Sivan, R. 1991. *Modern Signals and Systems*, Chapters 3 and 9. Englewood Cliffs, NJ: Prentice-Hall.

Lindquist, C. 1999. *Adaptive and Digital Signal Processing*, Chapter 10. Miami: Steward & Sons.

Oppenheim, A. V. and Schafer, R. W. 1999. *Discrete-Time Signal Processing*, 2nd ed., Chapter 7. Englewood Cliffs, NJ: Prentice-Hall.

CHAPTER 9
Oscillators

Oscillators find several uses in audio effects and plug-ins. The obvious use is as an audio test signal like the one RackAFX provides on the main interface. Additive synthesis of musical sounds uses multiple sinusoidal oscillators at harmonic frequencies to create complex waveforms. Wavetable synthesis stores a periodic waveform in a table for interpolation and playback when the musician strikes a key or a MIDI message is sent. Low-frequency oscillators (LFOs) are used in the design of modulated delay lines and modulated filters. Oscillators broadly fall into two categories: direct calculation and table lookup. We desire oscillators that have several important features:

- Stability over a wide range of frequencies
- No aliasing
- Purity of sinusoid (low THD+N) for sinusoidal oscillators
- Quadrature phase outputs
- Simplicity of calculation

9.1 Direct Form Oscillator

We can make a sinusoidal oscillator by placing a pair of poles directly on the unit circle in the z-plane. This produces a sinusoid at the pole angle (or frequency). The radius of the pole is always 1.0. Using the fundamental digital signal processing (DSP) z-plane equations, we can directly write the transfer function and difference equations:

$$H(z) = a_0 \left[\frac{1}{1 + b_1 z^{-1} + b_2 z^{-2}} \right]$$
$$= a_0 \left[\frac{1}{1 - 2R\cos(\Theta)z^{-1} + R^2 z^{-2}} \right] \quad (9.1)$$
$$y(n) = 2\cos(\Theta)y(n-1) - y(n-2)$$

Since the pole radius is 1.0, then the b_2 coefficient is 1.0 as well. The b_1 coefficient is then $-2\cos(\theta)$, where θ is the pole frequency from 0 to π.

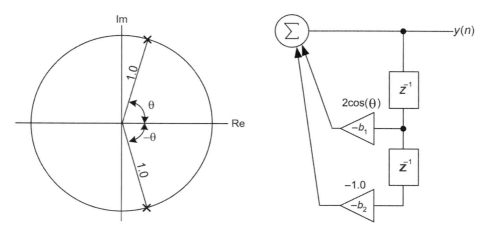

Figure 9.1: The direct form sinusoidal oscillator z-plane and block diagram.

You can see that the block diagram in Figure 9.1 has no input. Oscillators do not have inputs; instead they have *initial conditions* which, once started, will cause eternal oscillation. DSP theory suggests that kick-starting the system with an impulse will cause it to self-oscillate, which it does. However, the amplitude and phase are dependent on the frequency, θ. For an oscillator, we would like to precisely control the amplitude and starting phase so that they are exact. If we are trying to create a sinusoid with an initial phase offset of 0 degrees, then the first sample out of our oscillator at $n = 0$ would be 0.0, and the second sample out would be $\sin(\omega nT)$ where $n = 1$ and $T = 1/f_s$.

9.1.1 Initial Conditions

Suppose we want to generate the above sinusoid, oscillating at 500 Hz. If we want the first sample out to be 0.0, then continue rising after that, we need to initialize the delays as if a sinusoid had been oscillating, up to the point when we turn on the oscillator. So we can set our initial state by preloading those last two samples into the delays in our oscillator/filter, as shown in Figure 9.2.

The sinusoid is oscillating at $\sin(\omega nT)$ where $n = 0, 1, 2, 3\ldots$ after the oscillator starts and T is the sample period. The previous two samples with $f_o = 500$ Hz and $f_s = 44{,}100$ Hz would be $\sin(\omega(-1T))$ and $\sin(\omega(-2T))$, which are –0.0712 and –0.1532, respectively. The direct form oscillator block diagram is shown in Figure 9.3.

The difference equation is as follows:

$$y(n) = -b_1 y(n-1) - b_2 y(n-2) \tag{9.2}$$

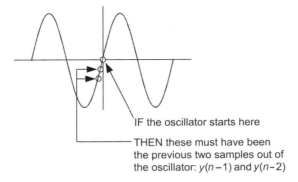

Figure 9.2: Initial conditions that would have produced a sinusoid whose first output sample is 0.0.

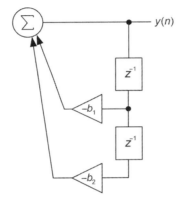

Figure 9.3: The direct form oscillator is really the feedback side of the bi-quad structure.

Specify:

- f_o, desired oscillation frequency

The design equations are as follows:

$$\theta = \frac{2\pi f_o}{f_s}$$
$$b_1 = -2\cos(\theta)$$
$$b_2 = 1.0$$

Initial conditions:
$$y(n-1) = \sin(-1\theta)$$
$$y(n-2) = \sin(-2\theta)$$

(9.3)

9.2 Design a Direct Form Oscillator Plug-In

In our first version of the direct form oscillator we are going to make it as simple as possible by restarting the oscillator when the user changes the oscillator frequency. This means we are going to recalculate the initial conditions as if the oscillator was starting from a phase of 0 degrees and the first sample out would have a value of 0.0. After we have that up and running, we will modify it to change frequency on the fly, automatically back-calculating the initial conditions for any given output sample. Here are the oscillator's specifications:

- Monophonic sinusoidal oscillator.
- We will need to implement a second-order feed-back block.
- We will need a slider for the user to control the oscillation frequency in Hz.
- We will need a cookFrequency() function to calculate our coefficients and reset our initial conditions when the user moves the frequency slider.
- The oscillator's range will be 200 Hz to 6 kHz.
- We need to assign Start and Stop buttons.

9.2.1 Project: DirectOscillator

Create a new project in RackAFX. On the new project dialog, make sure you check the box "Output Only Synthesizer Plug-In"—you must select this for any synthesis plug-in; you will not be able to play a wave file through it (Figure 9.4).

9.2.2 DirectOscillator GUI

Add a frequency slider (Figure 9.5) to the user interface (UI) with the parameters in Table 9.1. Next, set up the assignable buttons to create start and stop functions. Right-click inside the box labeled "Assignable Buttons" but do not click on a button, only in the area in between. A dialog pops up (Figure 9.6) that lets you set the button names. Choose "Start," "Stop," and make the last one blank with " " by entering the strings in the edit boxes. The buttons will automatically use the text you enter. Do not check the "Latching" buttons. See the website for notes on how to use latching buttons.

Table 9.1: The GUI control for the DirectOscillator plug-in

Slider Property	Value
Control Name	Frequency
Units	Hz
Variable Type	float
Variable Name	m_fFrequency_Hz
Low Limit	200
High Limit	6000
Initial Value	1000

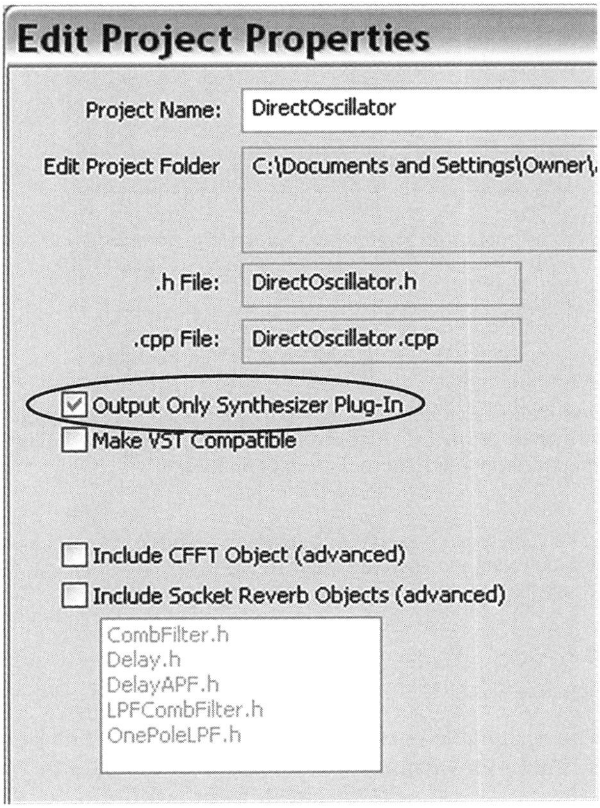

Figure 9.4: Make sure you check the "Output Only Synthesizer Plug-In" box for your oscillators.

Figure 9.5: The DirectOscillator GUI.

These assignable buttons will trigger your userInterfaceChange() function with their nControlIndex values of 50, 51, and 52. See the userInterfaceChange() function for more details.

Figure 9.6: The assignable button setup; notice that the last button is blank. RackAFX will hide it when you load the plug-in.

9.2.3 DirectOscillator.h File

In the .h file, declare the variables you need to implement the oscillator:

```
// Add your code here: ------------------------------------------------- //
//
// coefficients, 2nd Order FB
float m_f_b1;
float m_f_b2;

// delay elements, 2nd Order FB
float m_f_y_z1;
float m_f_y_z2;

// flag to start/stop oscillator
bool m_bRunOscillator;

// function to cook the Frequency and set initial conditions
void cookFrequency();
//
// END OF USER CODE --------------------------------------------------- //
```

9.2.4 DirectOscillator.cpp File

Constructor

- Turn off the oscillator by setting the flag you declared in the .h file. You will need to do these two steps any time you make an oscillator or other output-only plug-in.

```
CDirectOscillator::CDirectOscillator()
{
        <SNIP SNIP SNIP>

        // Finish initializations here
        // turn off
        m_bRunOscillator = false;

        // set 0
        m_f_b1 = 0;
        m_f_b2 = 0;

        // flush memory
        m_f_y_z1 = 0;
        m_f_y_z2 = 0;

        // call the cooking function to calc coeffs
        // and set initial states
        cookFrequency();
}
```

Implement the cooking function according to the design equations:

```
void CDirectOscillator::cookFrequency()
{
        // Oscillation Rate = theta = wT = w/fs
        float f_wT = (2.0*pi*m_fFrequency_Hz)/(float)m_nSampleRate;

        // coefficients according to design equations
        m_f_b1 = -2.0*cos(f_wT);
        m_f_b2 = 1.0;

        // set initial conditions so that first sample out is 0.0
        m_f_y_z1 = sin(-1.0*f_wT);    // sin(wnT) = sin(w(-1)T)
        m_f_y_z2 = sin(-2.0*f_wT);    // sin(wnT) = sin(w(-2)T)
}
```

prepareForPlay()

- Call the cooking functions to initialize the oscillator:

```
bool __stdcall CDirectOscillator::prepareForPlay()
{
        // Add your code here:
```

```
    // calc coeffs and initial conditions
    cookFrequency();

    retur true;
}
```

processAudioFrame()

- Implement the second-order feed-back filter.
- Check the oscillator flag. If the oscillator is not running you *must* send 0.0 samples to the output stream and then return without further processing. This is done at the very beginning of the function.

```
bool __stdcall CDirectOscillator::processAudioFrame(float* pInputBuffer, float*
                                                   pOutputBuffer, UINT uNumChannels)
{
    //
    // output = input -- change this for meaningful processing
    //
    // Do LEFT (MONO) Channel
    //
    // if not running, write 0s and bail
    if(!m_bRunOscillator)
    {
        // zero it all out
        pOutputBuffer[0] = 0.0;

        // Mono-In, Stereo-Out (AUX Effect)
        if(uNumInputChannels == 1 && uNumOutputChannels == 2)
            pOutputBuffer[1] = 0.0;

        // Stereo-In, Stereo-Out (INSERT Effect)
        if(uNumInputChannels == 2 && uNumOutputChannels == 2)
            pOutputBuffer[1] = 0.0;

        return true;
    }

    // otherwise, do the oscillator
    // do difference equation y(n) = -b1y(n-2) - b2y(n-2)
    pOutputBuffer[0] = -m_f_b1*m_f_y_z1 - m_f_b2*m_f_y_z2;

    // Mono-In, Stereo-Out (AUX Effect)
    if(uNumInputChannels == 1 && uNumOutputChannels == 2)
        pOutputBuffer[1] = pOutputBuffer[0];

    // Stereo-In, Stereo-Out (INSERT Effect)
    if(uNumInputChannels == 2 && uNumOutputChannels == 2)
        pOutputBuffer[1] = pOutputBuffer[0];
```

```
                // shuffle memory
                m_f_y_z2 = m_f_y_z1;
                m_f_y_z1 = pOutputBuffer[0];

                return true;
        }
```

userInterfaceChange()

- Call the frequency cooking function and implement the start and stop functions.
- Make sure your nControlIndex values match your switch/case statement.

```
        bool __stdcall CDirectOscillator::userInterfaceChange(int nControlIndex)
        {
                // cook variables
                switch(nControlIndex)
                {
                case 0: // freq
                        cookFrequency();
                        break;

                case 50: // start
                        m_bRunOscillator = true;
                        break;

                case 51: // stop
                        m_bRunOscillator = false;
                        break;
                default:
                        break;
                }
                return true;
        }
```

Build and load the dynamic link library (DLL). Then, hit the Start and Stop buttons to control your oscillator. Play with the frequency slider to change the value. You will notice a lot of clicking due to the discontinuities that occur each time the oscillator is reset to re-trigger from the 0.0 value. Other than that problem, the output should be a pure sinusoid with very low distortion and noise.

9.2.5 Improving the Oscillator Design

One of the drawbacks of oscillators is their initial conditions usually must be met (there are noise generators and other devices that can start up in a random state). In this case, we must preload the delay elements with samples that would have occurred before the initial output sample. If the user changes the frequency of oscillation, the initial states must change too. The problem is that the oscillator can be in any state when the user changes the controls. Suppose the user decreased the frequency of oscillation a bit and that the oscillator had some values in

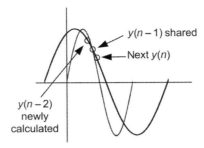

Figure 9.7: The new initial state samples, $y(n - 1)$ and $y(n - 2)$; we share the $y(n - 1)$ sample as if it was from the new frequency and alter the $y(n - 2)$ value to relocate it on the new sinusoid.

$y(n - 1)$ and $y(n - 2)$. Then to set up the new initial state, you can share the $y(n - 1)$ value between the two sinusoids and calculate a new $y(n - 2)$ value. The next sample out of the oscillator is at the new rate, using the new initial state you created (Figure 9.7).

The initial conditions from the design equations are in Equation 9.4:

$$\begin{aligned}\text{Initial conditions:} \\ y(n - 1) &= \sin(\omega n T) \\ y(n - 2) &= \sin(\omega(n - 1)T)\end{aligned} \qquad (9.4)$$

Our problem is that we do know the new frequency, ω, and since the sample rate is also known, then we also know ωT, but we don't know what sample interval n we are on. However, we can figure it out as follows:

- Take the inverse sin of the $y(n - 1)$ delay element.
- Find the value of n by dividing it by ωT.
- arcsine() returns a value between $-\pi/2$ to $+\pi/2$ which is the rising edge of the sinusoid like the original initial conditions.
- If we are on the rising edge, calculate the new $y(n - 2)$ value as $\sin((n - 1)\omega T)$.
- If we are on the falling edge, calculate the new $y(n - 2)$ value as $\sin((n + 1)\omega T)$.

It is simple to change in the cookFrequency() function as follows, so make the change in your code. Note the use of commenting out the original initial condition code in case you want to refer to it later. Also note the logic to see if we are rising or falling. We need this because the arcsine() function only returns the angle for half a sinusoid. If you think about it, this makes sense. If you give it a value of 0.707, it can't tell if the angle is 45 degrees or 135 degrees, both of which produce a sine of 0.707.

```
void CDirectOscillator::cookFrequency()
{
        // Oscillation Rate = theta = wT = w/fs
        float f_wT = (2.0*pi*m_fFrequency_Hz)/(float)m_nSampleRate;
```

```
        // coefficients according to design equations
        m_f_b1 = -2.0*cos(f_wT);
        m_f_b2 = 1.0;

        // set initial conditions so that first sample out is 0.0
//      m_f_y_z1 = sin(-1.0*f_wT);   // sin(wnT) = sin(w(-1)T)
//      m_f_y_z2 = sin(-2.0*f_wT);   // sin(wnT) = sin(w(-2)T)

        // re calculate the new initial conditions
        // arcsine of y(n-1) gives us wnT
        double wnT1 = asin(m_f_y_z1);
        // find n by dividing wnT by wT
        float n = wnT1/f_wT;

        // re calculate the new initial conditions
        // asin returns values from -pi/2 to +pi/2 where the sinusoid
        // moves from -1 to +1 -- the leading (rising) edge of the
        // sinewave. If we are on that leading edge (increasing)
        // then we use the value 1T behind.
        //
        // If we are on the falling edge, we use the value 1T ahead
        // because it mimics the value that would be 1T behind
        if(m_f_y_z1 > m_f_y_z2)
                n-=1;
        else
                n+=1;

        // calculate the new (old) sample
        m_f_y_z2 = sin((n)*f_wT);
}
```

Build and load the DLL. Then, hit the Start and Stop buttons to control your oscillator. Play with the frequency slider to change the value. If you change the slider too much you will still hear a click on some of the discontinuities due to the fact that the frequencies are so far apart that it produces a noticeable shift in the waveform.

Now that you understand how to code a direct calculation oscillator in RackAFX, next are a few more example oscillator designs for you to try out. Remember that you must set the flag m_bOutputOnlyPlugIn to make the oscillators work properly in RackAFX. You must provide a start/stop mechanism as well. Sample code is included to help you along.

9.3 The Gordon–Smith Oscillator

The Gordon–Smith oscillator uses a pair of delay elements arranged in a circular configuration to produce both sin and cos oscillations. The two outputs are almost perfectly in quadrature phase (90 degrees apart) only differing by half a sample period. Its sinusoidal oscillation is pure enough to be used in audio test equipment. And, because each output only uses one delay element, there are no initial states to update when the frequency is changed; the single $y(n-1)$ sample is shared with the new frequency the same way as in the direct form oscillator. Only the coefficient ε needs to be updated. A small amplitude variation is observed when the oscillation frequency changes, but it is small enough to not cause clicks or pops in the output. It sounds just as smooth as the direct form oscillator when the frequency is adjusted slowly.

The two outputs are labeled $y(n)$ and $y_q(n)$ where the "q" stands for quadrature. Therefore, there are two difference equations. The difference equation for $y_q(n)$ must be solved first because $y(n)$ is dependent on it. A Gordon–Smith oscillator block diagram is shown in Figure 9.8.

The difference equations are as follows:

$$y_q(n) = y_q(n-1) - \epsilon y(n-1)$$
$$y(n) = \epsilon y_q + y(n-1) \tag{9.5}$$

Specify:

- f_o, desired oscillation frequency

The design equations are as follows:

$$\theta = \frac{2\pi f_o}{f_s}$$
$$\epsilon = 2\sin(\theta/2) \tag{9.6}$$

Initial conditions:
$$y(n-1) = \sin(-1\theta)$$
$$y_q(n-1) = \cos(-1\theta)$$

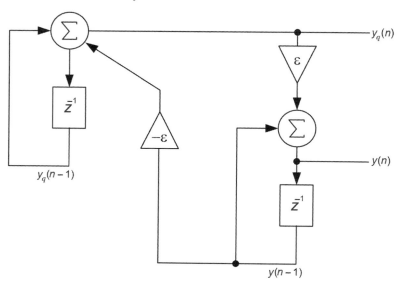

Figure 9.8: The Gordon–Smith Oscillator.

The C++ code for the Gordon–Smith oscillator looks as follows (two memory elements have been declared, m_f_yn_z and m_f_yq_z, as well as a coefficient m_fGorSmithEpsilon). In cookFrequency():

```
// calculate HS Epsilon
float f_wT = (2.0*pi*m_fFrequency_Hz)/(float)m_nSampleRate;
m_fGorSmithEpsilon = 2.0*sin(f_wT/2.0);
```

In processAudioFrame():

```
// form yq(n) first
float f_yqn = m_f_yq_z - m_fGorSmithEpsilon*m_f_yn_z;

// y(n)
float f_yn = m_fGorSmithEpsilon*f_yqn + m_f_yn_z;

// shuffle delays
m_f_yq_z = f_yqn;
m_f_yn_z = f_yn;

// write out
pOutputBuffer[0] = f_yn;

etc…
```

9.4 Wave Table Oscillators

A wave table oscillator is, as its name implies, a table-based system for creating periodic signals. A circular buffer is preloaded with one cycle, minus one sample, of a waveform. The waveform may be mathematical (sin, cos, pulse, square, saw, triangle, etc.) or it might be arbitrary or a recorded signal. The basic idea is that you read through the table and output samples from it. Consider a table of 1024 samples consisting of one cycle of a sinusoid, minus one sample (Figure 9.9).

Suppose you start at $i = 0$ and during each sample period, you read out one value and advance to the next. At the end of the buffer, you wrap around and start all over. If you did read out one value per sample period, what would be the resulting frequency of the waveform?

The answer is $f_{table} = f_s/L$ when the index increment is exactly 1.0 through the table. For a 1024-point wave table at a 44,100 Hz sample rate, the table frequency is 43.06 Hz. If you happen to really need a super precise sinusoid at exactly 43.06 Hz, then this method will produce nearly perfect results. The only factor is the precision of the sinusoid loaded into the table. If you had a saw-tooth waveform stored in the table, it too would have a fundamental frequency of 43.06 Hz.

Most likely, you are going to want to make any arbitrary frequency you like, perhaps with an emphasis on musical pitch frequencies. With the exception of the note A, these are going to be

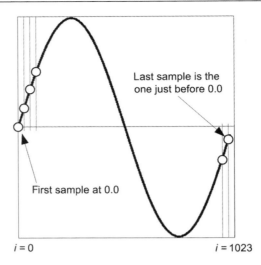

Figure 9.9: One cycle minus one sample of a sinusoid. The table is indexed with the value *i* which starts at 0; the last entry in the table is at *i* = 1023 and it is the sample just before the waveform starts all over again.

floating-point numbers with fractional parts. To make any frequency, you calculate the *inc* value with Equation 9.7:

$$inc = L\frac{f_{desired}}{f_s} \tag{9.7}$$

L is the table length and $f_{desired}$ is the target frequency. The increment value you get back will be used to skip through the table, moving forward by *inc* during each sample interval. If *inc* is less than 1.0, then the desired frequency is below the table frequency, and if it is above 1.0 it must be greater than the table frequency. If the *inc* is 2.0, then the resulting frequency is twice the table frequency. Most likely, the *inc* value is going to be noninteger and will therefore consist of an integer part and a fractional part. For example, if $inc = 24.9836$ then the integer part would be 24 and the fractional part 0.9836. Here, the integer part is called *int* and the fractional part is called *frac*.

As with the delay module you built in Chapter 8, there are several options for dealing with the fractional part of the increment value. You could:

- Truncate the value, and forget *frac*.
- Linearly interpolate the table *frac* distance between *int* and *int* + 1.
- Use polynomial interpolation or another interpolation method instead of linear interpolation.

If you truncate the *inc* value, then you have multiple problems—the note you synthesize won't be exactly in tune. Additionally, it will be distorted because of the inaccuracy in the transcription out of the table. Linear and polynomial interpolation both overcome these problems, though there is still distortion in the output. The industry standard is a fourth-order Lagrange interpolation on the wave table, where the neighboring four points (two to the left and two to the right) of the target interpolated value are used.

Before we code a wave table oscillator, there is one thing you need to be aware of: the danger of a wave table is that the cycle of data stored inside might be full of aliasing components. In other words, you can create a table of data that could have never made it past the input low-pass filter (LPF) if it were an analog signal being sampled. We will be doing just that and creating some signals that do alias, on purpose. Then, we'll examine a way to synthesize common signals without aliasing. The only meaningful signal that won't alias is a sinusoid, so we can start there.

9.5 Design a Wave Table Oscillator Plug-In

To code a wave table oscillator you need to create a table of data and initialize it in the constructor. A floating-point index value is used for *inc* and it will keep track of the current read location. Linear interpolation will be used to extract samples from the table. The circular buffer will run as long as the oscillator is engaged. Thus, we need to start with a plug-in design similar to the direct form oscillator. We can use the assignable buttons to trigger the oscillator to simulate MIDI note-on and note-off messages, and we can provide a slider for frequency control.

9.5.1 Project: WTOscillator

Create a new RackAFX project and make sure to check the synthesizer plug-in box (if you forget, you can always change it later).

9.5.2 WTOscillator GUI

The final GUI will feature a frequency control and multiple radio button controls for waveform, mode, and polarity. Ultimately, it will have the controls shown in Figure 9.10.

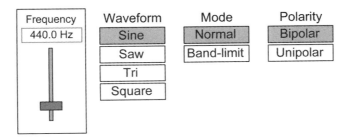

Figure 9.10: The final WTOscillator GUI.

Table 9.2: GUI controls for the wave table oscillator plug-in.

Slider Property	Value
Control Name	Frequency
Units	Hz
Variable Type	float
Variable Name	m_fFrequency_Hz
Low Limit	25
High Limit	4200
Initial Value	440

First, add a frequency slider to the UI and connect it to a variable named m_fFrequency_Hz with the limits 25 Hz to 4.2 kHz and an initial setting of 440 Hz (Table 9.2). The limits are chosen as such because they are close to the lower and upper fundamental frequencies of the notes on a standard (88 key) piano or synthesizer. Next, set up the assignable buttons to create start and stop functions just like you did for the DirectOscillator plug-in.

9.5.3 WTOscillator.h File

Declare the variables you need to implement the oscillator:

```
// Add your code here: ------------------------------------------------- //
//
// Array for the Table
float m_SinArray[1024];        // 1024 Point Sinusoid

// current read location
float m_fReadIndex;            // NOTE its a FLOAT!

// reset the read index
void reset()
{
    m_fReadIndex = 0.0;
}

// our inc value
float m_f_inc;

// our cooking function
void cookFrequency();

// our note on/off message
bool m_bNoteOn;

// END OF USER CODE ---------------------------------------------------- //
```

You can see that we've added the necessary ingredients (array, read index, *inc*, note-on/off), as well as two functions:

- cookFrequency() to update the *inc* value when the frequency changes.
- reset(), which just relocates the read index to the top of the buffer.

9.5.4 WTOscillator.cpp File

Constructor

- Turn off the oscillator by setting the flag you declared in the .h file.

```
CWTOscillator::CWTOscillator()
{
        <SNIP SNIP SNIP>
        // Finish initializations here
        // setup array
        for(int i = 0; i < 1024; i++)
        {
                // sample the sinusoid, 1024 points
                // sin(wnT) = sin(2pi*i/1024)
                m_SinArray[i] = sin( ((float)i/1024.0)*(2*pi) );
        }

        // clear variables
        m_fReadIndex = 0.0;
        m_f_inc = 0.0;

        // silent
        m_bNoteOn = false;

        // initialize inc
        cookFrequency();
}
```

In the constructor, you set the flags and set up the wave table by sampling it for 1024 points. This version is set up to produce the one-cycle-minus-one-sample waveform we desire. Write the cooking function:

```
void CWTOscillator::cookFrequency()
{
        // inc = L*fd/fs
        m_f_inc = 1024.0*m_fFrequency_Hz/(float)m_nSampleRate;
}
```

prepareForPlay()

- Reset the oscillator.
- Cook the variables.

```
bool __stdcall CWTOscillator::prepareForPlay()
{
        // Add your code here:
        // reset the index
        reset();

        // cook curent frequency
        cookFrequency();

        return true;
}
```

userInterfaceChange()

- Handle the slider control to cook the variables.
- Handle the start/stop buttons to turn the flag on/off and cook variables if turning on.
- Make sure your nControlIndex variables match with your GUI controls.

```
bool __stdcall CWTOscillator::userInterfaceChange(int nControlIndex)
{
        // add your code here
        switch(nControlIndex)
        {
        case 0:
                cookFrequency();
                break;

        // note on
        case 50:
                reset();
                cookFrequency();
                m_bNoteOn = true;
                break;
        // note off
        case 51:
                m_bNoteOn = false;
                break;

        default:
                break;
        }
        return true;
}
```

processAudioFrame()

- Implement the table look up.
- Notice that as with the delay line, we shift the frame of reference of the interpolated points to be between $n = 0$ and $n = 1$, then just use the *frac* component to find the value between the samples.

```cpp
bool __stdcall CWTOscillator::processAudioFrame(float* pInputBuffer, float*
                    pOutputBuffer, UINT uNumChannels, UINT uNumOutputChannels)
{
    // Do LEFT (MONO) Channel
    // if not running, write 0s and bail
    if(!m_bNoteOn)
    {
        pOutputBuffer[0] = 0.0;

        // Mono-In, Stereo-Out (AUX Effect)
        if(uNumInputChannels == 1 && uNumOutputChannels == 2)
            pOutputBuffer[1] = 0.0;

        // Stereo-In, Stereo-Out (INSERT Effect)
        if(uNumInputChannels == 2 && uNumOutputChannels == 2)
            pOutputBuffer[1] = 0.0;

        return true;
    }
    // our output value for this cycle
    float fOutSample = 0;

    // get INT part
    int nReadIndex = (int)m_fReadIndex;

    // get FRAC part
    float fFrac = m_fReadIndex - nReadIndex;

    // setup second index for interpolation; wrap the buffer if needed
    int nReadIndexNext = nReadIndex + 1 > 1023 ? 0 : nReadIndex + 1;

    // interpolate tht output (x1,x2,y1,y2,frac) - notice the way we set x1 and x2
    // to 0 and 1, then only use the frac value while nReadIndex and nReadIndexNext
    // acquire the values from the table
    fOutSample = dLinTerp(0, 1, m_SinArray[nReadIndex], m_SinArray[nReadIndexNext],
                    fFrac);

    // add the increment for next time
    m_fReadIndex += m_f_inc;

    // check for wrap
    if(m_fReadIndex > 1024)
        m_fReadIndex = m_fReadIndex - 1024;

    // write out
    pOutputBuffer[0] = fOutSample;

    // Mono-In, Stereo-Out (AUX Effect)
    if(uNumInputChannels == 1 && uNumOutputChannels == 2)
        pOutputBuffer[1] = fOutSample;

    // Stereo-In, Stereo-Out (INSERT Effect)
```

```
            if(uNumInputChannels == 2 && uNumOutputChannels == 2)
                pOutputBuffer[1] = fOutSample;

            return true;
}
```

Build and run the oscillator and check the pitch using the analyzer window. Adjust the oscillation frequency with the slider. Add a volume control slider if you wish.

9.6 Adding More Wave Tables

The sinusoidal oscillator is interesting, but you can implement any kind of table you want. You will see that adding more tables is actually easy since they all use the same frequency of oscillation equation. In this section we will add the following table types:

- Saw-tooth
- Triangle
- Square

These three waveforms will be mathematically as close as you can get in a discrete time system, but they will alias when played back. However, we still want to implement them because we can use the wave table oscillator as an LFO for upcoming effects like chorus and flanger. To add more tables we need to do the following:

- Add more arrays to store the tables.
- Initialize the arrays in the constructor.
- Provide the user with a way to switch oscillator types.

9.6.1 WTOscillator.h File

Start with the new arrays in the .h file:

```
// Add your code here: -------------------------------------------------
//
// Array for the Table
float m_SinArray[1024];            // 1024 Point Sinusoid
float m_SawtoothArray[1024];       // saw
float m_TriangleArray[1024];       // tri
float m_SquareArray[1024];         // sqr

// current read location
float m_fReadIndex;                // NOTE its a FLOAT!
```

9.6.2 WTOscillator.cpp File

Constructor

- These three new wave tables will be bipolar like the sinusoid table and swing between −1.0 and +1.0, and they will all start with the first sample = 0.0.

```
CWTOscillator::CWTOscillator()
{
        m_PlugInName = "WTOscillator";

        // set our oscillator flag:
        this->m_bOutputOnlyPlugIn = true;

        // slope and y-intercept values for the
        // Triangle Wave
        // rising edge1:
        float mt1 = 1.0/256.0;
        float bt1 = 0.0;

        // rising edge2:
        float mt2 = 1.0/256.0;
        float bt2 = -1.0;

        // falling edge:
        float mtf2 = -2.0/512.0;
        float btf2 = +1.0;

        // Sawtooth
        // rising edge1:
        float ms1 = 1.0/512.0;
        float bs1 = 0.0;

        // rising edge2:
        float ms2 = 1.0/512.0;
        float bs2 = -1.0;

        // setup arrays
        for(int i = 0; i < 1024; i++)
        {
                // sample the sinusoid, 1024 points
                // sin(wnT) = sin(2pi*i/1024)
                m_SinArray[i] = sin( ((float)i/1024.0)*(2*pi) );

                // saw, triangle and square are just logic mechanisms
                // can you figure them out?

                // Sawtooth
                m_SawtoothArray[i] = i < 512 ? ms1*i + bs1 : ms2*(i-511) + bs2;
```

310   Chapter 9

```
            // Triangle
            if(i < 256)
                    m_TriangleArray[i] = mt1*i + bt1; // mx + b; rising edge 1
            else if (i >= 256 && i < 768)
                    m_TriangleArray[i] = mtf2*(i-256) + btf2; // mx + b; falling edge
            else
                    m_TriangleArray[i] = mt2*(i-768) + bt2; // mx + b; rising edge 2

            // square:



            m_SquareArray[i] = i < 512 ? +1.0 : -1.0;
        }
        etc…
```

9.6.3 WTOscillator GUI

In order to let the user adjust the table, you can use the enumerated unsigned integer type (UINT) variable in the slider or the radio buttons. I will use the radio buttons and set them up here—right-click in the area around the buttons you want to label and fill in the properties with Table 9.3. The enumerated UINT variable lets you make the switch between oscillators. You set up the UINT variable name and the string of labels for the radio buttons here.

Table 9.3: The button properties for the oscillator type control.

Slider Property	Value
Control Name	Osc Type
Units	
Variable Type	enum
Variable Name	m_uOscType
Enum String	sine,saw,tri,square

Note: Here, we use a radio button list with an enumerated string.

9.6.4 WTOscillator.h File

In the .h file, you can see where RackAFX added the variable:

```
    // ADDED BY RACKAFX -- DO NOT EDIT THIS CODE!!! ------------------------------//
    // **--0x07FD--**

    float m_fFrequency_Hz;
    UINT m_uOscType;
    enum{sine,saw,tri,square};

    // **--0x1A7F--**
    // --------------------------------------------------------------------------- //
```

The m_uOscType is the switch variable and the enumerated list {sin, saw, tri, square} represents the integer set {0, 1, 2, 3}, which maintains the state of the switch.

9.6.5 WTOscillator.cpp File

processAudioFrame()

- A switch/case statement is used to choose the table to interpolate. Every other facet of the operation is the same as before, only the table is different.

```
bool __stdcall CDirectOscillator::processAudioFrame(float* pInputBuffer, float*
                  pOutputBuffer, UINT uNumChannels, UINT uNumOutputChannels)
{
        // Do LEFT (MONO) Channel
        <SNIP SNIP SNIP>
        // setup second index for interpolation; wrap the buffer if needed
        int nReadIndexNext = nReadIndex + 1 > 1023 ? 0 : nReadIndex + 1;

        // interpolate the output
        switch(m_uOscType)
        {
                case sine:
                        fOutSample = dLinTerp(0, 1, m_SinArray[nReadIndex],
                                              m_SinArray[nReadIndexNext], fFrac);
                        break;

                case saw:
                        fOutSample = dLinTerp(0, 1, m_SawtoothArray[nReadIndex],
                                              m_SawtoothArray[nReadIndexNext], fFrac);
                        break;
                case tri:
                        fOutSample = dLinTerp(0, 1, m_TriangleArray[nReadIndex],
                                              m_TriangleArray[nReadIndexNext], fFrac);
                        break;
                case square:
                        fOutSample = dLinTerp(0, 1, m_SquareArray[nReadIndex],
                                              m_SquareArray[nReadIndexNext], fFrac);
                        break;
                // always need a default
                default:
                        fOutSample = dLinTerp(0, 1, m_SinArray[nReadIndex],
                                              m_SinArray [nReadIndexNext], fFrac);
                        break;
        }
        // add the increment for next time
        m_fReadIndex += m_f_inc;
        <SNIP SNIP SNIP>
        return true;
}
```

Build and test the oscillator. Use the analyzer window to view the waveforms in time and frequency. Notice the large amount of upper harmonics in the non-sinusoidal waveforms—

some of those harmonics are actually aliased backwards. We'll need a way to get around the aliasing if we want to synthesize these waveforms for listening; however, we can still use this oscillator as a very effective LFO if we want.

9.7 Band-Limited Additive Wave Tables

If you looked at the analyzer's fast Fourier transform (FFT) window you saw an abundance of upper harmonics in the saw-tooth, triangle, and square waves. That is because mathematically, these waveforms contain harmonics out to infinite frequency. Equations 9.8 through 9.10 show the Fourier series equations for these waveforms.

$$y(n)_{SAW} = \sum_{k=1}^{\infty} (-1)^{k+1} \frac{1}{k} \sin(k\omega nT)$$
$$= \left[\sin(\omega nT) - \frac{1}{2}\sin(2\omega nT) + \frac{1}{3}\sin(3\omega nT) - \frac{1}{4}\sin(4\omega nT) + \ldots \right]$$
(9.8)

The saw-tooth waveform has both even and odd harmonics scaled according to $(1/k)$:

$$y(n)_{TRI} = \sum_{k=0}^{\infty} (-1)^k \frac{1}{(2k+1)^2} \sin((2k+1)\omega nT)$$
$$= \left[\sin(\omega nT) - \frac{1}{9}\sin(3\omega nT) + \frac{1}{25}\sin(5\omega nT) - \frac{1}{49}\sin(7\omega nT) + \ldots \right]$$
(9.9)

The triangle waveform has only odd harmonics. The $(-1)^k$ term alternates the signs of the harmonics. The harmonic amplitudes drop off at a rate given by $1/(2k+1)^2$ which is exponential in nature.

$$y(n)_{SQUARE} = \sum_{k=0}^{\infty} \frac{1}{2k+1} \sin((2k+1)\omega nT)$$
$$= \left[\sin(\omega nT) + \frac{1}{3}\sin(3\omega nT) + \frac{1}{5}\sin(5\omega nT) + \frac{1}{7}\sin(7\omega nT) + \ldots \right]$$
(9.10)

The square wave is also composed of odd harmonics like the triangle wave. The harmonic amplitudes drop off at a rate of $1/(2k+1)$, which is not as severe as the triangle wave. Therefore, the square wave has higher-sounding harmonics and a more gritty texture.

Using these Fourier series equations, you can implement Fourier synthesis (or additive synthesis) to build up a waveform that is band-limited to Nyquist. Let's modify the existing wave table oscillator to optionally use band-limited tables instead. The first version will be somewhat primitive, but will be a beginning source that you can use to make much more advanced synthesizer oscillators. This version will also span the frequencies of the piano.

The highest note on a piano is C8, vibrating at about 4186 Hz. Five times this value is 20,930 Hz, which is just barely below Nyquist. This means that we can synthesize our tables

Oscillators

with the fundamental plus the next five harmonics (also called partials) of the series given in Equations 9.8 through 9.10 and create our band-limited tables. The tables will therefore potentially contain the fundamental, 2nd, 3rd, 4th, 5th, and 6th harmonics. In order to do this, you need to modify the plug-in to allow for another mode: normal or band-limited. You can do this with another enumerated UINT data type, using either a slider or radio-button bank. I will use another radio button bank in this example.

9.7.1 WTOscillator GUI

Start with the UI and add another enumerated UINT by right-clicking inside the second bank of radio buttons and filling out the properties in Table 9.4.

Table 9.4: The button properties for the mode switches between "normal" and "bandlimited" on the UI.

Button Property	Value
Control Name	Mode
Units	
Variable Type	enum
Variable Name	m_uTableMode
Enum String	normal,bandlimit

9.7.2 WTOscillator.h File

This new information appears in the .h file:

```
// ADDED BY RACKAFX -- DO NOT EDIT THIS CODE!!! ----------------------------- //
// **--0x07FD--**

float m_fFrequency_Hz;
UINT m_uOscType;
enum{sine,saw,tri,square};
UINT m_uTableMode;
enum{normal,bandlimit};

// **--0x1A7F--**
// ------------------------------------------------------------------------- //
```

As long as we are in the .h file, we need to add more arrays for our band-limited tables. We want to keep these separate from the other tables to provide both modes of operation.

```
// Add your code here: ----------------------------------------------------- //
//
// Array for the Table
```

```
float m_SinArray[1024];              // 1024 Point Sinusoid
float m_SawtoothArray[1024];         // saw
float m_TriangleArray[1024];         // tri
float m_SquareArray[1024];           // sqr

// band limited to 5 partials
float m_SawtoothArray_BL5[1024];     // saw, BL = 5
float m_TriangleArray_BL5[1024];     // tri, BL = 5
float m_SquareArray_BL5[1024];       // sqr, BL = 5

// current read location
float m_fReadIndex;                  // NOTE its a FLOAT!
```

9.7.3 WTOscillator.cpp File

Constructor

- Initialize the tables according to the Fourier series equations. One of the problems with the additive method is that the final tables may be slightly outside the range of (-1.0 to $+1.0$), so we need to make normalized tables by saving the maximum value, then dividing each table by its maximum.

Here is the code for the five-partial tables:

```
CWTOscillator::CWTOscillator()
{
    <SNIP SNIP SNIP>

    // rising edge2:
    float ms2 = 1.0/512.0;
    float bs2 = -1.0;

    // setup arrays
    // to keep track of max-es for normalization later
    float fMaxTri = 0;
    float fMaxSaw = 0;
    float fMaxSqr = 0;

    for(int i = 0; i < 1024; i++)
    {
        // sample the sinusoid, 1024 points
        // sin(wnT) = sin(2pi*i/1024)
        m_SinArray[i] = sin( ((float)i/1024.0)*(2*pi) );

        <SNIP SNIP SNIP>

        // square: 1st sample is zero!
        if(i==1)
            m_SquareArray[i] = 0.0;
```

```
                else
                        m_SquareArray[i] = i < 512 ? +1.0 : -1.0;
        // zero to start, then loops build the rest
        m_SawtoothArray_BL5[i] = 0.0;
        m_SquareArray_BL5[i] = 0.0;
        m_TriangleArray_BL5[i] = 0.0;

        // sawtooth: += (-1)^g+1(1/g)sin(wnT)
        for(int g=1; g<=6; g++)
        {
                double n = double(g);
                m_SawtoothArray_BL5[i] += pow((float)-1.0,(float)(g+1))*
                                        (1.0/n)*sin(2.0*pi*i*n/1024.0);
        }

        // triangle: += (-1)^g(1/(2g+1+^2)sin(w(2n+1)T)
        // NOTE: the limit is 3 here because of the way the sum is constructed
        // (look at the (2n+1) components
        for(int g=0; g=3; g++)
        {
                double n = double(g);
                m_TriangleArray_BL5[i] += pow((float)-1.0, (float)n)*
                                        (1.0/pow((float)(2*n + 1),
                                        (float)2.0))*
                                        sin(2.0*pi*(2.0*n + 1)*i/1024.0);
        }

        // square: += (1/g)sin(wnT)
        for(int g=1; g<=5; g+=2)
        {
                double n = double(g);
                m_SquareArray_BL5[i] += (1.0/n)*sin(2.0*pi*i*n/1024.0);
        }
        // store the max values
        if(i == 0)
        {
                fMaxSaw = m_SawtoothArray_BL5[i];
                fMaxTri = m_TriangleArray_BL5[i];
                fMaxSqr = m_SquareArray_BL5[i];
        }
        else
        {
                // test and store
                if(m_SawtoothArray_BL5[i] > fMaxSaw)
                        fMaxSaw = m_SawtoothArray_BL5[i];

                if(m_TriangleArray_BL5[i] > fMaxTri)
                        fMaxTri = m_TriangleArray_BL5[i];
```

```
                    if(m_SquareArray_BL5[i] > fMaxSqr)
                        fMaxSqr = m_SquareArray_BL5[i];
            }
        }
        // normalize the bandlimited tables
        for(int i = 0; i < 1024; i++)
        {
            // normalize it
            m_SawtoothArray_BL5[i] /= fMaxSaw;
            m_TriangleArray_BL5[i] /= fMaxTri;
            m_SquareArray_BL5[i] /= fMaxSqr;
        }
        etc…
}
```

Because each sample in the table is built additively, you first zero out the value, then enter the loops to accumulate the additive signal in the sample slot. To increase the number of partials, you just increase the limit in the for() loops above.

processAudioFrame()

- A switch/case statement is used to change between normal and band-limited versions.

```
bool __stdcall CWTOscillator::processAudioFrame(float* pInputBuffer, float*
                                                pOutputBuffer, UINT uNumInputChannels,
                                                UINT uNumOutputChannels)
{
    // SNIP SNIP SNIP>

    // setup second index for interpolation; wrap the buffer if needed
    int nReadIndexNext = nReadIndex + 1 > 1023 ? 0 : nReadIndex + 1;

    // interpolate the output
    switch(m_uOscType)
    {
        case sine:
            fOutSample = dLinTerp(0, 1, m_SinArray[nReadIndex],
                                  m_SinArray[nReadIndexNext], fFrac);
            break;

        case saw:
            if(m_uTableMode == normal)       // normal
                fOutSample = dLinTerp(0, 1, m_SawtoothArray[nReadIndex],
                                      m_SawtoothArray[nReadIndexNext],
                                      fFrac);
            else                              // bandlimited
                fOutSample = dLinTerp(0, 1,
                                      m_SawtoothArray_BL5[nReadIndex],
                                      m_SawtoothArray_BL5[nReadIndexNext],
```

```
                                        fFrac);
                    break;

            case tri:
                    if(m_uTableMode == normal)    // normal
                            fOutSample = dLinTerp(0, 1, m_TriangleArray[nReadIndex],
                                                   m_TriangleArray[nReadIndexNext],
                                                   fFrac);
                    else                            // bandlimited
                            fOutSample = dLinTerp(0, 1,
                                                   m_TriangleArray_BL5[nReadIndex],
                                                   m_TriangleArray_BL5[nReadIndexNext],
                                                   fFrac);
                    break;
            case square:
                    if(m_uTableMode == normal)    // normal
                            fOutSample = dLinTerp(0, 1, m_SquareArray[nReadIndex],
                                                   m_SquareArray[nReadIndexNext],
                                                   fFrac);
                    else                            // bandlimited
                            fOutSample = dLinTerp(0, 1, m_SquareArray_BL5[nReadIndex],
                                                   m_SquareArray_BL5[nReadIndexNext],
                                                   fFrac);
                    break;

            // always need a default
            default:
                    fOutSample = dLinTerp(0, 1, m_SinArray[nReadIndex],
                                           m_SawtoothArray[nReadIndexNext], fFrac);
                    break;
    }
    // add the increment for next time
    m_fReadIndex += m_f_inc;
    etc…
}
```

Build and test the oscillator. Switch between the different modes and listen to the results, then use the analyzer to compare the oscillators in frequency and time. Next are some screenshots showing the differences for the saw-tooth and square waves.

9.7.4 Saw-Tooth

Figure 9.11 shows a normal saw-tooth waveform. Figure 9.12 shows a band-limited saw-tooth waveform.

9.7.5 Square Wave

Figure 9.13 shows a normal square wave. Figure 9.14 shows a band-limited square wave.

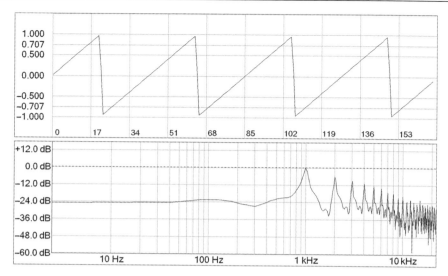

Figure 9.11: The mathematically perfect saw-tooth waveform with $f = 1$ kHz in the time domain (top) and frequency domain (bottom), which shows aliasing in the frequencies between 10 kHz and Nyquist.

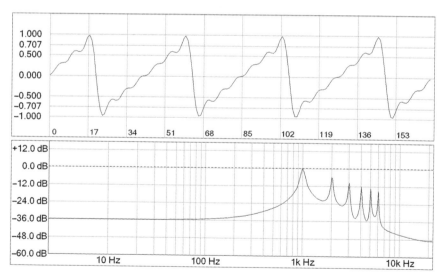

Figure 9.12: The five-harmonic band-limited saw-tooth waveform with $f = 1$ kHz in the time domain (top) and frequency domain (bottom). The aliasing is gone for this 1 kHz signal; it would reappear when the frequency was raised to the point that the highest harmonic went outside the Nyquist boundary. The fundamental plus the five harmonic peaks are clearly visible.

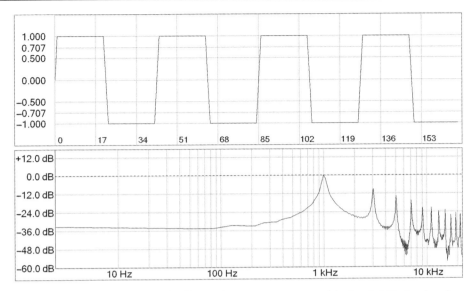

Figure 9.13: The mathematically perfect square wave with $f = 1$ kHz in the time domain (top) and frequency domain (bottom), which shows aliasing in the frequencies between 10 kHz and Nyquist. The square wave contains only the odd harmonics.

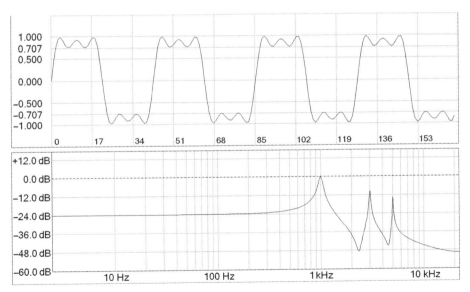

Figure 9.14: The five-harmonic band-limited square wave with $f = 1$ kHz in the time domain (top) and frequency domain (bottom). The aliasing is gone for this 1 kHz signal. There are only two harmonics (the third and fifth) because the next harmonic would be outside our limit.

9.8 Additional Oscillator Features (LFO)

For the modulated delay lines in the next chapters, we will need to use LFOs with a couple of additional properties. Specifically, we need

- A quadrature phase output
- Unipolar or bipolar operation
- Option to invert the output by 180 degrees
- A doOscillate() function that returns both outputs

All of these are easy to add to the current project. To make a quadrature phase output, we only need to declare a second read index which samples the table along with the current one. The quad phase read index will be initialized one-quarter of the way through the table to set up the quad phase output (Figure 9.15).

9.8.1 WTOscillator.h File

- Add a new variable for the inverted output option.
- Add a new variable for the quad phase read index; initialize it in reset().

```
// Add your code here: -------------------------------------------------

<SNIP SNIP SNIP>
// current read location
float m_fReadIndex;                      // NOTE its a FLOAT!
float m_fQuadPhaseReadIndex;             // NOTE its a FLOAT!
```

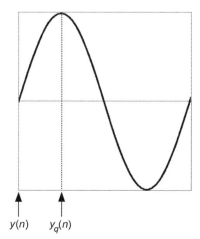

Figure 9.15: The concept of two read index values separated by one-quarter of the buffer; as long as they advance together by the same distance they will stay in quad phase. We currently have a bipolar oscillator which can easily be converted to a unipolar output by dividing by 2, then shifting the whole waveform up by 0.5.

```
// invert output
bool m_bInvert;

// reset the read index
void reset()
{
    m_fReadIndex = 0.0;                         // top of buffer
    m_fQuadPhaseReadIndex = 256.0;              // 1/4 of our 1024 point buffer
}
```

9.8.2 WTOscillator.cpp File

Constructor

- Initialize the new index variable in the constructor; it will be reset to the proper value later.

```
CWTOscillator::CWTOscillator()
{
    <SNIP SNIP SNIP>
    // clear variables
    m_fReadIndex = 0.0;
    m_fQuadPhaseReadIndex = 0.0;
    m_f_inc = 0.0;

    // silent
    m_bNoteOn = false;

    // norm phase
    m_bInvert = false;
    etc…
}
```

9.8.3 WTOscillator.h File

Add the doOscillate() function. It needs to be called externally in future projects. It also needs to provide both quad phase outputs. It will be a mono function. It requires no input argument and we can test it from within the plug-in by sending the two phases out to the left and right channels, then view it with the analyzer. In the .h file, add the following:

```
// Add your code here: -----------------------------------------------//
//
// funciton to do the Oscillator; may be called by an external parent Plug-In too
//   pYn is the normal output
//   pYqn is the quad phase output
void doOscillate(float* pYn, float* pYqn);

// Array for the Table
```

322 Chapter 9

```
float m_SinArray[1024];             // 1024 Point Sinusoid
float m_SawtoothArray[1024];        // saw
float m_TriangleArray[1024];        // tri
float m_SquareArray[1024];          // sqr
```

Etc...

9.8.4 WTOscillator.cpp File

Add the new function doOscillate() to the .cpp file. Cut and paste the contents of the processAudioFrame() for the left channel only. Just about everything is duplicated for the quad phase output. Of course, you can go in and streamline this code for more efficiency, but you can clearly see the operation of the two outputs here. Add the stuff in bold to the code you copied from processAudioFrame():

```
void CWTOscillator::doOscillate(float* pYn, float* pYqn)
{
    // IMPORTANT: for future modules, bail out if no note-on event!
    if(!m_bNoteOn)
    {
        *pYn = 0.0;
        *pYqn = 0.0;
        return;
    }
    // our output value for this cycle
    float fOutSample = 0;
    float fQuadPhaseOutSample = 0;

    // get INT part
    int nReadIndex = (int)m_fReadIndex;
    int nQuadPhaseReadIndex = (int)m_fQuadPhaseReadIndex;

    // get FRAC part - NOTE no Quad Phase frac is needed because it will be the same!
    float fFrac = m_fReadIndex - nReadIndex;

    // setup second index for interpolation; wrap the buffer if needed
    int nReadIndexNext = nReadIndex + 1 > 1023 ? 0 : nReadIndex + 1;
    int nQuadPhaseReadIndexNext = nQuadPhaseReadIndex + 1 > 1023 ? 0 :
                                    nQuadPhaseReadIndex + 1;

    // interpolate the output
    switch(m_uOscType)
    {
        case sine:
            fOutSample = dLinTerp(0, 1, m_SinArray[nReadIndex],
                            m_SinArray[nReadIndexNext], fFrac);
            fQuadPhaseOutSample = dLinTerp(0, 1,
                            m_SinArray[nQuadPhaseReadIndex],
                            m_SinArray[nQuadPhaseReadIndexNext], fFrac);
            break;
```

```
            case saw:
                if(m_uTableMode == normal)       // normal
                {
                        fOutSample = dLinTerp(0, 1, m_sawtoothArray[nReadIndex],
                                m_SawtoothArray[nReadIndexNext], fFrac);
                        fQuadPhaseOutSample = dLinTerp(0, 1,
                                m_SawtoothArray[nQuadPhaseReadIndex],
                                m_SawtoothArray[nQuadPhaseReadIndexNext], fFrac);
                }
                else                                             // bandlimited
                {
                        fOutSample = dLinTerp(0, 1,
                                m_SawtoothArray_BL5[nReadIndex],
                                m_SawtoothArray_BL5[nReadIndexNext], fFrac);
                        fQuadPhaseOutSample = dLinTerp(0, 1,
                                m_SawtoothArray_BL5[nQuadPhaseReadIndex],
                                m_SawtoothArray_BL5[nQuadPhaseReadIndexNext],
                        fFrac);
                }
                break;
        case tri:
        // ETC… FOLLOW THE SAME LOGIC AS THE OTHER TABLES //

        case SQUARE:
        // ETC… FOLLOW THE SAME LOGIC AS THE OTHER TABLES //
        // always need a default
        default:
                        fOutSample = dLinTerp(0, 1, m_SinArray[nReadIndex],
                                m_SinArray [nReadIndexNext], fFrac);
                        fQuadPhaseOutSample = dLinTerp(0, 1,
                                m_SinArray[nQuadPhaseReadIndex], m_SinArray
                                [nQuadPhaseReadIndexNext], fFrac);
                break;
}

// add the increment for next time
m_fReadIndex += m_f_inc;
m_fQuadPhaseReadIndex += m_f_inc;

// check for wrap
if(m_fReadIndex > 1024)
        m_fReadIndex = m_fReadIndex - 1024;

if(m_fQuadPhaseReadIndex > 1024)
        m_fQuadPhaseReadIndex = m_fQuadPhaseReadIndex - 1024;

// invert?
if(m_bInvert)
{
        fOutSample *= -1.0;
        fQuadPhaseOutSample *= -1.0;
}
```

```
            // write out
            *pYn = fOutSample;
            *pYqn = fQuadPhaseOutSample;
}
```

processAudioFrame()

- Use the doOscillate() function.

```
bool __stdcall CWTOscillator::processAudioFrame(float* pInputBuffer, float*
                        pOutputBuffer, UINT uNumInputChannels, UINT uNumOutputChannels)
{
        // Do LEFT (MONO) Channel
        // if not running, write 0s and bail
        if(!m_bNoteOn)
        {
                pOutputBuffer[0] = 0.0;
                if(uNumOutputChannels == 2)
                        pOutputBuffer[1] = 0.0;
                return true;
        }

        // some intermediate variables for return
        float fY = 0;
        float fYq = 0;

        // call the oscilator function, return values into fY and fYq
        doOscillate(&fY, &fYq);

        // write fY to the Left
        pOutputBuffer[0] = fY;

        // write fYq to the Right
        if(uNumOutputChannels == 2)
        {
                pOutputBuffer[1] = fYq;
        }
        return true;
}
```

Build and test the code. Figure 9.16 clearly shows the quad phase output: the sin() in the left channel, cos() is in the right channel.

9.9 Bipolar/Unipolar Functionality
9.9.1 WTOscillator GUI

Add the bipolar/unipolar switch to the UI using the next bank of radio buttons with the properties in Table 9.5. Edit it like you did with the previous buttons and give it the enumerations. I named my variable "m_uPolarity" and used "bipolar" and "unipolar" as my string/enums. The default will be bipolar.

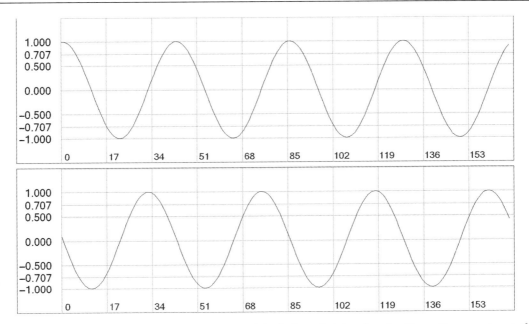

Figure 9.16: The quadrature-phase outputs of the left and right signals. Note: You must put the scope in free running mode to see the phase differences between channels.

Table 9.5: The button properties for the polarity control.

Slider Property	Value
Control Name	Polarity
Units	
Variable Type	enum
Variable Name	m_uPolarity
Enum String	bipolar,unipolar

9.9.2 WTOscillator.cpp File

Add the bipolar/unipolar functionality to the very last part of doOscillate() to divide by 2 then shift by 0.5 as follows:

```
void CWTOscillator::doOscillate(float* pYn, float* pYqn)
{
        <SNIP SNIP SNIP>
        // write out
        *pYn = fOutSample;
        *pYqn = fQuadPhaseOutSample;

        // create unipolar; div 2 then shift up 0.5
        if(m_uPolarity == unipolar)
```

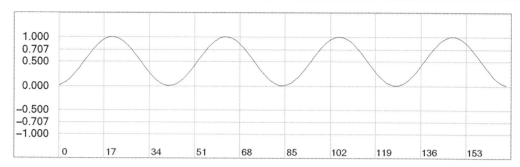

Figure 9.17: Unipolar operation, sinusoid.

```
        {
                *pYn  /= 2.0;
                *pYn  += 0.5;

                *pYqn /= 2.0;
                *pYqn += 0.5;
        }
}
```

Build and test the code; Figure 9.17 shows the output for a unipolar sinusoid.

Here are some projects to try:

- Higher-resolution tables; each octave of the piano fingerboard gets its own table with as many harmonics as the upper note of the octave permits.
- A two-oscillator synth module with a detuning control for one of the oscillators.
- An oscillator that only plays musical pitches.

In the next chapter, we'll use our new wave table oscillator to make modulated delay effects.

Bibliography

Dattorro, J. 2003. Effect design, Part 3, oscillators: Sinusoidal and pseudonoise. *Journal of the Audio Engineering Society*, 50(3): 114–146.

Gordon, J. W. and Smith, J. O. 1985. A sine generation algorithm for VLSI applications. *Proceedings from the International Computer Music Conference*, pp. 165–168.

Moore, R. 1990. *Elements of Computer Music*, Chapter 3. Englewood Cliffs, NJ: Prentice-Hall.

Roads, C. 1996. *The Computer Music Tutorial*, Chapter 2. Cambridge, MA: The MIT Press.

CHAPTER 10
Modulated Delay Effects

Perhaps the most interesting kinds of audio effects, from both listening and engineering standpoints, are the modulated delay effects. They include the chorus and flanger/vibrato; additionally, modulated delays are also found in some reverb algorithms. These time-varying filters are actually quite sophisticated and fun to implement, and we have gone to great lengths to create useful sub-modules. These include the digital delay effect or DDL (digital delay line) module and the wave table oscillator. If you can design, build, and implement modulated delay effects, then you are well on your way to proficiency in audio effects coding. These effects have parameters and coefficients which change on every sample period.

Figure 10.1a shows the standard delay line or DDL. The output is fed back via the *fb* coefficient. The delay is constant until the user changes it to another fixed value. In the modulated delay line (MDL) in Figure 10.1b the amount of delay is constantly changing over time. The system only uses a portion of the total available delay amount.

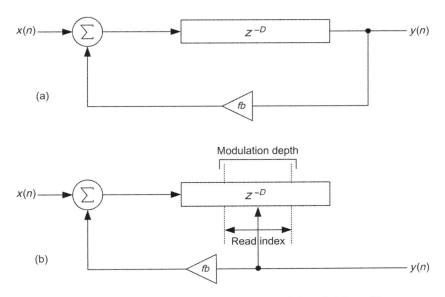

Figure 10.1: (a) A static delay and (b) modulated delay effect.

The *modulation depth* relates to the size of the portion of the delay line that is being read. The *modulation rate* is the speed at which the read index moves back and forth over the modulated delay section. A low-frequency oscillator (LFO) is used to modulate the delay amount and the LFO can be just about any kind of waveform, even noise. The most common waveforms are triangle and sinusoid. In order to make a modulated delay effect, you first need to make a delay line, and then modify it to constantly change and update the read location on each sample interval.

10.1 The Flanger/Vibrato Effect

The flanger effect gets its name from the original analog method of running two tape machines slightly out of sync with each other by placing a thumb on the flange (the ring that holds the tape) of one machine's reel and applying pressure to slow down the tape ever so slightly. Other variations include rubbing the flange circularly. The effect creates a time-varying delay that falls out of sync, then back in sync with the first tape machine. The effect has been compared to the whoosh of a jet engine, or the sound of a phaser effect. No matter how you define it, the flanger has a unique sound. The effect is heard on numerous recordings.

In order to simulate the flanger effect, you need a delay line whose read index moves away from, then back to the top or starting point of the delay. When the read pointer is back to the top, there is no delay and as it moves away from the write index, the pitch is modulated down. When the index turns around and moves back, the pitch is modulated up—this is the Doppler effect in practice. When the output of the modulated delay is sent out 100% wet, with no dry signal, the effect is that of vibrato—the pitch shifts up and down as the pointer moves back and forth. When the dry signal is mixed with the wet signal, the pitch shifting changes into the flanging effect. The amount of delay time varies between 0 and about 7–10 mSec; at a 44,100 Hz sample rate, this corresponds to around 308 to 410 samples. Feedback is usually added to increase the effect by creating resonances or anti-resonances, depending on whether the feedback is normal or inverted.

Figure 10.2 shows the two states of the modulator with increasing and decreasing pitch shifting due to the Doppler effect. Figure 10.3 shows alternate ways of diagramming the same modulator. The vibrato and flanger are the same effect underneath the hood—it's only the mix ratio that determines which effect you have. Because the delay is constantly moving around we are once again faced with options on how to handle the fractional delay that is required; after you calculate the number of samples of delay (which will most likely have a fractional component) you can

- Truncate the value and just use the integer part.
- Round the value up or down.

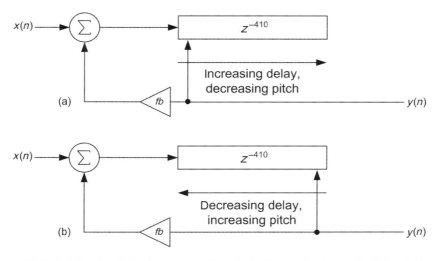

Figure 10.2: (a) As the delay increases, the pitch drops. At the end of the delay, the index turns around and begins heading back toward the write index (b), increasing the pitch; the maximum delay here is about 10 mSec at 44,100 Hz.

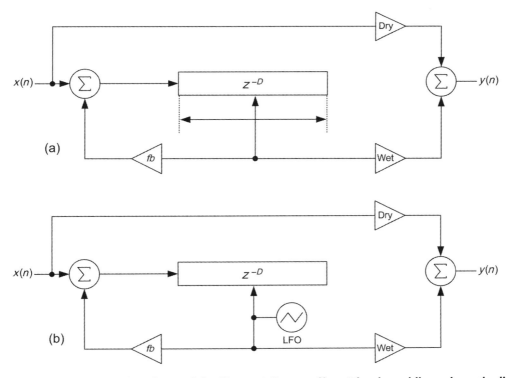

Figure 10.3: (a) The simplest form of the flanger/vibrato effect. The dotted lines show the limits of the delay modulation, from 0.0 (no delay) to the full range, z^{-D} samples. (b) An alternate version shows the LFO connection to the modulation index to the delay line.

- Interpolate the delayed value using linear or polynomial interpolation.
- Use an all-pass filter to make the fractional delay (see the Bibliography).

The flanger/vibrato controls consist of:

- Modulation depth: how much of the available range is used.
- Modulation rate: the rate of the LFO driving the modulation.
- LFO type: usually sinusoidal or triangular, but may be anything.
- Feedback: as with the normal delay, the feedback has a big effect on the final product.
- Wet/dry control: 100% wet = vibrato; 50/50 = flanger.

As shown in Figure 10.4, the flanger/vibrato technically always starts with 0.0 samples of delay. Because of this, we'd like to have a LFO that can generate a unipolar output from 0.0 to +1.0, which we can then map to a range of delays, from 0 to about 410 samples or so. Figure 10.5 shows a stereo version of the flanger. A common LFO modulates both delay lines, but the phases are off by 90 degrees, putting them in quadrature or quad-phase. The effect is interesting as the two LFOs seem to chase each other, one on the left and the other on the right.

The flanger technically modulates the delay right down to 0 samples. For analog tape flanging, this occurs when the two tape machines come back into perfect synchronization. This is called *through-zero flanging* or TZF. This means that for an instant, the output is double the input value. For low frequencies and/or high feedback values, this can cause a problem when the flanger delay time is between 0 and about 0.1 mSec and the waveform is at a peak location. In this case, large valued samples near each other on the waveform sum together, producing an increased and sometimes distorted bass response.

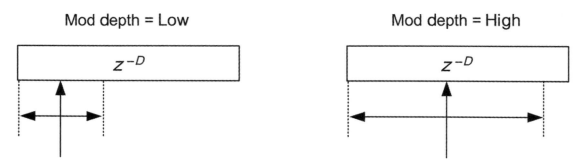

Figure 10.4: Two modulation depth settings; zero is always the minimum delay.

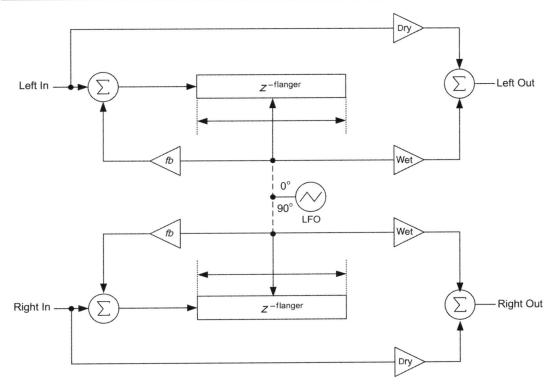

Figure 10.5: A stereo quadrature flanger; the dotted lines at the LFO show its 90-degree offset modulation. The max depth spans the entire delay line.

Table 10.1: Delay times for the flanger and chorus effects.

	Min Delay (mSec)	Max Delay (mSec)	Feedback (%)
Flanger	0.0	7–10	−99 to +99
Chorus	7	20–40	Typically 0

10.2 The Chorus Effect

The chorus effect appeared in the late 1970s and was used on countless recordings during the 1980s, becoming a signature sound effect of that decade. The chorusing effect can produce a range of effects from a subtle moving sheen to a swirling sea-sick de-tuned madness. Like the flanger, it is also based on a modulated delay line, driven by an LFO. Although different manufacturers adopted different designs, the basic algorithm for a single chorusing unit is the same. The differences between a flanger and chorus are shown in Table 10.1.

- Analog chorus units typically do not use any feedback (because our DDL module already has feedback, we can keep it for experimentation purposes).
- The range of delay times over which the device operates is the most significant difference.

The read index is modulated about the center of the pointer but the center is set to avoid the keep-out zone (Figures 10.6 and 10.7). Also, the modulation depth and the location of the center of the delay area vary across manufacturers (Figure 10.8).

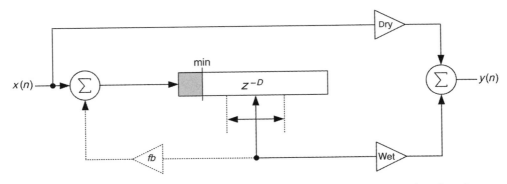

Figure 10.6: The basic chorus module. Note the feedback path is optional and not found in typical analog chorus units. The gray area is the flanger keep-out zone.

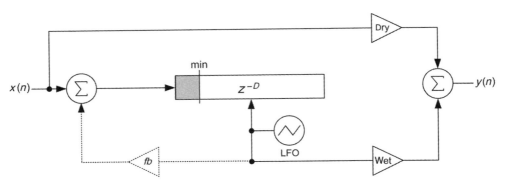

Figure 10.7: An alternative block diagram that describes the same chorus module.

Figure 10.8: (a) and (b) are two different chorus modules with the same maximum depth (dotted lines) but a different center of operation.

Many variations on the basic chorus module exist, including:

- Multiple chorus modules with different centers of operation
- Different LFO frequency or LFO phases applied to different modules
- Series modules
- Parallel modules
- Bass modules with a low-pass filter/high-pass filter on the front end

Figure 10.9 shows the stereo quad chorus with the left and right LFOs out of sync by 90 degrees. In this permutation, the maximum depths of the left and right channels are independent, as well as the ability to adjust them. The flowchart for the basic modulated delay effect for each processAudioFrame() is shown in Figure 10.10.

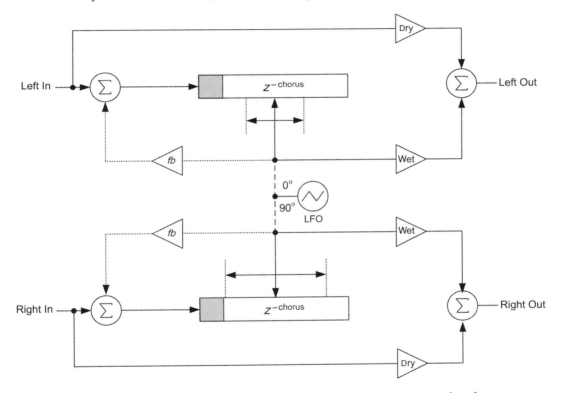

Figure 10.9: A stereo quadrature chorus; feedback paths are optional.

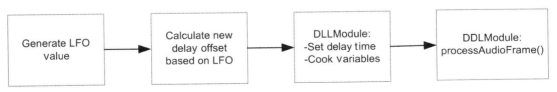

Figure 10.10: The flowchart for modulated delay effects.

You will use the DDL module to do most of the low-level work and generate the delayed values. The LFO will generate values that will be used to calculate the current delay offset. Since the LFO is varying, the delay offset will need to be recalculated on each sample interval.

10.3 Design a Flanger/Vibrato/Chorus Plug-In

Now we get down to the business of coding these effects. Because of the large number of variations, we should start by coding a module that can be used in larger parent plug-in objects. Figure 10.11 shows the block diagram.

We will provide several user interface (UI) components to control the plug-in and their variables are available for a parent plug-in to use as well. We will use the existing DDL module and wave table oscillator plug-ins. Since the DDL module has feedback path control, there will be many ways to combine these modules into sophisticated units. The controls we will provide for the user are discussed next. The plan of operation is as follows:

- LFO: needs to be set up to run unipolar and with the non-band-limited tables.
- The chorus delay offset control will effectively move the center of operation for the modulated delay.
- The vibrato mode will force the feedback value to 0.0% since it does not sound good with feedback.
- prepareForPlay()
  - Reset the LFO and start it by setting the m_bNoteOn flag; also set up the LFO and DDL modules.
  - Set the DDL m_nSampleRate variable; it needs this value to initialize its buffer properly.
  - Forward the prepareForPlay() call to the child objects, LFO and DDL.

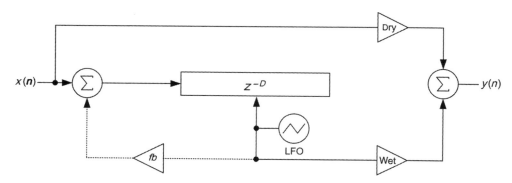

Figure 10.11: Generic modulated delay block diagram;
it can be any type of flanger/vibrato/chorus.

- processAudioFrame()
  - Call the doOscillate() function on the LFO.
  - Use the LFO value to calculate a delay offset in milliseconds for the DDL module.
  - Write the new delay time to the DDL module and call its cooking function.
  - Call processAudioFrame() on the DDL module to get the current output sample.
- userInterfaceChange()
  - Call the update functions for the LFO and DDL.
  - Call the type switching function to change from flanger to vibrato and chorus.

We'll go ahead and plan on some helper functions to make this process easier. This is not going to be the most streamlined or efficient code, but once you have it mastered you can always go back and improve it. Although this is a difficult project, you only need two member variables and four functions to accomplish the design.

The member objects are as follows:

```
CWTOscillator m_LFO;    // our LFO
CDDLModule m_DDL;       // our delay line module
```

The member variables are as follows:

```
float m_fMinDelay_mSec;  // min delay time
float m_fMaxDelay_mSec;  // max delay time
```

The member functions are as follows:

```
float updateLFO();     // change/set the type and rate
float updateDDL();     // change/set the wet/dry and feedack
float cookModType();   // set up the min/max delay times
float calculateDelayOffset(float fLFOSample);  // convert the LFO to a delay value
```

10.3.1 Project: ModDelayModule

Start a new project and make sure to include your WTOscillator and DDLModule objects (Figure 10.12).

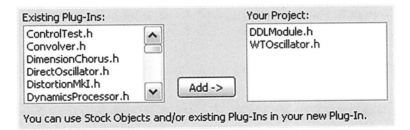

Figure 10.12: Project settings for the ModDelayModule.

Table 10.2: The GUI components, variable names, and values.

Slider Property	Value
Control Name	Depth
Units	%
Variable Type	float
Variable Name	m_fModDepth_pct
Low Limit	0
High Limit	100
Initial Value	50

Slider Property	Value
Control Name	Rate
Units	Hz
Variable Type	float
Variable Name	m_fModFrequency_Hz
Low Limit	0.02
High Limit	5
Initial Value	0.18

Slider Property	Value
Control Name	Feedback
Units	%
Variable Type	float
Variable Name	m_fFeedback_pct
Low Limit	-100
High Limit	100
Initial Value	0

Slider Property	Value
Control Name	Chorus Offset
Units	mSec
Variable Type	float
Variable Name	m_fChorusOffset
Low Limit	0
High Limit	30
Initial Value	0

Button Property	Value
Control Name	Mod Type
Units	
Variable Type	enum
Variable Name	m_uModType
Enum String	Flanger, Vibrato, Chorus

Button Property	Value
Control Name	LFO
Units	
Variable Type	enum
Variable Name	m_uLFOType
Enum String	tri,sin

10.3.2 ModDelayModule GUI

Set up the graphical user interface (GUI) in RackAFX with Table 10.2.

10.3.3 ModDelayModule.h File

Add the member objects, variables, and function declarations to the .h file:

```
// Add your code here: ------------------------------------------------ //
CWTOscillator m_LFO;    // our LFO
CDDLModule m_DDL;       // our delay line module

// these will depend on the type of mod
float m_fMinDelay_mSec;
float m_fMaxDelay_mSec;
```

```
// functions to update the member objects
void updateLFO();
void updateDDL();

// cooking function for mod type
void cookModType();

// convert a LFO value to a delay offset value
float calculateDelayOffset(float fLFOSample);

// END OF USER CODE ------------------------------------------------- //
```

10.3.4 ModDelayModule.cpp File

Constructor

- Initialize all variables.

```
CModDelayModule::CModDelayModule()
{
    <SNIP SNIP SNIP>

    // Finish initializations here
    m_fMinDelay_mSec = 0.0;
    m_fMaxDelay_mSec = 0.0;
    m_fChorusOffset = 0.0;

    m_LFO.m_fFrequency_Hz = 0;
    m_LFO.m_uOscType = m_uLFOType; // triangle  enum{tri,sin};

    m_DDL.m_bUseExternalFeedback = false;
    m_DDL.m_fDelay_ms = 0;
}
```

Write the update() functions:

```
// LFO function to set:
//                  - the LFO Frequency
//                  - the oscillator type
void CModDelayModule::updateLFO()
{
    // set raw data
    m_LFO.m_fFrequency_Hz = m_fModFrequency_Hz;
    m_LFO.m_uOscType = m_uLFOType;

    // cook it
    m_LFO.cookFrequency();
}
```

```
// DDL function to set:
//                  - the DDL Feedback amount (disabled for Vibrato)
void CModDelayModule::updateDDL()
{
    // test and set if needed
    if(m_uModType != Vibrato)
        m_DDL.m_fFeedback_pct = m_fFeedback_pct;

    // cook it
    m_DDL.cookVariables();
}
```

Write the cookModType() function using the information in Table 10.3—this method is called when the user changes the effect type.

```
// cookMod() function:
/*
              Min Delay (mSec)    Max delay (mSec)    Wet/Dry(%)    Feedback(%)
Flanger       0                   7-10                50/50         -100 to +100
Vibrato       0                   7-10                100/0         0
Chorus        5                   20-40               50/50         -100 to +100
*/
void CModDelayModule::cookModType()
{
    switch(m_uModType)
    {
        case Flanger:
        {
            m_fMinDelay_mSec = 0;
            m_fMaxDelay_mSec = 7;
            m_DDL.m_fWetLevel_pct = 50.0;
            m_DDL.m_fFeedback_pct = m_fFeedback_pct;
            break;
        }
        case Vibrato:
        {
            m_fMinDelay_mSec = 0;
            m_fMaxDelay_mSec = 7;
            m_DDL.m_fWetLevel_pct = 100.0;
            m_DDL.m_fFeedback_pct = 0.0; // NOTE! no FB for vibrato
            break;
        }
```

Table 10.3: The default delay line settings for the various effects.

	Min Delay (mSec)	Max Delay (mSec)	Wet/Dry (%)	Feedback (%)
Flanger	0	7-10	50/50	-100 to +100
Vibrato	0	7-10	100/0	0
Chorus	7	20-40	50/50	Typically 0

```
                case Chorus:
                {
                        m_fMinDelay_mSec = 5;
                        m_fMaxDelay_mSec = 30;
                        m_DDL.m_fWetLevel_pct = 50.0;
                        m_DDL.m_fInputAttenuation = 1.0;
                        break;
                }

                default: // is Flanger
                {
                        m_fMinDelay_mSec = 0;
                        m_fMaxDelay_mSec = 7;
                        m_DDL.m_fWetLevel_pct = 50.0;
                        m_DDL.m_fFeedback_pct = m_fFeedback_pct;
                        break;
                }
        }
}
```

Write the calculateDelayOffset() function. It works by taking the LFO value from 0 to 1.0 and mapping it on to the range from min to max delay. Then, this value is scaled by the depth percent value so that at 100% the maximum range is used. For the chorus, we effectively move the operating range and chorus center by adding a delay offset to the minimum value, then using that as the starting (reference) point in the calculation.

```
// calculateDelayOffset():
/*
        fLFOSample: a value from 0.0 to 1.0 from the LFO object

        returns: the calculated delay time in mSec for each effect

        NOTES: - the range for the flanger/vibrato is simply mapped from min to max
                 starting at min: fLFOSample*(m_fMaxDelay_mSec - m_fMinDelay_mSec)) +
                 m_fMinDelay_mSec

               - the range for the Chorus includes the starting offset
                 fStart = m_fMinDelay_mSec + m_fChorusOffset;
*/
if(m_uModType == Flanger || m_uModType == Vibrato)
{
        // flanger 0->1 gets mapped to 0->maxdelay
        return (m_fModDepth_pct/100.0)*(fLFOSample*(m_fMaxDelay_mSec -
                m_fMinDelay_mSec)) + m_fMinDelay_mSec;
}
else if(m_uModType == Chorus)
{
        // chorus adds starting offset to move delay range
        float fStart = m_fMinDelay_mSec + m_fChorusOffset;
```

```
            return (m_fModDepth_pct/100.0)*(fLFOSample*(m_fMaxDelay_mSec -
                    m_fMinDelay_mSec)) + fStart;
        }
}
```

10.3.5 PrepareForPlay()

- Forward the call to prepareForPlay() on the member objects.
- Reset and start up the LFO in unipolar mode.

```
bool __stdcall CModDelayModule::prepareForPlay()
{
        // Add your code here:
        m_LFO.prepareForPlay();

        // DDL needs to know sample rate to initialize its buffer
        m_DDL.m_nSampleRate = m_nSampleRate;
        m_DDL.prepareForPlay();

        m_LFO.m_uPolarity = 1;    // 0 = bipolar, 1 = unipolar
        m_LFO.m_uTableMode = 0;   // normal, no band limiting
        m_LFO.reset();            // reset it

        // initialize
        cookModType();
        updateLFO();
        updateDDL();

        // start the LFO!
        m_LFO.m_bNoteOn = true;

        return true;
}
```

userInterfaceChange()

- Call the update functions and cook the mod type. We won't bother with the switch/case statement since changes to any slider will require updates of the LFO and DDL.
- Make sure your nControlIndex values match your GUI components.

```
bool __stdcall CModDelayModule::userInterfaceChange(int nControlIndex)
{
        // change the min/max limits; set wet/dry and Feedback
        if(nControlIndex == 41) // 41 is mod type switch
                cookModType();

        // else just call the other updates which handle all the rest
        //
        // frequency and LFO type
        updateLFO();
```

```
        // Wet/Dry and Feedback
        updateDDL();

        return true;
}
```

processAudioFrame()

- Call the doOscillate() function on the LFO.
- Use the LFO value to calculate a delay offset in milliseconds for the DDL module.
- Write the new delay time to the DDL module and call its cooking function.
- Call processAudioFrame() on the DDL module to get the current output sample.

```
bool __stdcall CModDelayModule::processAudioFrame(float* pInputBuffer, float*
                                                  pOutputBuffer, UINT uNumInputChannels,
                                                  UINT uNumOutputChannels)
{
        // Do LEFT (MONO) Channel
        //
        // 1. Get LFO Values, normal and quad phase
        float fYn = 0;
        float fYqn = 0;
        m_LFO.doOscillate(&fYn, &fYqn);

        // 2. calculate delay offset
        float fDelay = 0.0;
        if(m_uLFOPhase == quad)
                fDelay = calculateDelayOffset(fYqn); // quadrature LFO
        else
                fDelay = calculateDelayOffset(fYn);  // normal LFO

        // 3. set the delay & cook
        m_DDL.m_fDelay_ms = fDelay;
        m_DDL.cookVariables();

        // 4. get the delay output one channel in/one channel out
        m_DDL.processAudioFrame(&pInputBuffer[0], &pOutputBuffer[0], 1, 1);

        // Mono-In, Stereo-Out (AUX Effect)
        if(uNumInputChannels == 1 && uNumOutputChannels == 2)
                pOutputBuffer[1] = pOutputBuffer[0];

        // Stereo-In, Stereo-Out (INSERT Effect)
        if(uNumInputChannels == 2 && uNumOutputChannels == 2)
                pOutputBuffer[1] = pOutputBuffer[0];

        return true;
}
```

Build and test the module. Be sure to listen to the flanger as the delay hits the zero point.

10.3.6 Challenge

Add another radio button switch to turn on or off TZF that we employ. To turn off TZF, set the minimum delay time for flanging to a nonzero value (try 0.1 mSec to begin with) in the cookModType() method. You now have a working modulated delay module that can be combined in many ways to create a wide variety of modulated delay effects. Let's try a couple of them now.

10.4 Design a Stereo Quadrature Flanger Plug-In

The block diagram of the stereo quad flanger is shown in Figure 10.5 and it consists of two identical flanger delay lines running off of LFO values that are in quadrature phase. Since our mod delay module has a built-in LFO and delay, we can assemble this plug-in quickly.

10.4.1 Project: StereoQuadFlanger

Make sure to add all the existing modules when you create the new StereoQuadFlanger project:

- ModDelayModule.h.
- DDLModule.h (because ModDelayModule #includes it).
- WTOscillator.h (because DDLModule #includes it).

10.4.2 StereoQuadFlanger GUI

We will use a simpler UI consisting of the controls shown in Table 10.4.

10.4.3 StereoQuadFlanger.h File

In the .h file, declare two instances of the ModDelayModule, one for the left and one for the right. Also add one helper function to initialize and update the modules. These two member variables and the one function are all that you need.

```
// Add your code here: ----------------------------------------- //
CModDelayModule m_ModDelayLeft;
CModDelayModule m_ModDelayRight;

void updateModDelays();
// END OF USER CODE -------------------------------------------- //
```

Table 10.4: The GUI elements for the StereoQuadFlanger.

Slider Property	Value		Slider Property	Value
Control Name	Depth		Control Name	Feedback
Units	%		Units	%
Variable Type	float		Variable Type	float
Variable Name	m_fModDepth_pct		Variable Name	m_fFeedback_pct
Low Limit	0		Low Limit	−100
High Limit	100		High Limit	100
Initial Value	50		Initial Value	50

Slider Property	Value		Button Property	Value
Control Name	Rate		Control Name	LFO
Units	Hz		Units	
Variable Type	float		Variable Type	enum
Variable Name	m_fModFrequency_Hz		Variable Name	m_uLFOType
Low Limit	0.02		Enum String	tri,sin
High Limit	5			
Initial Value	0.18			

10.4.4 StereoQuadFlanger.cpp File

Add the one helper function updateModDelays(); this is also the function that forces the two mod lines into quad phase:

```
// update all parameters of the ModDelayModules:
/*
        - setup for stereo quad phase
        - mod depth
        - mod rate
        - feedback
        - mod type = 0 = flanger
        - LFO type
        - call the update() functions to cook everything
*/
void CStereoQuadFlanger::updateModDelays()
{
        // setup quad phase
        m_ModDelayLeft.m_uLFOPhase = 0; // 0: normal
        m_ModDelayRight.m_uLFOPhase = 1; // 1: quad phase

        m_ModDelayLeft.m_fModDepth_pct = m_fModDepth_pct;
        m_ModDelayRight.m_fModDepth_pct = m_fModDepth_pct;
```

```
            m_ModDelayLeft.m_fModFrequency_Hz = m_fModFrequency_Hz;
            m_ModDelayRight.m_fModFrequency_Hz = m_fModFrequency_Hz;

            m_ModDelayLeft.m_fFeedback_pct = m_fFeedback_pct;
            m_ModDelayRight.m_fFeedback_pct = m_fFeedback_pct;

            // FLANGER!
            m_ModDelayLeft.m_uModType = 0;
            m_ModDelayRight.m_uModType = 0;

            m_ModDelayLeft.m_uLFOType = m_uLFOType;
            m_ModDelayRight.m_uLFOType = m_uLFOType;

            // cook them
            m_ModDelayLeft.updateLFO();
            m_ModDelayLeft.updateDDL();

            m_ModDelayRight.updateLFO();
            m_ModDelayRight.updateDDL();
    }
```

prepareForPlay()

- Forward the call to the member objects.
- updateModDelays().

```
    bool __stdcall CStereoQuadFlanger::prepareForPlay()
    {
            // Add your code here:
            // call forwarding!
            m_ModDelayLeft.prepareForPlay();
            m_ModDelayRight.prepareForPlay();

            // don't leave this out - it inits and cooks
            updateModDelays();

            return true;
    }
```

userInterfaceChange()

- updateModDelays().

```
    bool __stdcall CStereoQuadFlanger::userInterfaceChange(int nControlIndex)
    {
            // just do a brute force update of all
            updateModDelays();
    }
```

processAudioFrame()

- Forward the processAudioFrame() function to the member objects to do the processing.

```
bool __stdcall CStereoQuadFlanger:: processAudioFrame(float* pInputBuffer, float*
                                                     pOutputBuffer, UINT uNumInputChannels,
                                                     UINT uNumOutputChannels)
{
    // Do LEFT (MONO) Channel
    m_ModDelayLeft.processAudioFrame(&pInputBuffer[0], &pOutputBuffer[0], 1, 1);

    // Mono-In, Stereo-Out (AUX Effect)
    if(uNumInputChannels == 1 && uNumOutputChannels == 2)
        pOutputBuffer[1] = pOutputBuffer[0];

    // Stereo-In, Stereo-Out (INSERT Effect)
    if(uNumInputChannels == 2 && uNumOutputChannels == 2)
        m_ModDelayRight.processAudioFrame(&pInputBuffer[1], &pOutputBuffer[1],
        1, 1);

    return true;
}
```

Build and test the project. The code is so sparse because we already did most of the work. The only tricky part about this is making sure you have the objects properly initialized and cooked and that you forward the calls correctly. If you have errors or crashes, it's probably because your objects are not properly initialized.

10.4.5 Challenges

- Add a control to toggle TZF mode.
- Add a control to let you turn on and off quad phase (it's currently hard-coded "on").
- Add more LFO shapes.

10.5 Design a Multi-Unit LCR Chorus Plug-In

This LCR or left-center-right chorus (based on the Korg Triton®) uses three chorusing modules run by three independent LFOs, each with its own depth and rate controls (Figure 10.13). If we play it right, we can code this with a minimum of effort, but we have to be very careful about book-keeping since we have many variables here. The UI will also be more complicated, with three sliders per chorus module: depth, rate, and feedback. The LFO type is fixed as a triangle for all units. L is in quad phase, R is in inverse-quad phase, and C is normal.

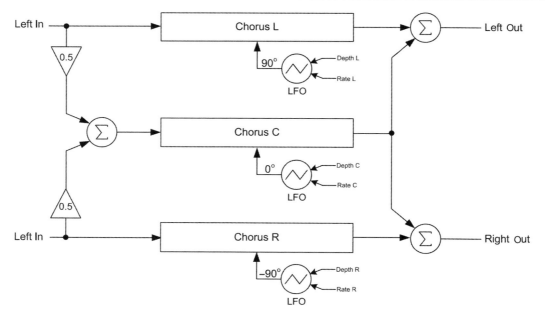

Figure 10.13: The LCR chorus.

10.5.1 Project: StereoLCRChorus

Make sure to add all the existing modules when you create the new project:

- ModDelayModule.h.
- DDLModule.h (because ModDelayModule #includes it).
- WTOscillator.h (because DDLModule #includes it).

10.5.2 StereoLCRChorus GUI

This project contains repeated sets of controls. There are depth, rate, and feedback values for each of the three modules. The sets of controls share the same min and max values. You will have nine slider controls in all. Use Table 10.5 to set up the controls.

10.5.3 StereoLCRChorus.h File

Declare your member objects in the .h file. Also, declare an update() function to modify the parameters all at once.

```
// Add your code here: ----------------------------------------- //
//
// One module for each LCR
```

Table 10.5: GUI elements for the LCR chorus.

Depth (%)	Rate (Hz)	Feedback (%)
Min = 0	Min = 0.02	Min = −100
Max = 100	Max = 5	Max = +100
Initial L = 50	Initial L = 0.18	Initial L = 0
Initial C = 50	Initial C = 0.27	Initial C = 0
Initial R = 50	Initial R = 0.49	Initial R = 0
Variable names:	**Variable names:**	**Variable names:**
m_fModDepth_pct_L	m_fModFrequency_Hz_L	m_fFeedback_pct_L
m_fModDepth_pct_C	m_fModFrequency_Hz_C	m_fFeedback_pct_C
m_fModDepth_pct_R	m_fModFrequency_Hz_R	m_fFeedback_pct_R

```
CModDelayModule m_ModDelayLeft;
CModDelayModule m_ModDelayCenter;
CModDelayModule m_ModDelayRight;

// function to transfer out variables to it and cook
void updateModules(); // you could split this out into smaller functions
// END OF USER CODE ------------------------------------------------- //
```

10.5.4 StereoLCRChorus.cpp File

Constructor

There is nothing to initialize in the constructor because we have no variables; the member objects will initialize themselves at construction time. Implement the updateModules() function next. We need to set up the LFO phases and invert flags according to the block diagram. We also need to put the modules in chorus mode and call the internal update functions:

```
// updateModules()
/*
        - set LFO Phases
        - set LFO inversion (right only)
        - Mod Depths
        - Mod Freqs
        - Feedback
        - Mod Types = chorus (2)
        - LFO Types = triangle (0)
        - call updaters
*/
void CStereoLCRChorus::updateModules()
{
        // setup quad phase
```

```
            m_ModDelayLeft.m_uLFOPhase = 1;   // 1: quad phase
            m_ModDelayCenter.m_uLFOPhase = 0; // 0: normal
            m_ModDelayRight.m_uLFOPhase = 1;  // 1: quad phase

            m_ModDelayLeft.m_LFO.m_bInvert = false;
            m_ModDelayCenter.m_LFO.m_bInvert = false;
            m_ModDelayRight.m_LFO.m_bInvert = true; // this one is inverted

            m_ModDelayLeft.m_fModDepth_pct = m_fModDepth_pct_L;
            m_ModDelayCenter.m_fModDepth_pct = m_fModDepth_pct_C;
            m_ModDelayRight.m_fModDepth_pct = m_fModDepth_pct_R;

            m_ModDelayLeft.m_fModFrequency_Hz = m_fModFrequency_Hz_L;
            m_ModDelayCenter.m_fModFrequency_Hz = m_fModFrequency_Hz_C;
            m_ModDelayRight.m_fModFrequency_Hz = m_fModFrequency_Hz_R;

            m_ModDelayLeft.m_fFeedback_pct = m_fFeedback_pct_L;
            m_ModDelayCenter.m_fFeedback_pct = m_fFeedback_pct_C;
            m_ModDelayRight.m_fFeedback_pct = m_fFeedback_pct_R;

            // CHORUS!
            m_ModDelayLeft.m_uModType = 2;
            m_ModDelayCenter.m_uModType = 2;
            m_ModDelayRight.m_uModType = 2;

            m_ModDelayLeft.m_uLFOType = 0;   // triangle
            m_ModDelayCenter.m_uLFOType = 0; // triangle
            m_ModDelayRight.m_uLFOType = 0;  // triangle

            // cook them
            m_ModDelayLeft.updateLFO();
            m_ModDelayCenter.updateLFO();
            m_ModDelayRight.updateLFO();

            m_ModDelayLeft.updateDDL();
            m_ModDelayCenter.updateDDL();
            m_ModDelayRight.updateDDL();
    }
```

prepareForPlay()

- Forward the calls to prepareForPlay() to the member objects.

```
    bool __stdcall CStereoLCRChorus::prepareForPlay()
    {
            // Add your code here:
            // call forwarding!
            m_ModDelayLeft.prepareForPlay();
            m_ModDelayCenter.prepareForPlay();
            m_ModDelayRight.prepareForPlay();
```

```
            // don't leave this out - it inits and cooks
            updateModules();

            return true;
}
```

UserInterfaceChange()

- Just do a brute force update of all modules (you can streamline this later):

```
bool __stdcall CStereoLCRChorus::userInterfaceChange(int nControlIndex)
{
            // add your code here
            // brute force update
            updateModules();

            return true;
}
```

ProcessAudioFrame()

- Split and sum the L + R to feed the center chorus.
- Split and sum the outputs (L = L + C) and (R = R + C).

```
bool __stdcall CStereoLCRChorus:: processAudioFrame(float* pInputBuffer, float*
                                                    pOutputBuffer, UINT
                                                    uNumInputChannels, UINT
                                                    uNumOutputChannels)
{
        // declare some output variables
        float fChorusOut_L = 0;
        float fChorusOut_C = 0;
        float fChorusOut_R = 0;

        // get the left and right inputs; note the setup for mono
        float fLeftIn = pInputBuffer[0];
        float fRightIn = pInputBuffer[0];

        // if stereo
        if(uNumInputChannels == 2)
              fRightIn = pInputBuffer[1];

        // form the center channel
        float fCenterIn = 0.5*fLeftIn + 0.5*fRightIn;

        // call the processAudioFrame()
        m_ModDelayLeft.processAudioFrame(&fLeftIn, &fChorusOut_L, 1);
        m_ModDelayCenter.processAudioFrame(&fCenterIn, &fChorusOut_C, 1);
        m_ModDelayRight.processAudioFrame(&fRightIn, &fChorusOut_R, 1);
```

```
              // sum to create Left Out
              pOutputBuffer[0] = fChorusOut_L + fChorusOut_C;

              // Mono-In, Stereo-Out (AUX Effect)
              if(uNumInputChannels == 1 && uNumOutputChannels == 2)
                    pOutputBuffer[1] = pOutputBuffer[0];

              // Stereo-In, Stereo-Out (INSERT Effect)
              if(uNumInputChannels == 2 && uNumOutputChannels == 2)
                    pOutputBuffer[1] = fChorusOut_R + fChorusOut_C;

              return true;
      }
```

Build and test the code. Try to find the most interesting settings for different instruments then go back and set your own initial values accordingly. Be sure to save lots of presets—this plug-in can produce a wide variety of sounds.

10.6 More Modulated Delay Algorithms

In addition to the extra algorithms in Sections 10.1 and 10.2, next are some more designs for you to experiment with. Remember to figure out the UI first then code it using the modules we've already built.

10.6.1 Stereo Cross-Flanger/Chorus (Korg Triton®)

Like the quadrature version, the only difference between the flanger and chorus is in the module setting—the same block diagram is used for both. The LFO can be engaged in normal or quadrature phase (Figure 10.14).

10.6.2 Multi-Flanger (Sony DPS-M7®)

The DPS-M7 has some intensely thick modulation algorithms. This one has two flanger circuits that can be combined in parallel or series on either channel. The channels are also cross mixable. Each module marked "flanger" contains a complete flanger module: depth, rate, feedback, and wet/dry. Note also the use of pre-delays with feedback too. All LFOs and pre-delays are independent and fully adjustable (Figure 10.15).

10.6.3 Bass Chorus

The bass chorus in Figure 10.16 splits the signal into low-frequency and high-frequency components and then only processes the high-frequency component. This leaves the fundamental intact. The comb filtering of the chorus effect smears time according to how

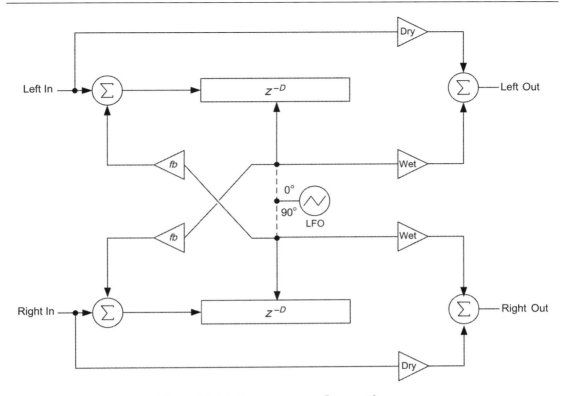

Figure 10.14: A stereo cross-flanger/chorus.

much delay is being used. For the bass guitar, this results in a decreased fundamental with an ambiguous transient edge or starting point. Because bass players need to provide a defined beat, the bass chorus will preserve this aspect of the instrument. If you want to implement this effect, use the Linkwitz–Riley low-pass and high-pass filters to split the signal. Invert the output of one of the filters—it doesn't matter which one—so that their phase responses sum properly.

10.6.4 Dimension-Style (Roland Dimension D®)

This chorus unit (Figure 10.17) is based on the Roland Dimension D® chorus. Known for its subtle transparent sound, it features a shared but inverted LFO and an interesting output section where each output is a combination of three signals:

1. Dry
2. Chorus output
3. Opposite channel chorus output, inverted and high-pass filtered

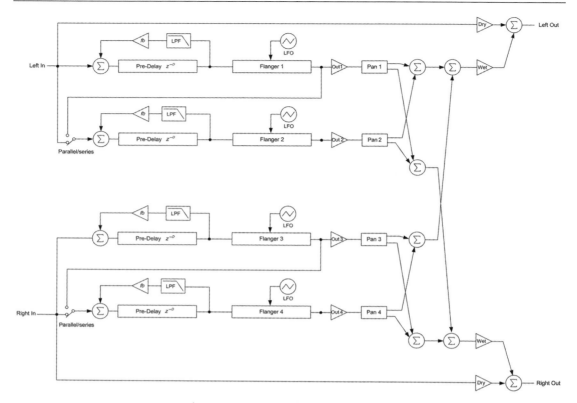

Figure 10.15: DPS M7 multi-flanger.

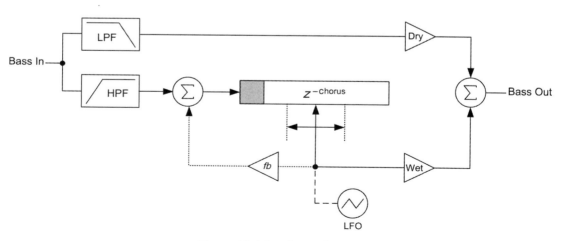

Figure 10.16: A bass chorus.

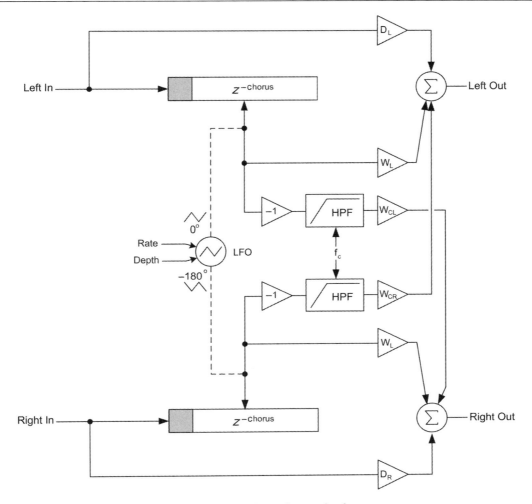

Figure 10.17: A dimension-style chorus.

The controls on the original unit consisted of four switches only; these were hardwired presets and changed

- The LFO rate (either 0.25 Hz or 0.5 Hz)
- The depth
- For one preset, the wet/dry mix ratios for the two channels (W_L vs D_L, W_R vs D_R)

Here I have added two more level controls for experimentation, W_{CL} and W_{CR}, which are the filtered, inverted chorused signals from the opposite (crossed) channels. Experiment with various rate, depth, and output mixture combinations as well as different high-pass filter cut-off frequencies.

10.6.5 Deca-Chorus (Sony DPS-M7®)

The deca-chorus has 10 (deca) chorus units, 5 per channel. Each chorus has its own pre-delay, LFO, gain, and pan control. It can also run in a mono mode (10 chorus units in parallel) (Figure 10.18).

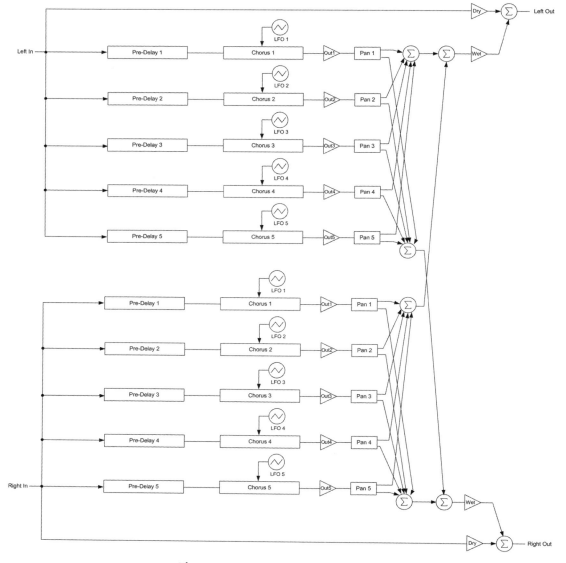

Figure 10.18: DPS-M7 deca chorus.

Bibliography

Cole, M. 2007. "Roland dimension C clone for Eventide 7000, Orville and H8000." Accessed August 2012 from http://www.eventidestompboxes.com/forummedia/PATCHES/Orville/ProFXalgorithms/Roland%20Dimension%20C.doc.

Coulter, D. 2000. *Digital Audio Processing*, Chapter 11. Lawrence, KS: R&D Books.

Dattorro, J. 1997. Effect design part 2: Delay line modulation and chorus. *Journal of the Audio Engineering Society*, 45(10): 764–786.

Korg, Inc. 2000. *Triton Rack Parameter Guide*. Tokyo: Korg Inc.

Phillips, D. 1991. *Wavestation SR Reference Guide*. Tokyo: Korg Inc.

Roads, C. 1996. *The Computer Music Tutorial*, Chapter 3. Cambridge, MA: The MIT Press.

Roland Corporation. 1990. "Boss CH-1 schematics." Accessed August 2012 from http://superchorusmods.blogspot.com/2009/08/schematics.html.

Sony, Corp. 1995. *DPS-M7 Manual*. Tokyo: Sony Corp.

White, P. 2001. "Understanding and emulating vintage dffects." *Sound on Sound Magazine*. Accessed August 2012 from http://www.soundonsound.com/sos/jan01/articles/vintage.asp.

CHAPTER 11
Reverb Algorithms

Reverb algorithms might represent the Holy Grail of audio signal processing. They have an appeal that seems universal, perhaps because we live in a reverberant world. Our ears are time-integrating devices that use time domain cues and transients for information, so we are sensitive to anything that manipulates these cues. In this chapter, we discuss reverb algorithms as applied mainly to room simulation. There are two general ways to create the reverberation effect:

- Reverb by direct convolution—the physical approach.
- Reverb by simulation—the perceptual approach.

In the physical approach, the impulse response of a room is convolved with the input signal in a large finite impulse response (FIR) filter. For large rooms, these impulse responses might be 5 to 10 seconds. In the mid 1990s, Gardner developed a hybrid system for fast convolution that combined direct convolution with block fast Fourier transform (FFT) processing (Gardner 1995). Around the same time, Reilly and McGrath (1995) described a new commercially available system that could process 262,144-tap FIR filters for convolving impulses over 5 seconds in length. The required processing power is sometimes too large to be practical in a plug-in. Aside from the computing expense, another drawback to this approach is that an impulse response is frozen in time, and measures the room at one location only, under a certain set of conditions. To create a reverb unit (or *reverberator*) that is general enough to provide many different reverbs of different spaces would require a large library of impulse response files. In addition, its parameters can't be adjusted in real time.

The perceptual approach aims to simulate the reverberation with enough quality to fool the ears and give the same perception of real reverb but with much less processing power. The advantages are numerous, from the minimal processing required to the ability to vary many parameters in real time. Browne (2001) proposed a hybrid system combining a short impulse response convolution along with recursive filtering, combining both approaches. There are several key engineers who developed much of the theory still in use today. These include Schroeder's (1962) initial work in the early 1960s with continued research and contributions from Moorer, Griesinger, Gerzon, Gardner, Jot, Chaigne, Smith, Roscchesso, and others across the decades. Most of this chapter is owed to their work in the field. We will focus on the perceptual approach and try to find computationally efficient algorithms for interesting reverberation effects.

Griesinger (1989) states that it is impossible to perfectly emulate the natural reverberation of a real room and thus the algorithms will always be approximations. It seems that the area of reverberation design has the most empirically derived or trial-and-error research of just about any audio signal processing field. There is no single "correct" way to implement a reverb algorithm, so this chapter focuses on giving you many different reverberator modules to experiment with.

11.1 Anatomy of a Room Impulse Response

The first place to start is by examining impulse responses of actual rooms. There are several popular methods for capturing the impulse response, from cap pistols and balloons to deconvolution of chirp signals and pseudo random sequences. The resulting time domain plot is useful for investigating the properties of reverberation.

Figure 11.1 shows the impulse response plots for two very different spaces; a large concert hall and a cathedral. The initial impulse is followed by a brief delay called the *pre-delay*. As the impulse pressure wave expands, it comes into contact with the nearby structures—walls, floor, and ceiling—and the first echoes appear. These initial echoes, called *early reflections*, are important to the simulation of reverb because of the auditory cues we get from them. The pressure wave continues to expand and more reflections occur, with reflected signal upon reflected signal piling on top of each other while decaying in energy. The resulting reverb "tail" is called *late reverberation* or *subsequent reverberation*.

The top impulse response in Figure 11.1 is a modern concert hall designed for a pleasing reverberant quality. The initial reflections are from irregularly shaped back and side walls and they pile up in amplitude; they are also piling up in density, shown as the impulse gets "fatter" in the middle section. The dense reverberant tail follows, which decays in an exponential fashion, and the sound pressure energy is absorbed by the seats and acoustical room treatments. The cathedral is a different story. After a short pre-delay, a few large early reflections arrive from the nearby structures but they don't increase in amplitude or density in the same way as the hall. The reverb's decay is also much longer in time as there is little in the cathedral to absorb the sound pressure.

The block diagram in Figure 11.2 shows the three components we observe. However, there is debate as to whether or not this is a good way to break the problem down. Some feel that the reverb algorithm should not need to separate out the early reflections from the late reverberation. In other words, a good algorithm will create all of the reverberation aspects at once.

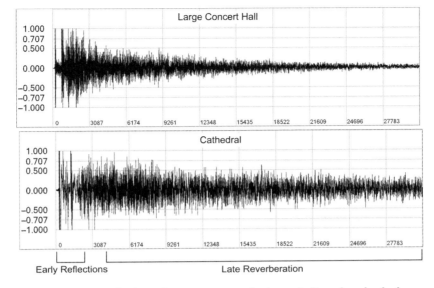

Figure 11.1: The impulse responses of a large hall and cathedral.

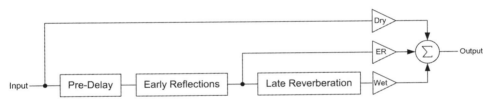

Figure 11.2: A generalized model of a reverb algorithm.

11.1.1 RT_{60}: The Reverb Time

The most common measurement for reverb is the RT_{60} reverb time. Reverb time is measured by first stimulating the room into a reverberant state, then turning off the source and plotting the pressure-squared level as a function of time. The amount of time it takes for this energy decay curve to drop by 60 dB is the *reverb time*, or RT_{60}. Sabine's (1973) pioneering work in this area leads to the following formula in Equation 11.1:

$$RT_{60} = 0.5 \frac{V_R}{S_R A_{RAve}} \tag{11.1}$$

where

$$V_R = \text{volume of room in ft}^3$$
$$S_R = \text{surface area of room in ft}^2$$
$$A_{RAve} = \text{average absorption coefficient}$$

Sabine measured and tabulated the absorption coefficients for various materials. The units are given in *Sabines*. A room made of several materials is first partitioned to find the partial surface area of each material type, then the average is found by weighting the areas with the absorption coefficients and summed. The reverb time is probably the most common control, found on just about every kind of reverb plug-in.

11.2 Echoes and Modes

In Schroeder's early work, he postulates that a natural sounding artificial reverberator has both a large echo density and a colorless frequency response. The echo density is simply the number of echoes per second that a listener experiences in a given position in the reverberant environment (Equation 11.2).

$$E_D = \frac{\text{echoes}}{\text{second}} \tag{11.2}$$

If the echo density is too low the ear discerns the individual echoes and a fluttering sound is the result. As the echo density increases, the echoes fuse together, producing a wash of sound. Schroeder (1962) postulated that for a natural-sounding reverb, the echo density needs to be 1000 echoes/second. Greisinger (1989) increases this to 10,000 echoes/second or more. Statistically, the echo density is shown in Equation 11.3, which reveals that the echoes build up as the square of time.

$$\text{Echo density} \propto \frac{4\pi c^3}{V_R} t^2 \tag{11.3}$$

where

$$c = \text{speed of sound}$$
$$t = \text{time}$$
$$V_R = \text{volume of room}$$

Reverb is often modeled statistically as decaying white noise, which implies that ideal reverberant rooms have flat frequency responses. A room's geometry can be used to predict its frequency response. The dimensions of the room can be used to calculate the frequencies that will naturally resonate as the wave bounces back and forth between parallel surfaces. An untreated rectangular room will have multiple resonances and anti-resonances.

Given a rectangular room with length, width, and height of l, w, and h, the resonances are found in Equation 11.4, a well-known and useful equation for predicting room resonances, also called *modes* (Beranek 1986). As the frequency increases, so do the number of resonances.

$$f_e = \frac{c}{2}\sqrt{\left(\frac{n_x}{l}\right)^2 + \left(\frac{n_y}{w}\right)^2 + \left(\frac{n_z}{h}\right)^2}$$

where (11.4)

n_x, n_y, n_z = half wave numbers 0, 1, 2, 3...
l, w, h = length(x), width(y) and height(z) of the room

Above a certain frequency the resonances overlap and fuse together. Each resonant frequency has its own envelope whose curve is bell (band-pass filter [BPF]) shaped, meaning that it has a peak resonance at the mode frequency, but is still excitable at frequencies around it and thus the quality factor (Q) of the curve relates to this excitability. The number of resonances increases with frequency. Figure 11.3 shows a fictitious room example with the bell-shaped resonances.

The *modal density* is the number of resonant peaks per Hz. Physicists call the resonant frequencies or modes *eigenfrequencies* (note this is not an acoustics-specific term; an eigenfrequency is the resonant frequency of any system). Schroeder's second postulation is that for a colorless frequency response, the modal density should be 0.15 eigenfrequencies/Hz or one eigenfrequency every 6.67 Hz or approximately 3000 resonances spread across the audio spectrum. Thus, it makes sense that good reverberant environments have interesting geometries with many nonparallel walls. The multitude of resonances is created by the many paths an impulse can take from the source to the listener. Kuttruff (1991) derived the approximation for

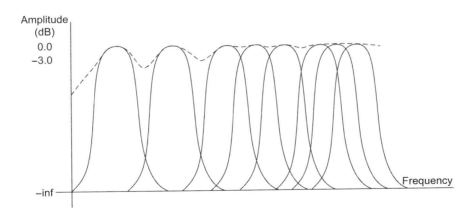

Figure 11.3: The resonances of a room all contribute to create the final frequency response. When the bandwidths of the modes overlap, they fuse together. The dotted line represents the combined frequency response.

> Schroeder's rules for natural reverb:
>
> - Echo density: At least 1000 echoes/sec (Greisinger: 10,000 echoes/sec)
> - Modal density: At least 0.15 eigenfrequencies/Hz
>
> In physical rooms we know that:
>
> - Echo density increases with the square of time.
> - Modal density increases with the square of frequency.

the modal density as it relates to the volume of the room and modal frequency. Equation 11.5 shows that the resonances build up as the square of frequency.

$$D_m = \frac{4\pi V_R}{c^3} f_m^2 \qquad (11.5)$$

where

V_R = volume of the room
f_m = modal frequency in question

The *energy decay relief* plot (or EDR) shows how the energy decays over both frequency and time for a given impulse response of a room. Figure 11.4 shows a very simple fictitious EDR.

In Figure 11.4, notice that the frequency axis comes out of the page; low frequencies are in the back. It also shows that this room has a resonant peak, which forms almost right away

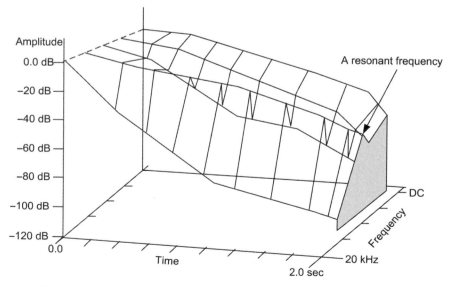

Figure 11.4: The EDR shows time (*x*-axis), frequency (*z*-axis) and amplitude (*y*-axis) of the energy decay of a room.

in time. In an EDR, the modal density is shown across the *z*-axis (frequency) while the echo density appears across the *x*-axis (time). This simplified plot shows just one resonance and no echo density build-up for ease of viewing. Figure 11.5 shows the EDR of an actual room, in this case a theater. The eigenfrequencies are visible as the ridges that run perpendicularly to the frequency axis.

Figure 11.6 shows the EDR for a cathedral. In this case, the echo pile-ups are clearly visible running perpendicular to the time axis. Comparing both EDRs shows that both rooms have high echo and modal densities; therefore, they should be good reverberant spaces. Both EDRs show an initial high frequency roll-off and, especially in the theater's case, the high frequencies decay faster than the low frequencies. The high-frequency decay is a property of the treatment of the room surfaces along with the fact that the air absorbs high frequencies more easily than low frequencies. The high-frequency energy decay in the theater is caused by the carpet, seats, and acoustic treatment which the cathedral lacks. Therefore, for a good quality reverb, we will need to take this high-frequency decay into account along with the echo and modal density.

Reverb algorithms are typically made of arrangement of modules called *reverberator modules*. From our basic observations we can tell that good reverberator modules are going to produce dense echoes along with a large number of resonances. If the resonances are distributed properly across the spectrum, the reverb will sound flat. If they are not, there will be metallic pinging and other annoyances that will color the frequency response in an

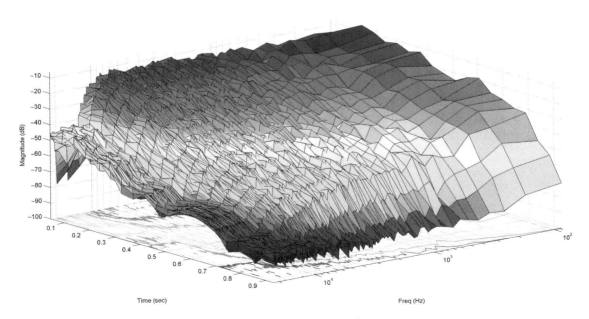

Figure 11.5: An EDR for a theater.

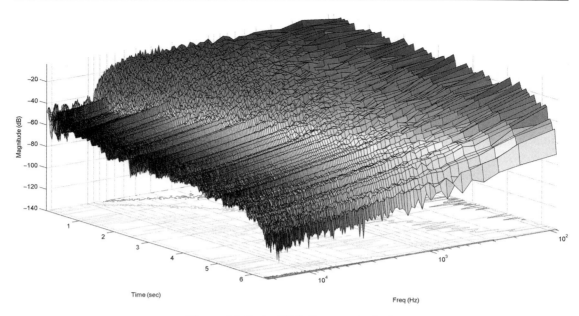

Figure 11.6: An EDR for a cathedral.

unnatural way. The majority of the rest of the chapter is devoted to revealing, analyzing, and explaining these building blocks and the algorithms that use them. The best place to start is with Schroeder's reverb modules.

11.3 The Comb Filter Reverberator

One of the reverberator modules that Schroeder proposed is a comb filter. Remember that we are looking for modules that produce echoes and the comb filter with feedback will do just that. The comb filter reverberator is familiar looking because it's also the delay with feedback algorithm you studied in Chapter 7. We derived the difference equation and transfer function for the comb filter in Figure 11.7 in Chapter 7:

$$y(n) = x(n-D) + gy(n-D)$$

and (11.6)

$$H(z) = \frac{z^{-D}}{1 - gz^{-D}}$$

We also performed a pole-zero analysis and generated frequency plots. We showed that the feedback path caused a series of poles evenly spaced around the inside of the unit circle. The resulting impulse response is easy to predict as the echo recirculates through the feedback path, being attenuated by multiplying by g each time through the loop.

While Figure 11.8 might look simple, the results certainly trend in the right direction. The frequency response plot shows a set of resonant peaks that could correspond to a room's eigenfrequencies and the impulse response shows a set of decaying impulses, mimicking the energy loss as the echoes bounce off of surfaces. The modal density for a comb filter is given by Jot and Chaigne (1991) in Equation 11.7.

$$M_d = \frac{D}{f_s}$$

$$\Delta f = \frac{f_s}{D}$$
(11.7)

where

D = the delay length
f_s = the sample rate

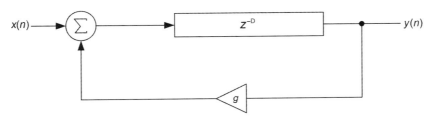

Figure 11.7: The basic comb filter.

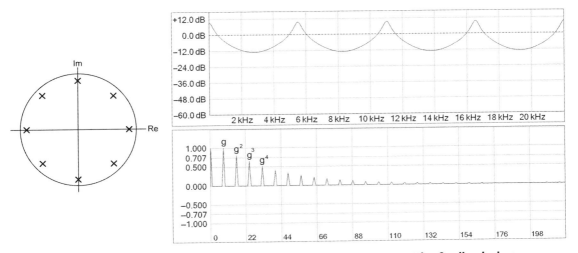

Figure 11.8: The poles in the z-plane produce the resonances. The feedback that creates those poles also recirculates and scales the echoes by g. In this example, D = 8 samples, feedback = 80%.

So, the comb filter produces D resonances across DC to f_s (or $D/2$ resonances from DC to Nyquist) with each resonance separated by $\Delta f = f_s/D$. This is exactly what we found when we analyzed the comb filter in Chapter 7. In order to get Schroeder's desired 0.15 eigenfrequencies/Hz from DC to 20 kHz at $f_s = 44{,}100$, you would need $D = 6622$. This also makes sense because we need about 3000 ($D/2$) resonances to cover the range from DC to 20 kHz. The problem is that while we will get the desired number of resonances, they will be linearly spaced rather than piling up as the square of frequency, as Equation 11.5 predicts. Additionally, it will take 6622 samples (~150 mSec with a 44.1 kHz sample rate) before the reverberator begins outputting samples. Schroeder was able to achieve high density by placing comb filters in parallel, then summing the results. This would then cause the comb filter responses to overlap, however, care must be taken when choosing the delay lengths—if they are mathematically related then resonances or anti-resonances will color the sound.

Figure 11.9 shows an example of four comb filters in parallel. Each has its own delay time ($D1$–$D4$) and gain ($g1$–$g4$). However, care must be taken with the gain values. As you remember from Chapter 7, the value of g alone determines the pole radii. If the poles of each filter have the same radii, their impulses will decay at the same rate. This is desirable since it mimics an uncolored reverberation pattern. If the pole radii are different for the different comb filters then the resulting reverberation will have undesirable coloration in it. The pole radii are given in Equation 11.8 by Gardner (Kahrs and Bradenberg 1998). The relationship between the pole radii, the delay D and the reverb time RT_{60} is given in Equation 11.9.

$$r = g^{1/D} \tag{11.8}$$

where

$$g = \text{the feedback gain}$$
$$D = \text{the delay length}$$

$$RT_{60} = \frac{3DT_s}{\log(1/g)}$$

or

$$1/g = 10^{\frac{3DT_s}{RT_{60}}}$$

or

$$g = 10^{\frac{-3DT_s}{RT_{60}}} \tag{11.9}$$

T_s is the sample period

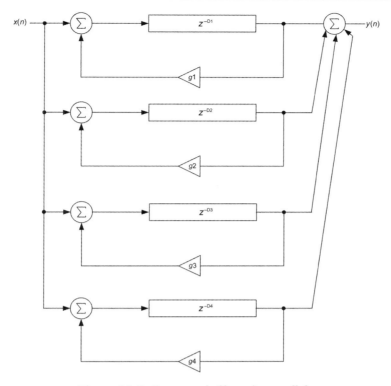

Figure 11.9: Four comb filters in parallel.

This means we can control the reverb time by using the gain factor *or* the delay length D. The tradeoff is that if we increase g to increase reverb time, the poles get very near the unit circle, causing resonances. If we increase D, then the echoes become distinctly audible rather than smearing together. This means that there is a tradeoff of the modal density versus the echo density in the design. The modal density for a parallel bank of N comb filters in Equation 11.10 is given by Jot (1991). From Equation 11.10 you can see that the modal density remains constant across all frequencies, which is not what happens in real rooms where the modal density increases with the square of the frequency (Equation 11.5).

$$M_d = \sum_{i=0}^{N} D_i T_s$$

where

D_i = the delay length of the ith comb filter
T_s = the sample period

(11.10)

or

$$M_d = N\overline{D}T_s$$

where

$\overline{D}$ = the mean delay length of all filters averaged

The echo density for the parallel combs with delay lengths close to each other is given in Equation 11.11 (Jot 1991).

$$E_d = \sum_{i=0}^{N} \frac{1}{D_i T_s}$$

where

D_i = the delay length of the ith comb filter   (11.11)

or

$$E_d = \frac{N}{\overline{D} T_s}$$

where

$\overline{D}$ = the mean delay length of all filters averaged

Knowing the desired M_d and E_d you can then calculate the number of parallel comb filters N with average delay time $\overline{D} T_s$ with Equation 11.12.

$$N = \sqrt{E_d M_d}$$
$$\overline{D} T_s = \sqrt{M_d / E_d} \qquad (11.12)$$

Plugging Schroeder's values of $M_d = 0.15$ eigenfrequencies/Hz and $E_d = 1000$ echoes/sec into Equation 11.12 yields $N = 12$ and the average delay time $DT_s = 12$ mSec.

11.4 The Delaying All-Pass Filter Reverberator

Schroeder also proposed the delaying all-pass filter (APF) as a reverberator unit. The impulse response is a decaying set of echoes but the frequency response is technically flat. Schroeder's APF is shown in Figure 11.10. In Figure 11.11, you can see that we get echoes that decay faster than the comb filter and at a different rate. It has a flat magnitude response due to the APF reciprocal zero/pole radii. However, as Gardner points out, our ears only perform a short-time integration, whereas the APF requires an infinite integration time to achieve the flat response, so we will still hear some timbral coloration in this unit.

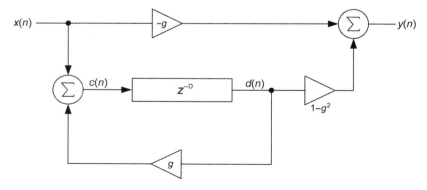

Figure 11.10: Schroeder's APF reverberator.

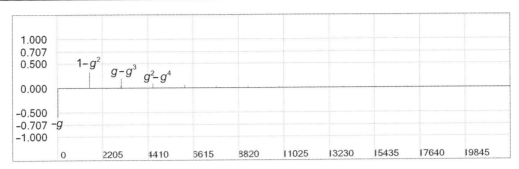

Figure 11.11: Impulse response of Schroeder's APF reverberator; the frequency response is flat. The delay time was $D = 35$ mSec with $g = 0.6$.

Inspection of the block diagram in Figure 11.10 reveals that this is a rather complex feedback/feed-forward structure. We need to extract the difference equation so we can synthesize the reverb unit. To start, we label the nodes $c(n)$ and $d(n)$ in the block diagram, then fashion the output $y(n)$ with respect to them:

$$c(n) = x(n) + gd(n)$$
$$d(n) = c(n)z^{-D} \qquad (11.13)$$
$$y(n) = -gx(n) + (1-g^2)d(n)$$

Now, we expand out the internal nodes and use the familiar time and frequency shifting properties of the z transform to continue Equation 11.14.

$$\begin{aligned} d(n) &= c(n)z^{-D} \\ &= x(n-D) + gd(n-D) \\ y(n) &= -gx(n) + (1-g^2)[x(n-D) + gd(n-D)] \end{aligned} \qquad (11.14)$$

Examining the last term in Equation 11.14, $d(n-D)$, we can rearrange Equation 11.13 to continue and get Equation 11.15:

$$d(n) = \frac{y(n) + gx(n)}{(1-g^2)}$$

so
$$d(n-D) = \frac{y(n-D) + gx(n-D)}{(1-g^2)} \qquad (11.15)$$

Substituting Equation 11.15 back into Equation 11.14 you get the difference equation in Equation 11.16:

$$\begin{aligned} y(n) &= -gx(n) + (1-g^2)\left[x(n-D) + g\left(\frac{y(n-D) + gx(n-D)}{(1-g^2)}\right)\right] \\ &= -gx(n) + (1-g^2)x(n-d) + gy(n-D) + g^2x(n-D) \\ &= -gx(n) + x(n-d) + gy(n-D) \end{aligned} \qquad (11.16)$$

11.5 More Delaying All-Pass Filter Reverberators

There are multiple ways to synthesize the delaying APF. Figure 11.12 shows another one. Deriving the difference equation is easier for this version; we only need to define one extra node, $w(n)$ in Equation 11.17:

$$\begin{aligned} w(n) &= x(n) + gw(n-D) \\ y(n) &= -gw(n) + w(n-D) \\ &= -g[x(n) + gw(n-D)] + [x(n-D) + gw(n-2D)] \\ &= -gx(n) - g^2 w(n-D) + x(n-D) + gw(n-2D) \end{aligned} \quad (11.17)$$

grouping:

$$= -gx(n) + x(n-D) - g^2 w(n-D) + gw(n-2D)$$

In order to finish, find $y(n-D)$ from the second line in Equation 11.18 and notice that this is the same as the last two terms in Equation 11.17.

$$y(n) = -gw(n) + w(n-D)$$

then

$$y(n-D) = -gw(n-D) + w(n-2D) \quad (11.18)$$

and

$$gy(n-D) = -g^2 w(n-D) + gw(n-2D)$$

Thus, the last two terms in Equation 11.17 can be replaced by $gy(n-D)$, and the final difference equation matches Schroeder's APF:

$$y(n) = -gx(n) + x(n-D) + gy(n-D) \quad (11.19)$$

Figure 11.13 shows another version of the same APF. The proof that it has the same difference equation is easy if you look at the node $w(n)$ in Equation 11.20:

$$\begin{aligned} w(n) &= x(n) + gw(n-D) \\ y(n) &= -gw(n) + w(n-D) \end{aligned} \quad (11.20)$$

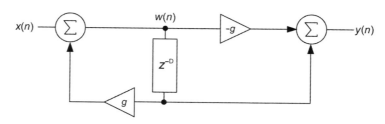

Figure 11.12: Another version of the delaying all-pass reverberator.

These are the first two lines in the derivation for the delaying APF above. Figure 11.14 shows yet another structure. To figure this one out, once again find the internal node $w(n)$ as it relates to the output in Equation 11.21:

$$w(n) = x(n) + gy(n)$$
$$y(n) = -gx(n) + w(n-D)$$

then

$$w(n-D) = x(n-D) + gy(n-D)$$

and

$$y(n) = -gx(n) + x(n-D) + gy(n-D)$$

(11.21)

Equation 11.21 is identical to Equation 11.20, but care must be taken in the synthesis of this in code—you must form the internal nodes first to avoid a zero-delay loop (the $x(n) + gy(n)$ term above). Inverted APFs simply swap signs on the g coefficients. This has the effect of inverting the impulse response while keeping the frequency response theoretically flat. By combining both inverting and regular delaying APFs, you can try to achieve a more uncorrelated echo density.

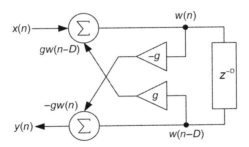

Figure 11.13: Alternate version of the delaying APF; the transfer function and difference equations are identical to the other structures.

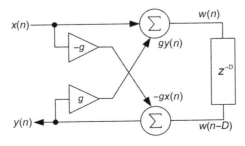

Figure 11.14: Another delaying APF structure.

11.6 Schroeder's Reverberator

Schroeder combined a parallel comb filter bank with two APFs to create his first design. The comb filters produce the long series of echoes and the APFs multiply the echoes, overlaying their own decaying impulse response on top of each comb echo. The resulting reverberation unit, shown in Figure 11.15, sounds marginal but it is very simple to implement.

Schroeder suggests that the ratio of smallest to largest delay in the comb filters should be about 1:1.5 and originally specified a range of 30–45 mSec in total. The 1:1.5 ratio rule turns out to be useful for just about any set of parallel comb filters.

The comb filters should have the following properties:

- Choose the delays to have the 1:1.5 ratio above.
- Choose delay times that have no common factors (e.g., 2191, 2971, 3253, 3307).
- Set the gain values according to Equation 11.9.

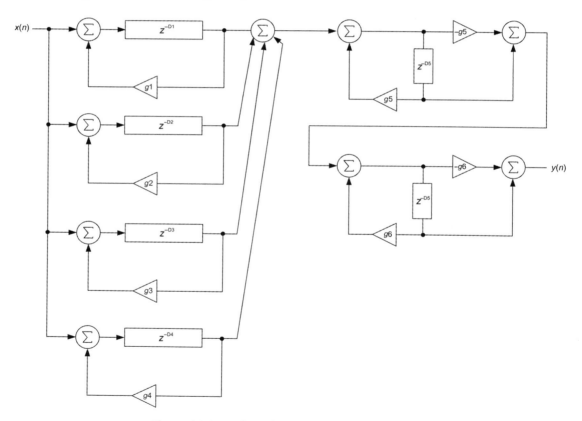

Figure 11.15: Schroeder's original reverb design.

The APFs should have the following properties:

- Choose delays that are much shorter than the comb filters, 1 mSec to 5 mSec.
- Set both gain values the same, between 0.5 and 0.707.

11.7 The Low-Pass Filter–Comb Reverberator

One of the things missing from the original Schroeder reverb is the fact that in a real room, high frequencies decay much more rapidly than low frequencies, as shown in the EDRs of Figure 11.6 (Moorer 1979). Placing a low-pass filter (LPF) in the comb filter's feedback path will roll off the high-frequency content of successive echoes exponentially, which is what we want. The LPF–comb reverberator block diagram is shown in Figure 11.16.

The LPF is chosen to be a one-pole feedback type (infinite impulse response [IIR]) so that it will introduce not only low-pass filtering, but also its own impulse response into the echoes going through the feedback path. In Figure 11.17 the LPF is shown in the dotted box; notice that it is turned around backward to follow the flow of the feedback path.

To find the difference equation, it is easier to start in the frequency domain with the z transforms of the comb and first-order feed-forward filter, since we are already familiar with them by now (Equation 11.22).

$$H_C(z) = \frac{z^{-D}}{1 - g_1 z^{-D}}$$
$$H_{LP}(z) = \frac{1}{1 - g_2 z^{-1}}$$
(11.22)

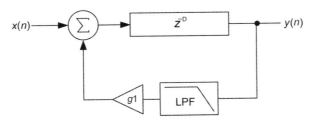

Figure 11.16: The LPF and comb filter combination.

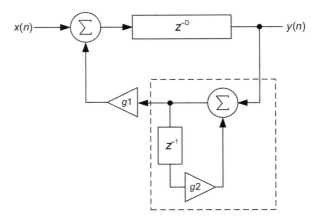

Figure 11.17: The LPF–comb filter expanded.

Filtering the feedback loop in the frequency domain is done by simply multiplying the LPF transfer function by the feedback term $g_1 z^{-D}$ in Equation 11.23.

$$H(z) = \frac{z^{-D}}{1 - H_{LP}(z) g_1 z^{-D}}$$

$$= \frac{z^{-D}}{1 - g_1 z^{-D} \frac{1}{1 - g_2 z^{-1}}}$$

$$= \frac{z^{-D}}{1 - \frac{g_1 z^{-D}}{1 - g_2 z^{-1}}} \qquad (11.23)$$

$$\frac{Y(z)}{X(z)} = \frac{z^{-D}}{1 - \frac{g_1 z^{-D}}{1 - g_2 z^{-1}}}$$

The next step is to separate variables and multiply out the denominator:

$$Y(z)\left[1 - \frac{g_1 z^{-D}}{1 - g_2 z^{-1}}\right] = X(z) z^{-D}$$

$$Y(z)[1 - g_2 z^{-1} - g_1 Y(z) z^{-D}] = X(z) z^{-D}(1 - g_2 z^{-1})$$

$$Y(z) - g_2 Y(z) z^{-1} - g_1 Y(z) z^{-D} = X(z) z^{-D} - g_2 X(z) z^{-D} z^{-1} \qquad (11.24)$$

$$Y(z) - g_2 Y(z) z^{-1} - g_1 Y(z) z^{-D} = X(z) z^{-D} - g_2 X(z) z^{-D-1}$$

Lastly, take the inverse z transform by inspection:

$$Y(z) - g_2Y(z)z^{-1} - g_1Y(z)z^{-D} = X(z)z^{-D} - g_2X(z)z^{-D-1}$$

then

$$y(n) - g_2y(n-1) - g_1y(n-D) = x(n-D) - g_2x(n-D-1) \quad (11.25)$$

and

$$y(n) = x(n-D) - g_2x(n-D-1) + g_2y(n-1) + g_1y(n-D)$$

In order for the filter combination to remain stable (after all, it is a massive feedback system) g_1 and g_2 can be related as in Equation 11.26:

$$g = \frac{g_2}{1-g_1}$$

where

$$g < 1.0$$

(11.26)

Because the RT_{60} determines the pole radii, and therefore the value of g_1, you can rearrange Equation 11.26 as Equation 11.27:

$$g_2 = g(1-g_1)$$

where

(11.27)

$$g < 1.0$$

11.8 Moorer's Reverberator

Moorer proposed a similar design to Schroeder's reverberator which uses a parallel bank of LPF–comb filters. Because the LPFs remove energy from the system, more units are needed in parallel for a given reverb time. Moorer's reverb sounds better because it mimics the high-frequency losses of a real room.

In Figure 11.18 you can see the differences from Schroeder's reverb—there are more comb units and only one all-pass on the output. The same care must be taken to ensure the pole radii of the comb filters are still the same, using Equation 11.9 to set them according to the desired reverb time. Table 11.1 shows Moorer's preset values for a reverb time of about 2 seconds.

Table 11.1: Some settings for Moorer's reverb.

Comb Filter	Delay (mSec)	g_{COMB} (48 kHz)	g_{LPF} (48 kHz)
1	50	0.46	0.4482
2	56	0.47	0.4399
3	61	0.475	0.4350
4	68	0.48	0.4316
5	72	0.49	0.4233
6	78	0.50	0.3735
All-Pass Filter	Delay (mSec)	g	–
1	6	0.7	–

Note: RT_{60} 2 seconds, total $g = 0.83$

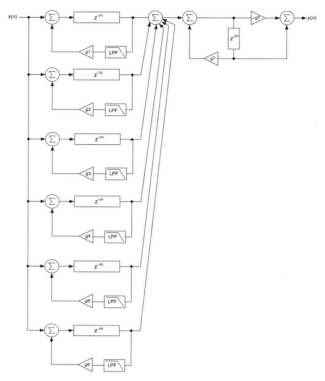

Figure 11.18: Moorer's reverb.

11.9 Stereo Reverberation

Conducting listening tests, Schroeder (1984) found that listeners overwhelmingly preferred a stereo reverb to a mono version of the same algorithm. Both Schroeder and Moorer's reverbs can be adapted for stereo usage. In fact, this scheme can be used with just about any reverb consisting of comb and APFs. The first thing to note is that mathematically, there is no reason why you can't place the APFs before the comb filters since their transfer functions multiply in the z domain. Then, the individual outputs of the comb filters can be combined through a mixing matrix to provide the left and right outputs. The mixing matrix is an array of weighting values for each comb filter's output. Jot proposed that the matrix should be orthogonal for the most uncorrelated output. The mixing matrix is shown in Equation 11.28. The rows are the outputs of the comb filters and the columns are left and right. Figure 11.19 shows a mixing matrix for the left channel of a Schroeder reverb unit.

$$g \begin{bmatrix} +1 & +1 \\ +1 & -1 \\ +1 & +1 \\ +1 & -1 \end{bmatrix} \quad (11.28)$$

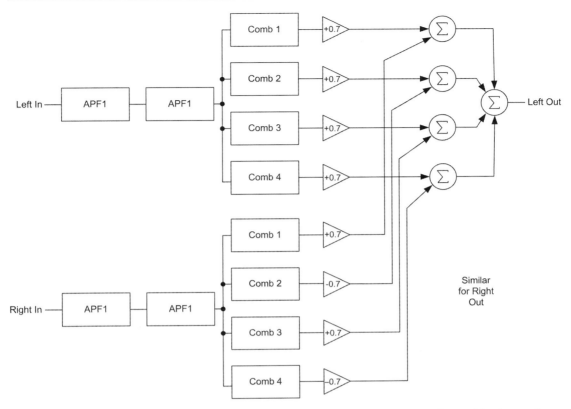

Figure 11.19: A stereo implementation for a Schroeder reverb—only one channel is shown for clarity; the right output is fashioned in a similar way. Notice the mix matrix values do not have to be ±1.0 but need to follow the orthogonality of alternating signs.

11.10 Gardner's Nested APF Reverberators

Schroeder experimented with reverbs made only from delaying APF modules in series. The abundant time smearing suggested that this might be a viable option. Unfortunately the consensus was that this reverb sounded metallic and that it didn't build up the echo density fast enough. Schroeder's original series APFs started with the maximum delay in the first module (about 100 mSec) and then decreased by 0.3 for each successive module, through five modules. Gardner (1992) noted that Schroeder (1962), Gerzon (1972), and Moorer (1979) all experimented with *nesting* multiple all-pass structures, but without success. He suggested a strategy for nesting delaying APF modules along a single transversal delay line. In order to understand how this works, re-fashion the delaying APF structure using the transversal delay line structure you saw in Chapter 8. Figure 11.20 shows the same delaying APF structure as shown in Figure 11.13 but with the delay replaced by a set of single delay elements.

The sequence of operation—reading before writing—is key for implementing this design. Specifically, you need to

1. Read the output of the last delay cell, $x(n - D)$.
2. Read the value of the first delay cell, $x(n)$.
3. Form the new value for the last cell $-gx(n) + x(n - D)$ and write it back into the cell as $y(n)$.
4. Form the new value for the first cell $x(n) = x(n) + gy(n)$.

Gardner's idea was to nest multiple APF structures inside each other so that they *shared* the same delay line. This would produce layers of embedded echoes. Figure 11.20 shows a delaying APF with a total delay time of z^{-5}. Nesting another APF with a delay of z^{-3} inside it results in Figure 11.21.

Gardner also devised a new schematic representation of his nested filter structures that removes the clutter of the delay cells, gain, and summation components. Figure 11.22 shows a nested APF structure. The outer APF has a delay time $D2$ and gain g_2 while the inner APF uses $D1$ and g_1 as its values. Additionally, pre- and post-delays may be added to the transversal delay line before and after the nested structure as shown in Figure 11.23.

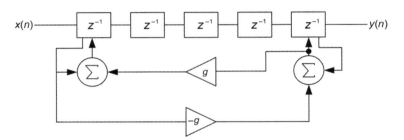

Figure 11.20: The delaying APF structure sitting across a transversal delay line.

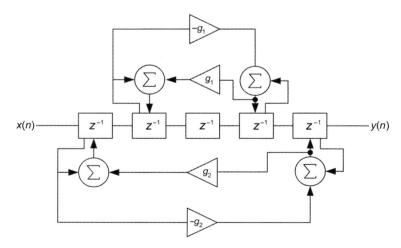

Figure 11.21: Two delaying APF structures sharing the same delay line.

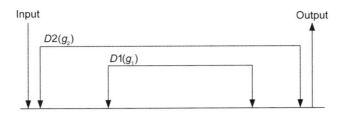

Figure 11.22: A nested APF schematic; Gardner gave the delay times in mSec.

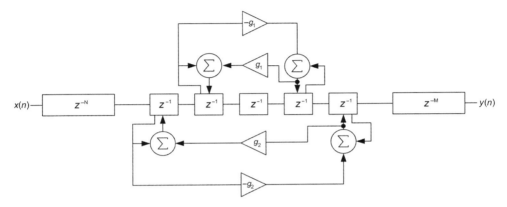

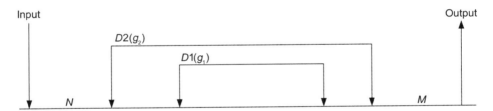

Figure 11.23: Pre- and post-delays of length N and M have been added to the nested APF structure. The second diagram shows how Gardner would lay this out schematically.

Consider the nested APF structure in Figure 11.24. It has an outer APF with a delay time of 35 mSec with $g = 0.3$ and two inner APFs with delay and gain values of 22 mSec ($g = 0.4$) and 8.3 mSec ($g = 0.6$). This is actually the first part of one of Gardner's larger designs. Figure 11.25 shows the impulse responses as each APF is added to the structure.

You can certainly see how the echo density begins to grow as you add each APF unit. We note several things about the nesting system:

- In the example, the first nested APF is 22 mSec in length; because of the commutative property of the delay operator, it doesn't matter where the 22 mSec delay is placed within the 35 mSec outer element.
- The 8.3 mSec APF is not nested inside the 22 mSec APF; it comes anywhere after it but still inside the outer APF.

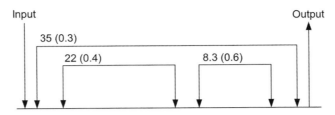

Figure 11.24: Three APFs—an outer filter and two nested units.

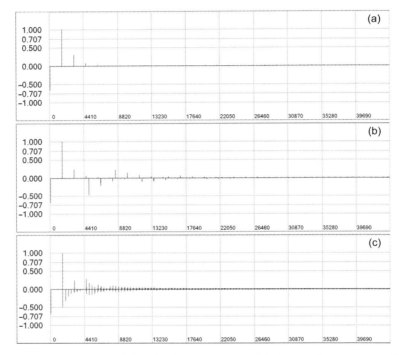

Figure 11.25: (a) The output of the single outer APF with delay of 35 mSec and gain of 0.3 shows the typical APF impulse response. (b) The first nested APF with delay of 22 mSec has been added. (c) The second nested APF with delay of 8.3 mSec has been added to the first two.

- Echo density piles up as time increases.
- The system can produce ringing or instabilities.
- The system still sounds metallic.

Gardner's solution to the ringing or metallic sound was to create a feedback path around the entire system with an LPF in the loop, thus embedding the nested APFs in a massive comb filter. He also notes that his reverb designs were ultimately arrived at by trial-and-error. Figure 11.26 shows Gardner's three reverb designs for small, medium, and large room simulations. In each algorithm, the reverb time is controlled by the loop gain or "g" control (notice there are two of them in the medium room algorithm).

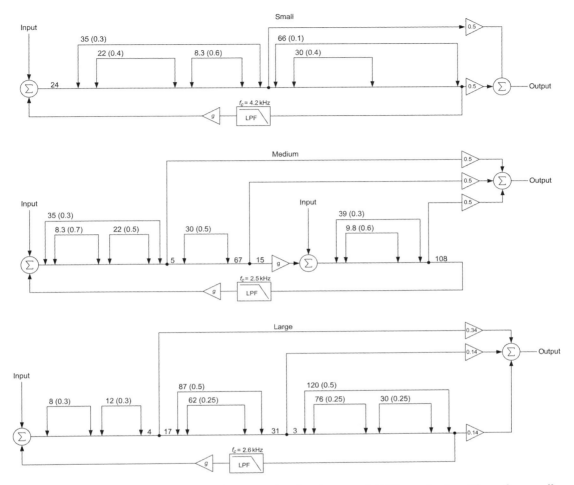

Figure 11.26: From top to bottom these are Gardner's nested APF reverb algorithms for small, medium, and large room simulations. Notice that the medium room algorithm in the middle has two input locations. Also notice the use of pre- and post-delays on some of the nested modules.

11.11 Modulated APF and Comb/APF Reverb

The modulated delay line can be used to further increase time smearing and echo density by using it in an APF module. The low-frequency oscillator (LFO) rate is kept low (<1 Hz) and the depth also very low (10–20 samples) to ensure that the detuning and chorusing effect is not overly obvious. In 1997, Dattorro presented a plate reverb algorithm "in the style of Greisinger," revealing the use of a modulated APF. He also notes that the modulated filter technically mimics a room whose walls are slowly moving back and forth.

Frenette (2000) showed that the use of modulated delay lines could reduce the computational requirements without a perceptual reduction in quality of reverb. Frenette used modulated comb filters (chorus modules) in addition to modulated APFs to further reduce the overall complexity and memory requirements. The modulated APF is shown in Figure 11.27.

The modulated APF in Figure 11.27 modulates only the very end of the delay line, producing a delay described by Equation 11.29:

$$Delay = z^{(-D + u(nT))} \qquad (11.29)$$

where $u(nT)$ is the discrete time modulation signal (LFO). The inverting version can be implemented with a swap of the g coefficients, as with the other APFs.

11.12 Dattorro's Plate Reverb

Dattorro's plate reverb algorithm in Figure 11.28 has a block diagram that reveals its figure-8 recirculating tank circuit. This figure-8 circuit could be applied to the other reverb block diagrams to generate countless variations on this theme.

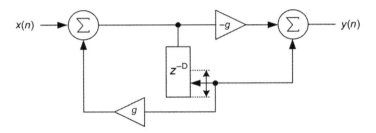

Figure 11.27: A modulated APF.

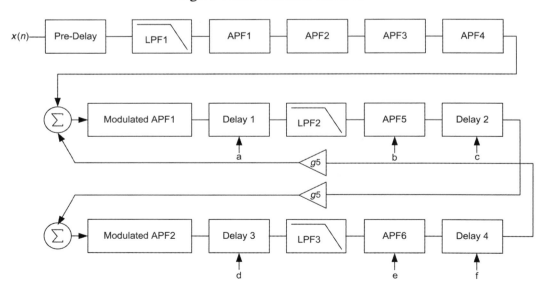

Figure 11.28: Block diagram of Dattorro's reverb.

You might notice something strange about Figure 11.29—it has no output y(n) node. In fact, the left and right outputs are taken from various locations within the delay lines, marked a–f in the diagram. This is a mono-in, stereo-out reverberator. The first LPF is marked "diffusion" while the second pair (LPF2 and LPF3) are designated "damping." The first filter controls the diffusion in the series APFs while the second pair controls the high-frequency roll-off in the tank circuit. These LPFs are all DC-normalized single pole feedback filters shown in Figure 11.30.

Table 11.2 gives the various values for the filters, followed by the equations that give the left and right outputs. The original design was for a sample rate of 29.8 kHz. Values for $f_s = 44.1$ kHz are calculated and given in the tables. Figure 11.30 shows the entire reverb algorithm block diagram. Table 11.3 lists the control ranges and defaults.

Table 11.2: Gain and delay values for Dattorro's plate reverb.

APF Dx	Delay (samples) $f_s = 29.8$ k	Delay (samples) $f_s = 44.1$ k	g (index)	g (value)
1	142	210	1	0.75
2	107	158	1	0.75
3	379	561	2	0.625
4	277	410	2	0.625
9	2656	3931	3	0.5
10	1800	2664	3	0.5
Fixed Delay Dx	Delay (samples) $f_s = 29.8$ k	Delay (samples) $f_s = 44.1$ k	–	–
7	4217	6241	–	–
8	4453	6590	–	–
11	3136	4641	–	–
12	3720	5505	–	–
Mod APF Dx	Delay (samples) $f_s = 29.8$ k	Delay (samples) $f_s = 44.1$ k	g (index)	g (value)
5	908 +/−8	1343+/−12	4	0.7
6	672 +/−8	995+/−12	4	0.7

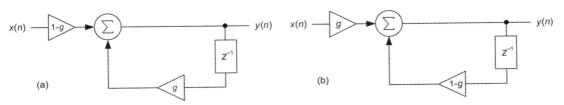

Figure 11.29: (a) DC-normalized one-pole filter, easy to use in reverb algorithms, including the comb filter and LPF type module. (b) This version merely reverses the effect of the slider or control.

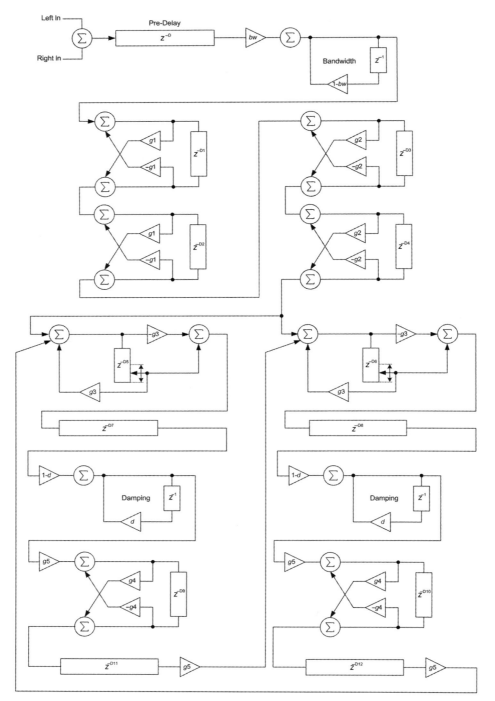

Figure 11.30: Dattorro's plate reverb algorithm.

Table 11.3: Control values for Dattorro's plate reverb.

Control	Range	Default
Decay (g5)	0.0–1.0	0.5
Bandwidth	0.0–1.0	0.9995
Damping	0.0–1.0	0.0005

The left and right outputs (Equation 11.30) are summed from points within the delay lines labeled a–f in Figure 11.28.

$f_s = 29.8$ kHz:
$$y_L = a[266] + a[2974] - b[1913] + c[1996] - d[1990] - e[187] - f[1066] \quad (11.30)$$
$$y_R = d[353] + d[3627] - e[1228] + f[2673] - a[2111] - b[335] - c[121]$$

$f_s = 44.1$ kHz:
$$y_L = a[394] + a[4401] - b[2831] + c[2954] - d[2945] - e[277] - f[1578]$$
$$y_R = d[522] + d[5368] - e[1817] + f[3956] - a[3124] - b[496] - c[179]$$

11.13 Generalized Feedback Delay Network Reverbs

Another approach to reverberation is to realize that the listener is experiencing dense echoes caused by multiple reflections off of surfaces along with potentially constructive or destructive interference. The generalized *feedback delay network* (FDN) approach began with Gerzon's work in 1972 on preserving energy in multichannel reverberators. It was continued by Stautner and Puckette (1982) as well as Jot and Chaigne (1991). Generally speaking, the idea is to model the room with some number of delay lines with potential feedback paths to and from every delay line in the system. The inputs and outputs of each delay line are also scaled by some values b and c. Consider a simple version that consists of two delay lines with feedback paths to and from each one shown in Figure 11.31.

If you look at Figure 11.31 and think about Schroeder's parallel comb filter bank, you can see that this is a variation, indeed a generalization, on the structure. In the general FDN, every possible feedback path is accounted for. Additionally, each delay line has input and output amplitude controls. If you let $b1$, $b2$, $c1$, and $c2$ all equal 1.0 and set $g12$ and $g22$ to 0.0, you get Schroeder's parallel comb filter network exactly. Notice how the feedback coefficients are labeled:

- $g11$: Feedback from delay line 1 into delay line 1
- $g12$: Feedback from delay line 1 into delay line 2
- $g21$: Feedback from delay line 2 into delay line 1
- $g22$: Feedback from delay line 2 into delay line 2

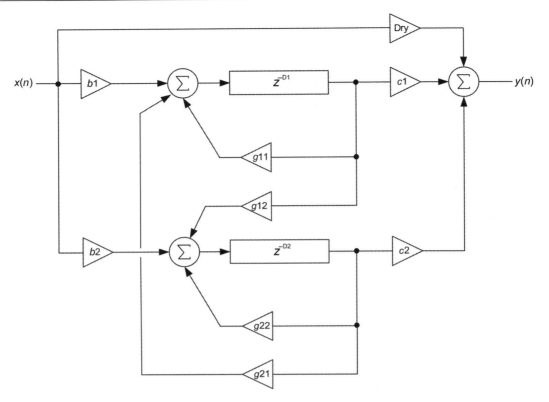

Figure 11.31: A two-delay-line feedback network.

An equivalent way to look at Figure 11.31 is shown in Figure 11.32. You can see that the feedback coefficients have been grouped together in what looks like a linear algebra matrix. The total energy is preserved if the matrix of coefficients is *unitary*. A unitary matrix multiplied by its *transpose matrix* results in the *identity matrix*. The transpose matrix is formed by turning the rows into columns and vice versa. These matrices are square. The identity matrix is a square matrix whose diagonal values are all 1. An FDN that uses a unitary matrix for its coefficients is called a *unitary feedback delay network* or UFDN.

Examine the gain matrix G in Equation 11.31:

$$G = \begin{bmatrix} 0 & 1 \\ -1 & 0 \end{bmatrix}$$

Then multiplying it by its transpose yields

$$GG^T = \begin{bmatrix} 0 & 1 \\ -1 & 0 \end{bmatrix} \begin{bmatrix} 0 & -1 \\ 1 & 0 \end{bmatrix} = \begin{bmatrix} 1 & 0 \\ 0 & 1 \end{bmatrix} \quad (11.31)$$

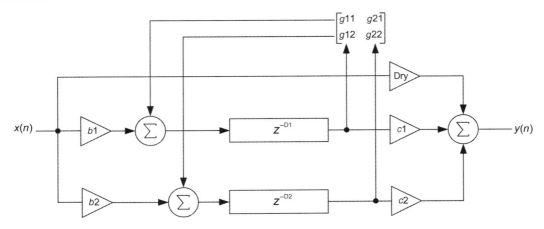

Figure 11.32: Another version of the two-delay-line FDN.

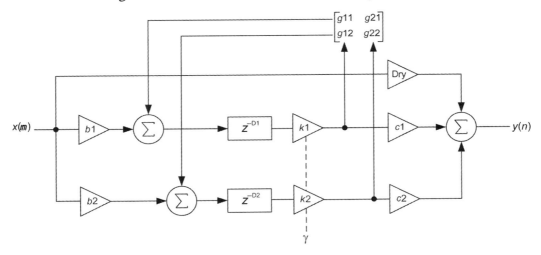

Figure 11.33: The FDN with decay factor control.

It is a unitary matrix. This reverberator would ring forever because of 1.0 gain values. Jot proposed adding an absorptive coefficient k to each delay line. For a colorless reverb, the value for k in Equation 11.32 is given by Jot.

$$k = \gamma^M$$

where (11.32)

M = the length of the delay
γ = the decay factor, set by the user

In Figure 11.33, each delay line undergoes the proper attenuation to keep the reverb colorless. However, it does not include the frequency dependent absorptive losses we noted from the energy decay relief diagrams. To accomplish this, Jot then suggested inserting LPFs instead of the static attenuators. The LPF magnitudes are cleverly chosen in relation to the frequency-dependent reverb time, $RT_{60}(\omega)$ in Equation 11.33:

$$20 \log(|h|) = \frac{-60 T_s}{RT_{60}(\omega)} M \tag{11.33}$$

where

M = delay length
h = the magnitude of the filter at some frequency ω

The problem with this setup is that the pole radii are no longer the same (circular in the z plane) and we know that this produces coloration in the signal. The solution (Jot 1992) was to add a correction filter at the end of the whole system, $T(z)$, whose magnitude is proportional to the inverse square root of the frequency-dependent reverb time, $RT_{60}(\omega)$. This is shown in Figure 11.34.

Finally, Figure 11.35 shows a generalized version of the FDN. It does not include the absorptive loss components (either static k values or H_{LPFs}) for clarity. The feedback matrix is always square, $N \times N$, where N is the number of delay lines. Because it is square, it can be made to be unitary.

As an experiment, you can try to implement a four delay line version using the unitary matrix Equation 11.34 and gain coefficients proposed by Stautner and Puckette (1982), who were researching multichannel reverb algorithms.

$$G = \frac{g}{\sqrt{2}} \begin{bmatrix} 0 & 1 & 1 & 0 \\ -1 & 0 & 0 & -1 \\ 1 & 0 & 0 & -1 \\ 0 & 1 & -1 & 0 \end{bmatrix} \tag{11.34}$$

where

$$|g| < 1.0$$

You can also try to de-correlate the four delay line outputs by using Jot's orthogonal matrix and set the coefficients c_N according to Equation 11.35:

$$C = g_2 \begin{bmatrix} 1 \\ -1 \\ 1 \\ -1 \end{bmatrix} \tag{11.35}$$

where

$$|g_2| < 1.0$$

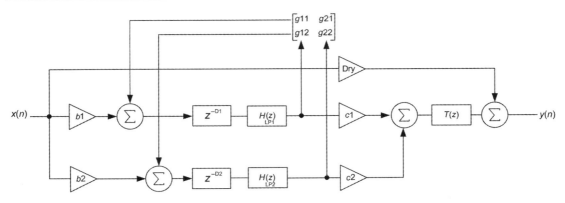

Figure 11.34: Addition of absorptive LPFs to the delay lines, plus a correction filter $T(z)$ at the end.

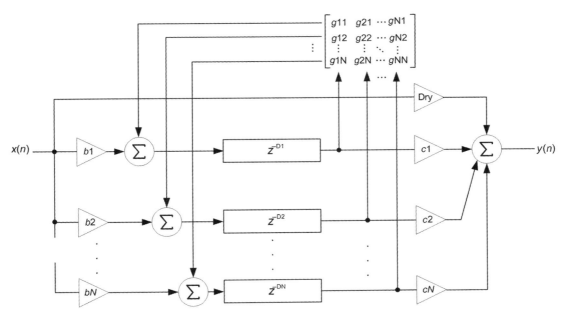

Figure 11.35: A generalized, N-delay-line FDN.

11.14 Other FDN Reverbs

Smith (1985) developed a variation on the FDN theme with his waveguide reverberators. Each waveguide consists of a two delay lines that move in opposite directions with coupling coefficients for connecting waveguides together. Any number of waveguides may be connected in any geometric pattern or shape, leading to a generalized set of scattering matrices. The scattering matrices are similar to the feedback matrices—they contain the coefficients that control the waveguide junction gains. The scattering matrices may also be made to be lossless by adhering to mathematical conditions involving the matrices' eigenvalues and eigenvectors (Smith and Rocchesso 1994).

Dahl and Jot (2000) proposed another UFDN type of reverb algorithm based on a structure they call the *absorbent all-pass filter* (AAPF). Figure 11.36 shows the block diagram of the AAPF which consists of a standard delaying APF with an LPF inserted in signal path. They combined an early reflections block that consisted of a multi-tapped delay line with a late reverberation block. In their late reverberation model, they use series AAPFs in a UFDN loop as shown in Figure 11.37. The block marked M is the unitary feedback matrix that mixes and creates the inputs for the filter loops.

Chemistruck, Marcolini, and Pirkle (2012) also experimented with FDNs for reverb algorithms. They used the genetic algorithm (GA) to generate coefficients for feedback delay and filtering blocks in their proposed system. A random unitary matrix generator was coded to seed the GA so that the starting point is a UFDN. Their overall algorithm block diagram consists of a four-delay line FDN followed by a diffusion block, as shown in Figure 11.38. The GA was only used to find the delay network coefficients while the diffusion block remained constant throughout.

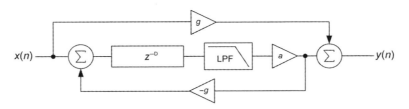

Figure 11.36: The absorbent APF features an LPF and attenuator in line with the delay element.

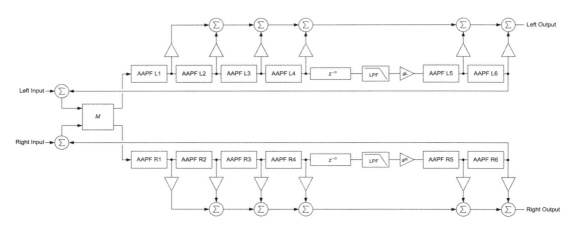

Figure 11.37: Dahl and Jot's late reverberator using a unitary feedback matrix M and absorbent APFs.

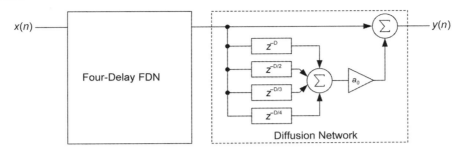

Figure 11.38: The FDN and diffusion network from the GA reverb.

Two different FDNs were used; the normal four-delay line and the delay and LPF in series (Figure 11.35, but without the correction filter on the output). An impulse response was taken for the target listening environment. The GA fitness function used the time domain envelope of the target impulse response for the matching criteria on the output of the system. During each iteration, the top 35% of offspring candidates were kept and inter-mated. Mutation rates of 10% for the delay coefficients and 5% for the LPF cutoff frequencies were used. More details can be found in the original reference.

We've analyzed many types of reverberator building blocks and complete algorithms. By now you should be able to see that reverb design is fairly open-ended. An example room reverb design follows.

11.15 An Example Room Reverb

For an example reverb, we have assembled a combination of reverberator modules to implement a mono-in, stereo-out reverb. The reverb is based on many of the modules and classic algorithms we've studied so far. The idea is to produce a realistic-sounding algorithm that uses several of the more common reverberator modules so that you can use it as a springboard for your own designs. It is a good place to start your experiments because it features a variety of reverberator modules to help you get a better understanding of how the parameters change the overall reverb sound. Figure 11.39 shows the stereo impulse response for the factory preset; notice the pre-delay, early reflections, and late reverb sections are all present. Figure 11.40 shows the block diagram and graphical user interface (GUI) components for the example reverb. The complete algorithm consists of the following:

- Pre-delay with output attenuation.
- Input LPF ("bandwidth").
- Two input APF diffusion modules.
- Two parallel comb filter banks; four comb filters per bank; one bank for left output, one for right.

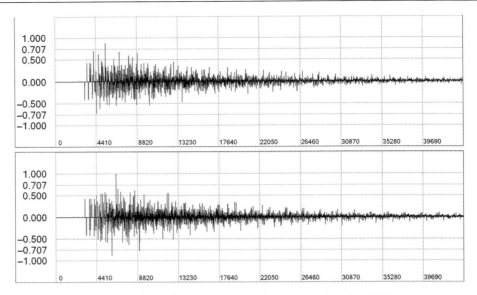

Figure 11.39: A stereo impulse response of the reverb plug-in, 100% wet output.

- Two comb filters in each bank are LPF–comb filters.
- Each parallel comb filter bank feeds an output LPF ("damping").
- The output LPF damping control also adjusts the LPF–comb filter g values.
- Each output LPF feeds an output APF diffusion module.
- A single RT_{60} Control for all comb filter gain variables.
- Individual APF and LPF controls for all parameters.

You can see many familiar components of this algorithm. The inputs are summed together to form a mono-signal and the left and right outputs are taken from two separate parallel comb filter echo generation units. The input bandwidth and output damping filters are the same as the Dattorro version but they also control the LPF–comb filter cutoffs. The parallel comb filter banks are reminiscent of the Schroeder and Moorer algorithms. These reverb algorithms often contain multiple instances of the same kinds of filters—APF, Comb, LPFComb, and so on. RackAFX provides you with several of these stock objects to use in your plug-ins. This will allow you to quickly assemble variations on the algorithms. Here are some key points for the design:

- We follow Schroeder's rule of about 1:1.5 ratio for shortest to longest comb filter delay (27.50 to 39.34 is our range, almost exactly 1:1.5).
- The RT_{60} time is used to calculate the gain variables for the comb filters for a less colored sound.
- The diffusion APFs alternate the sign of the g coefficient to create inverted/normal diffusion.

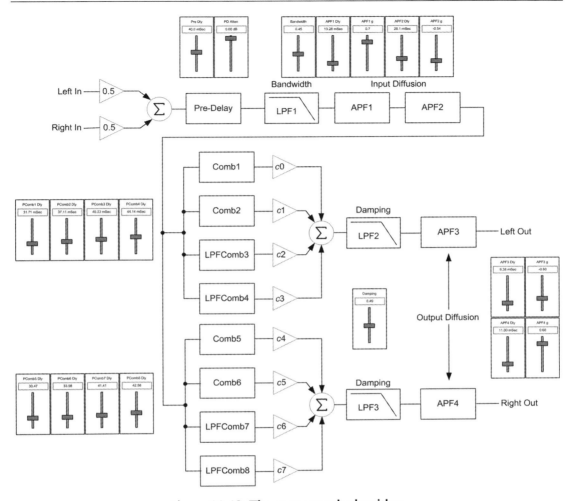

Figure 11.40: The room reverb algorithm.

- Adjusting the input APF delays has an effect on early reflections.
- Adjusting the output APF delays has a tonal coloration effect on the final reverb.
- The key to less ringing/metallic sound is getting the parallel comb filter delays just right, and not violating the 1:1.5 rule.
- The parallel comb filter attenuators, $c0$—$c7$ are hard-coded as 0.15, but there are several ways you could experiment with this, or allow the user to control it. They also have alternating signs, which was found to greatly reduce pinging.
- You can also provide unequal weighting such that the shorter (or longer) combs are more (or less) emphasized.

11.16 RackAFX Stock Objects

You can use any or all of the following stock objects in your plug-in—you just have to let RackAFX know when you create your project initially. The stock objects consist of the following:

- COnePoleLPF
- CDelay
- CCombFilter
- CLPFCombFilter
- CDelayAPF

When you tell RackAFX to add them to your project, RackAFX automatically copies the files, #includes them, and adds them to your C++ project file. Here's a quick reference guide to the objects. With these objects, coding our example reverb unit will be much easier than you think.

11.16.1 COnePoleLPF

Implements a one-pole LPF with a single coefficient, g in Figure 11.41. Table 11.4 shows object members.

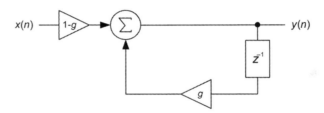

Figure 11.41: The one-pole LPF.

Table 11.4: The COnePoleLPF object.

Member Variables	Purpose
float m_fLPF_g	Implements the one and only gain coefficient g.
float m_fLPF_z1	Register to hold the single sample delay, z^{-1}.
Member Functions	
void setLPF_g(float fLPFg) Parameters: • float fLPFg	set() function for gain coefficient. Input: the new gain value.
void init()	Function to initialize and clear out the z^{-1} register.
bool processAudio(float* pInput, float* pOutput) Parameters: • float* pInput • float* pOutput	Process one input sample. Input: pointer to a float input sample. Output: pointer to the output sample destination.

11.16.2 CDelay

Implements a delay of *D* samples with an output attenuator. This is also the base class for other stock objects in Figure 11.42. Table 11.5 shows object members.

Table 11.5: The CDelay object.

Member Variables	Purpose
float m_fDelayInSamples	Implements the one and only gain coefficient g
float m_fOutputAttenuation	Output attenuation variable
float* m_pBuffer	Pointer to our dynamically declared buffer
int m_nReadIndex	Current read location
int m_nWriteIndex	Current write location
int m_nBufferSize	Max buffer size
int m_nSampleRate	Sample rate
float m_fDelay_ms	Delay in mSec, set by the parent plug-in
float m_fOutputAttenuation_dB	Output attenuation in dB, set by the parent plug-in
Member Functions	
void init(int nDelayLength) Parameters:	Declare the buffer and initialize it with 0.0s
• int nDelayLength	Input: the buffer length in samples
void cookVariables()	Function to cook delay (nSec) and attenuation (dB) to the variables actually used in the calculation
void resetDelay()	Reset the pointers to top, flush buffer
void setDelay_mSec(float fmSec) Parameters:	Called by parent to set the delay in mSec
• float fmSec	Input: the desired delay time
void setSampleRate(int nFs) Parameters:	Set the sample rate; called by parent
• int nFs	Input: the sample rate integer
void setOutputAttenuation_ dB(float fAttendB) Parameters:	Set the output attenuator; called by parent
• float fAttendB	Input: the attenuation in dB
float readDelay()	Read the delay with the user defined delay time without incrementing any pointers or writing input data
float readDelayAt(float fmSec) Parameters: • float fmSec	Read the delay at an arbitrary delay time without incrementing any pointers or writing input data; caller must make sure that this does not exceed max delay time Input: the delay time to read
void writeDelayAndInc(float fDelayInput) Parameters: • float fDelayInput	Write the input value and increment the pointer indices Input: the input audio sample
bool processAudio(float* pInput, float* pOutput) Parameters: • float* pInput • float* pOutput	Processes one input sample Input: pointer to a float input sample Output: pointer to the output sample destination

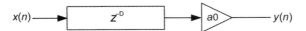

Figure 11.42: The delay with output attenuator.

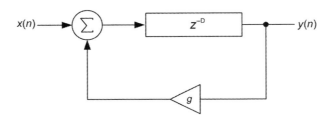

Figure 11.43: The comb filter.

11.16.3 CCombFilter

Implements a D-sample comb filter with feedback coefficient g in Figure 11.43. Table 11.6 shows object members.

Table 11.6: The CCombFilter object.

Member Variables	Purpose
float m_fComb_g	The one and only feedback gain coefficient g
Member Functions	
void setComb_g(float fCombg) Parameters: • float fCombg	set( ) function for feedback gain coefficient Input: the new feedback gain value
void setComb_g_with_RTSixty(float fRT) Parameters: • float fRT	Set the feedback gain using the RT60 time Input: *RT60* time in mSec
CDelay Base Class Overrides	
bool processAudio(float* pInput, float* pOutput) Parameters: • float* pInput • float* pOutput	Process one input sample Input: pointer to a float input sample Output: pointer to the output sample destination

11.16.4 CLPFCombFilter

Implements a D-sample comb filter with LPF in feedback loop (with coefficient $g2$) and feedback coefficient $g1$ in Figure 11.44. Table 11.7 shows object members.

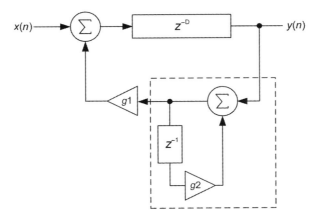

Figure 11.44: The LPF–comb filter.

Table 11.7: The CLPFCombFilter object.

Member Variables	Purpose
float m_fComb_g	The one and only feedback gain coefficient g_1
float m_fLPF_g	The LPF gain coefficient, g_2
float m_fLPF_z1	Register for one pole LPF
Member Functions	
void setComb_g(float fCombg) Parameters: • float fCombg	set( ) function for feedback gain coefficient Input: the new feedback gain value
void setComb_g_with_RTSixty(float fRT) Parameters: • float fRT	void setComb_g_with_RTSixty(float fRT) Parameters: float fRT
void setLPF_g(float fOverAllGain) Parameters: • float fOverAllGain	Set the LPF gain knowing an overall gain value; uses Equation 11.27 Input: the overall gain in Equation 11.27
CDelay Base Class Overrides	
void init(int nDelayLength) Parameters: • int nDelayLength	Dynamically create and init the buffer Input: delay length in samples
bool processAudio(float* pInput, float* pOutput) Parameters: • float* pInput • float* pOutput	Process one input sample Input: pointer to a float input sample Output: pointer to the output sample destination

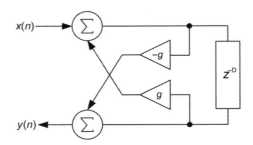

Figure 11.45: The delaying APF.

11.16.5 CDelayAPF

Implements a *D*-sample delaying APF in Figure 11.45. Table 11.8 shows object members.

Table 11.8: The CDelayAPF object.

Member Variables	Purpose
float m_fAPF_g	The one and only gain coefficient *g*
Member Functions	
void setAPF_g(float fAPFg) Parameters: • float fAPFg	set( ) function for gain coefficient Input: the new gain value
CDelay Base Class Overrides	
bool processAudio(float* pInput, float* pOutput) Parameters: • float* pInput • float* pOutput	Process one input sample Input: pointer to a float input sample Output: pointer to the output sample destination

11.17 Design the Room Reverb

11.17.1 Project: Reverb

Start a new project named "Reverb" and be sure to check the box to automatically include the stock reverb objects.

11.17.2 Reverb GUI

The first thing to do is decide on a user interface to control the reverb's modules. Since this is a learning tool as well, we will give the user many more controls than they would normally find on a reverb for the purpose of experimentation. Refer back to Figure 11.37, which has the block diagram and GUI components with it. There are going to be a lot of sliders to set up since this is a complex design (Table 11.9). However the use of the stock reverb objects is going to make the rest of the design easy.

Table 11.9: UI Control Properties for the Room Reverb: Pre-Delay

Slider Property	Value
Control Name	Pre Delay
Units	mSec
Variable Type	float
Variable Name	m_fPreDelay_mSec
Low Limit	0
High Limit	100
Initial Value	40

Slider Property	Value
Control Name	Pre Dly Atten
Units	dB
Variable Type	float
Variable Name	m_fPreDelayAtten_dB
Low Limit	−96
High Limit	0
Initial Value	0

Input Diffusion

Slider Property	Value
Control Name	Bandwidth
Units	
Variable Type	float
Variable Name	m_fInputLPF_g
Low Limit	0
High Limit	1
Initial Value	0.45

Slider Property	Value
Control Name	APF1 Dly
Units	mSec
Variable Type	float
Variable Name	m_fAPF_1_Delay_mSec
Low Limit	0
High Limit	100
Initial Value	13.28

Slider Property	Value
Control Name	APF1 g
Units	
Variable Type	float
Variable Name	m_fAPF_1_g
Low Limit	−1
High Limit	1
Initial Value	0.7

Slider Property	Value
Control Name	APF2 Dly
Units	mSec
Variable Type	float
Variable Name	m_fAPF_2_Delay_mSec
Low Limit	0
High Limit	100
Initial Value	28.13

Slider Property	Value
Control Name	APF2 g
Units	
Variable Type	float
Variable Name	m_fAPF_2_g
Low Limit	−1
High Limit	1
Initial Value	−0.7

Parallel Comb Filter Bank 1

Slider Property	Value
Control Name	PComb1 Dly
Units	mSec
Variable Type	float
Variable Name	m_fPComb_1_Delay_mSec
Low Limit	0
High Limit	100
Initial Value	31.71

Slider Property	Value
Control Name	PComb2 Dly
Units	mSec
Variable Type	float
Variable Name	m_fPComb_2_Delay_mSec
Low Limit	0
High Limit	100
Initial Value	37.11

Slider Property	Value
Control Name	PComb3 Dly
Units	mSec
Variable Type	float
Variable Name	m_fPComb_3_Delay_mSec
Low Limit	0
High Limit	100
Initial Value	40.23

Slider Property	Value
Control Name	PComb4 Dly
Units	mSec
Variable Type	float
Variable Name	m_fPComb_4_Delay_mSec
Low Limit	0
High Limit	100
Initial Value	44.14

Parallel Comb Filter Bank 2

Slider Property	Value
Control Name	PComb5 Dly
Units	mSec
Variable Type	float
Variable Name	m_fPComb_5_Delay_mSec
Low Limit	0
High Limit	100
Initial Value	30.47

Slider Property	Value
Control Name	PComb6 Dly
Units	mSec
Variable Type	float
Variable Name	m_fPComb_6_Delay_mSec
Low Limit	0
High Limit	100
Initial Value	33.98

Slider Property	Value
Control Name	PComb7 Dly
Units	mSec
Variable Type	float
Variable Name	m_fPComb_7_Delay_mSec
Low Limit	0
High Limit	100
Initial Value	41.41

Slider Property	Value
Control Name	PComb8 Dly
Units	mSec
Variable Type	float
Variable Name	m_fPComb_8_Delay_mSec
Low Limit	0
High Limit	100
Initial Value	42.58

Input Diffusion and Damping

Slider Property	Value
Control Name	Damping
Units	
Variable Type	float
Variable Name	m_fLPF2_g2
Low Limit	0
High Limit	1
Initial Value	0.5

Slider Property	Value
Control Name	APF3 Dly
Units	mSec
Variable Type	float
Variable Name	m_fAPF_3_Delay_mSec
Low Limit	0
High Limit	100
Initial Value	9.38

Slider Property	Value
Control Name	APF3 g
Units	
Variable Type	float
Variable Name	m_fAPF_3_g
Low Limit	−1
High Limit	1
Initial Value	−0.6

Slider Property	Value
Control Name	APF4 Dly
Units	mSec
Variable Type	float
Variable Name	m_fAPF_4_Delay_mSec
Low Limit	0
High Limit	100
Initial Value	11

Slider Property	Value
Control Name	APF4 g
Units	
Variable Type	float
Variable Name	m_fAPF_4_g
Low Limit	−1
High Limit	1
Initial Value	0.6

Reverb Output

Slider Property	Value
Control Name	Reverb Time
Units	mSec
Variable Type	float
Variable Name	m_fRT60
Low Limit	0
High Limit	5000
Initial Value	1000

Slider Property	Value
Control Name	Wet/Dry
Units	%
Variable Type	float
Variable Name	m_fWet_pct
Low Limit	0
High Limit	100
Initial Value	50

Build the UI using the list of controls above; the variable names I used in the project are listed in the table. You can use your own variable names but you will need to make sure they are correctly implemented. Figure 11.46 shows my GUI; I grouped sets of controls together, following the algorithm block diagram.

11.17.3 Reverb.h

In the .h file, declare the member objects plus one cooking function that will change all modules at once (it's not streamlined but it is simple to cook all at once). There is a stock object for each block in the algorithm.

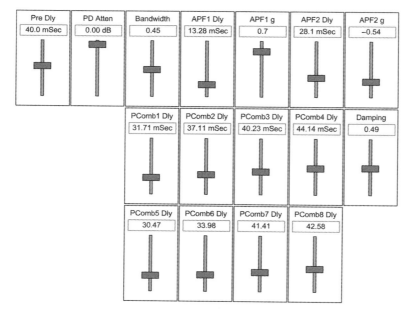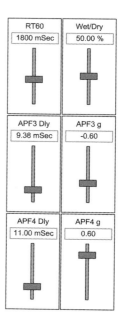

Figure 11.46: The completed reverb prototype GUI.

```
    // Add your code here: --------------------------- //
    //
    // Pre-Delay Block
    CDelay m_PreDelay;

    // input Diffusion
    COnePoleLPF m_InputLPF;
    CDelayAPF m_InputAPF_1;
    CDelayAPF m_InputAPF_2;

    // parallel Comb Bank 1
    CCombFilter m_ParallelCF_1;
    CCombFilter m_ParallelCF_2;
    CLPFCombFilter m_ParallelCF_3;
    CLPFCombFilter m_ParallelCF_4;

    // parallel Comb Bank 2
    CCombFilter m_ParallelCF_5;
    CCombFilter m_ParallelCF_6;
    CLPFCombFilter m_ParallelCF_7;
    CLPFCombFilter m_ParallelCF_8;

    // damping
    COnePoleLPF m_DampingLPF1;
    COnePoleLPF m_DampingLPF2;

    // output diffusion
    CDelayAPF m_OutputAPF_3;
    CDelayAPF m_OutputAPF_4;

    // function to cook all member object's variables at once
    void cookVariables();

    // END OF USER CODE --------------------------- //
```

11.17.4 Reverb.cpp

Write the only extra function, cookVariables():

```
// function to cook all variables at once
void CReverb::cookVariables()
{
    // Pre-Delay
    m_PreDelay.setDelay_mSec(m_fPreDelay_mSec);
    m_PreDelay.setOutputAttenuation_dB(m_fPreDelayAtten_dB);

    // input diffusion
    m_InputAPF_1.setDelay_mSec(m_fAPF_1_Delay_mSec);
    m_InputAPF_1.setAPF_g(m_fAPF_1_g);
```

```
    m_InputAPF_2.setDelay_mSec(m_fAPF_2_Delay_mSec);
    m_InputAPF_2.setAPF_g(m_fAPF_2_g);

    // output diffusion
    m_OutputAPF_3.setDelay_mSec(m_fAPF_3_Delay_mSec);
    m_OutputAPF_3.setAPF_g(m_fAPF_3_g);

    m_OutputAPF_4.setDelay_mSec(m_fAPF_4_Delay_mSec);
    m_OutputAPF_4.setAPF_g(m_fAPF_4_g);

    // comb filters
    // set delays first...
    m_ParallelCF_1.setDelay_mSec(m_fPComb_1_Delay_mSec);
    m_ParallelCF_2.setDelay_mSec(m_fPComb_2_Delay_mSec);
    m_ParallelCF_3.setDelay_mSec(m_fPComb_3_Delay_mSec);
    m_ParallelCF_4.setDelay_mSec(m_fPComb_4_Delay_mSec);
    m_ParallelCF_5.setDelay_mSec(m_fPComb_5_Delay_mSec);
    m_ParallelCF_6.setDelay_mSec(m_fPComb_6_Delay_mSec);
    m_ParallelCF_7.setDelay_mSec(m_fPComb_7_Delay_mSec);
    m_ParallelCF_8.setDelay_mSec(m_fPComb_8_Delay_mSec);

    // ...then calculate comb g's from RT60:
    m_ParallelCF_1.setComb_g_with_RTSixty(m_fRT60);
    m_ParallelCF_2.setComb_g_with_RTSixty(m_fRT60);
    m_ParallelCF_3.setComb_g_with_RTSixty(m_fRT60);
    m_ParallelCF_4.setComb_g_with_RTSixty(m_fRT60);
    m_ParallelCF_5.setComb_g_with_RTSixty(m_fRT60);
    m_ParallelCF_6.setComb_g_with_RTSixty(m_fRT60);
    m_ParallelCF_7.setComb_g_with_RTSixty(m_fRT60);
    m_ParallelCF_8.setComb_g_with_RTSixty(m_fRT60);

    // LPFs
    m_DampingLPF1.setLPF_g(m_fLPF2_g2);
    m_DampingLPF2.setLPF_g(m_fLPF2_g2);
    m_InputLPF.setLPF_g(m_fInputLPF_g);

    // LPF-comb filters
    m_ParallelCF_3.setLPF_g(m_fLPF2_g2);
    m_ParallelCF_4.setLPF_g(m_fLPF2_g2);
    m_ParallelCF_7.setLPF_g(m_fLPF2_g2);
    m_ParallelCF_8.setLPF_g(m_fLPF2_g2);

}
```

Constructor

There is nothing to do here because the child-object constructors handle these details.

prepareForPlay()

- Initialize all the objects with their max delay times; all delay times except the pre-delay have maximum values of 100 mSec.
- Initialize the pre-delay for its maximum of 2 seconds.
- Reset all delays.
- Set the sample rate for the delays.
- Cook all the variables at once.
- Init the delays to flush buffers.

```
bool __stdcall CReverb::prepareForPlay()
{
    // Add your code here:
    // up to 2 seconds predelay
    m_PreDelay.init(2.0*(m_nSampleRate));

    // init up to 100 mSec
    m_InputAPF_1.init(0.1*(m_nSampleRate));
    m_InputAPF_2.init(0.1*(m_nSampleRate));

    // 100 mSec each max
    m_ParallelCF_1.init(0.1*(m_nSampleRate));
    m_ParallelCF_2.init(0.1*(m_nSampleRate));
    m_ParallelCF_3.init(0.1*(m_nSampleRate));
    m_ParallelCF_4.init(0.1*(m_nSampleRate));
    m_ParallelCF_5.init(0.1*(m_nSampleRate));
    m_ParallelCF_6.init(0.1*(m_nSampleRate));
    m_ParallelCF_7.init(0.1*(m_nSampleRate));
    m_ParallelCF_8.init(0.1*(m_nSampleRate));

    // 100 mSec each max
    m_OutputAPF_3.init(0.1*(m_nSampleRate));
    m_OutputAPF_4.init(0.1*(m_nSampleRate));

    // init the three LPFs
    m_InputLPF.init();
    m_DampingLPF1.init();
    m_DampingLPF2.init();

    // Call all delay resets
    m_PreDelay.resetDelay();
    m_InputAPF_1.resetDelay();
    m_InputAPF_2.resetDelay();

    m_ParallelCF_1.resetDelay();
    m_ParallelCF_2.resetDelay();
    m_ParallelCF_3.resetDelay();
    m_ParallelCF_4.resetDelay();
    m_ParallelCF_5.resetDelay();
```

```
    m_ParallelCF_6.resetDelay();
    m_ParallelCF_7.resetDelay();
    m_ParallelCF_8.resetDelay();

    m_OutputAPF_3.resetDelay();
    m_OutputAPF_4.resetDelay();

    // set sample rates on combs (needed to calc g values)
    m_ParallelCF_1.setSampleRate(this->m_nSampleRate);
    m_ParallelCF_2.setSampleRate(this->m_nSampleRate);
    m_ParallelCF_3.setSampleRate(this->m_nSampleRate);
    m_ParallelCF_4.setSampleRate(this->m_nSampleRate);
    m_ParallelCF_5.setSampleRate(this->m_nSampleRate);
    m_ParallelCF_6.setSampleRate(this->m_nSampleRate);
    m_ParallelCF_7.setSampleRate(this->m_nSampleRate);
    m_ParallelCF_8.setSampleRate(this->m_nSampleRate);

    // flush buffers
    m_InputLPF.init();
    m_DampingLPF1.init();
    m_DampingLPF2.init();

    // cook everything
    cookVariables();
    return true;
}
```

userInterfaceChange()

- Call the one and only cooking function

```
bool __stdcall CReverb::userInterfaceChange(int nControlIndex)
{
    // add your code here
    cookVariables();
    return true;
}
```

processAudioFrame()

In this function, you tie all the member objects together by creating variables to pass into/out of their processAudio() functions. By stepping through the code, you can see just how the series and parallel elements are connected. You can always go back and make this code more streamlined by removing intermediate variables, and so on. The four outputs of each comb filter bank are mixed at 15% ratios with alternating signs.

```
bool __stdcall CReverb::processAudioFrame(float* pInputBuffer, float* pOutputBuffer, UINT
                                           uNumInputChannels, UINT uNumOutputChannels)
{
    //
    // output = input - change this for meaningful processing
```

```cpp
//
// Form our input = L + R (if there is a R)
//
float fInputSample = pInputBuffer[0];
if(uNumInputChannels == 2)
{
    // mix
    fInputSample += pInputBuffer[1];
    // attenuate by 0.5
    fInputSample *= 0.5;
}

// begin the series/parallel signal push
// Pre-Delay
float fPreDelayOut = 0;
m_PreDelay.processAudio(&fInputSample, &fPreDelayOut);

// Pre-Delay Out -> fAPF_1_Out
float fAPF_1_Out = 0;
m_InputAPF_1.processAudio(&fPreDelayOut, &fAPF_1_Out);

// fAPF_1_Out -> fAPF_2_Out
float fAPF_2_Out = 0;
m_InputAPF_2.processAudio(&fAPF_1_Out, &fAPF_2_Out);

// fAPF_2_Out -> fInputLPF
float fInputLPF = 0;
m_InputLPF.processAudio(&fAPF_2_Out, &fInputLPF);

// comb filter bank
// variables for each output
float fPC_1_Out = 0;
float fPC_2_Out = 0;
float fPC_3_Out = 0;
float fPC_4_Out = 0;
float fPC_5_Out = 0;
float fPC_6_Out = 0;
float fPC_7_Out = 0;
float fPC_8_Out = 0;
float fC1_Out = 0;
float fC2_Out = 0;

// fInputLPF -> fPC_1_Out, fPC_2_Out, fPC_3_Out, fPC_4_Out
m_ParallelCF_1.processAudio(&fInputLPF, &fPC_1_Out);
m_ParallelCF_2.processAudio(&fInputLPF, &fPC_2_Out);
m_ParallelCF_3.processAudio(&fInputLPF, &fPC_3_Out);
m_ParallelCF_4.processAudio(&fInputLPF, &fPC_4_Out);

// fInputLPF -> fPC_5_Out, fPC_6_Out, fPC_7_Out, fPC_8_Out
m_ParallelCF_5.processAudio(&fInputLPF, &fPC_5_Out);
m_ParallelCF_6.processAudio(&fInputLPF, &fPC_6_Out);
```

```
            m_ParallelCF_7.processAudio(&fInputLPF, &fPC_7_Out);
            m_ParallelCF_8.processAudio(&fInputLPF, &fPC_8_Out);

            // form outputs: note attenuation by 0.15 for each and alternating signs
            fC1_Out = 0.15*fPC_1_Out - 0.15*fPC_2_Out + 0.15*fPC_3_Out - 0.15*fPC_4_Out;
            fC2_Out = 0.15*fPC_5_Out - 0.15*fPC_6_Out + 0.15*fPC_7_Out - 0.15*fPC_8_Out;

            // fC1_Out -> fDamping_LPF_1_Out
            float fDamping_LPF_1_Out = 0;
            m_DampingLPF1.processAudio(&fC1_Out, &fDamping_LPF_1_Out);

            // fC2_Out -> fDamping_LPF_2_Out
            float fDamping_LPF_2_Out = 0;
            m_DampingLPF2.processAudio(&fC2_Out, &fDamping_LPF_2_Out);

            // fDamping_LPF_1_Out -> fAPF_3_Out
            float fAPF_3_Out = 0;
            m_OutputAPF_3.processAudio(&fDamping_LPF_1_Out, &fAPF_3_Out);

            // fDamping_LPF_2_Out -> fAPF_4_Out
            float fAPF_4_Out = 0;
            m_OutputAPF_4.processAudio(&fDamping_LPF_2_Out, &fAPF_4_Out);

            // form output = (100-Wet)/100*x(n) + (Wet/100)*fAPF_3_Out
            pOutputBuffer[0] = ((100.0 - m_fWet_pct)/100.0)*fInputSample +
                                (m_fWet_pct/100.0)*(fAPF_3_Out);

            // Do RIGHT Channel if there is one
            if(uNumOutputChannels == 2)
            {
                    // form output = (100-Wet)/100*x(n) + (Wet/100)*fAPF_4_Out
                    pOutputBuffer[1] = ((100.0 - m_fWet_pct)/100.0)*fInputSample +
                                        (m_fWet_pct/100.0)*(fAPF_4_Out);
            }
            return true;
    }
```

Build and test the example reverb module. Use RackAFX's impulse response tool in the analyzer to tweak your settings. Adjust the impulse response time according to your RT_{60} times. Notice how the impulse response can drastically change with only a slight modification of the parameters. Also, try adjusting the mix coefficients for the comb filter outputs by alternating signs or using different weightings. Visit the website www.willpirkle.com for more example reverb algorithms and code.

11.18 Challenge

Design your own reverb. Start by implementing some of the classics (Schroeder, Moorer) and some of the more recent versions (Gardner, Jot, Dattorro) combining different modules.

Or start with the reverb design here and modify it. For example, try replacing the APFs with nested all-pass modules or add or remove comb filters. You can easily identify ringing and oscillations using the impulse response tool, so keep it open as you experiment.

Bibliography

Griesinger, D. 1995. How loud is my reverberation? *Journal of the Audio Engineering Society*, preprint 3943, pp. 1–11.
Roads, C. 1996. *The Computer Music Tutorial*, Chapter 3. Cambridge, MA: The MIT Press.

References

Beranek, L. 1986. *Acoustics*. New York: American Institute of Physics.
Browne, S. 2001. "Hybrid reverberation algorithm using truncated impulse response convolution and recursive filtering." Master's diss., University of Miami, Miami, FL.
Chemistruck, M., Marcolini, K., and Pirkle, W. 2012. Generating matrix coefficients for feedback delay networks using genetic algorithms. *Journal of the Audio Engineering Society*, preprint N/A.
Dahl, L. and Jot, J-M. 2000. A reverberator based on absorbent all-pass filters. *Proceedings of the COST G-6 Convention on Digital Audio Effects*, DAFX-00, pp. 1–6.
Dattorro, J. 1997. Effect design part 1: Reverberators and other filters. *Journal of the Audio Engineering Society* 45(9): 660–665.
Frenette, J. 2000. "Reducing artificial reverberation requirements using time-variant feedback delay networks." Master's diss., University of Miami, Miami, FL.
Gardner, W. G. 1992. "The virtual acoustic room." Master's diss., Massachusetts Institute of Technology, Boston, MA.
Gardner, W. G. 1995. Efficient convolution without input-output delay. *Journal of the Audio Engineering Society* 43(3): 127–135.
Gerzon, M. A. Synthetic stereo reverberation. *Studio Sound Magazine*, January 14, 1972.
Griesinger, D. 1989. Practical processors and programs for digital reverberation. *7th International Conference of the Audio Engineering Society*, pp. 187–195.
Jot, J-M. and Chaigne, A. 1991. Digital delay networks for designing artificial reverberators. *Journal of the Audio Engineering Society*, preprint 3030, pp. 1–16.
Jot, J-M. 1992. "Design and implementation of a sound spatializer based on physical and perceptual models." PhD diss., Telecom, Paris.
Kahrs, M. and Brandenberg, K. 1998. *Applications of Digital Signal Processing to Audio and Acoustics*, Chapter 3. Boston: Kluwer Academic Publishers.
Kuttruff, H. 1991. *Room Acoustics*. New York: Elsevier.
Moorer, J. A. 1979. About this reverberation business. *Computer Music Journal* 3(2): 13–28.
Reilly, A. and McGrath, D. S. 1995. Convolution processing for realistic reverb. *Journal of the Audio Engineering Society*, preprint 3977, pp. 1–8.
Sabine, W. 1973 "Reverberation." Lindsay, R. B., ed., *Acoustics: Historical and Philosophical Development*. Stroudsburg, PA: Dowden, Hutchinson & Ross.
Schroeder, M. 1962. Natural-sounding artificial reverberation. *Journal of the Audio Engineering Society* 10(3): 209–213.
Schroeder, M. 1984. Progress in architectural acoustics and artificial reverberation: Concert hall acoustics and number theory. *Journal of the Audio Engineering Society* 32(4): 194–202.
Smith, J. O. 1985. A new approach to digital reverberation using closed waveguide networks. *Proceedings of the 1985 International Computer Music Conference*, pp. 47–53.
Smith, J. O. and Rocchesso, D. 1994. Connections between feedback delay networks and waveguide networks for digital reverberation. *Proceedings of the 1994 International Computer Music Conference*, pp. 376–377.
Stautner, J. and Puckette, M. 1982. Designing multi-channel reverberators. *Computer Music Journal* 6(1): 52–65.

CHAPTER 12
Modulated Filter Effects

Modulated filter effects alter one or more filter parameters with other signals called control signals. The filter parameters might include cutoff frequency, Q, bandwidth, filter type, or overall gain. The control signals are usually periodic low-frequency oscillators (LFOs), a signal envelope, an envelope generator (EG) or even another audio signal. In the modulated delay effects, the LFO control signal changed the delay amount of the signal. In modulated filter effects, the filter parameters are the ones that are changing. In this chapter you will design and implement the following effects:

- Mod filter using an LFO
- Envelope follower using an audio detector
- Phaser

The phaser is a specialized effect that uses a bank of all-pass filters (APFs) to try to brute force delay the signal to make a flanging effect. An LFO modulates the depth of the APFs. In Figure 12.1 you can see a simple modulated filter. The filter type is a low-pass filter (LPF). The control signal is an LFO. The control parameter is the cutoff frequency of the filter.

In Figure 12.2, an EG moves the filter parameter. An EG is triggered by some external event, such as a note-on or -off event, or possibly triggered when the input level crosses a threshold. A modulated filter might have multiple parameters controlled by multiple control signals. For example, Figure 12.3 shows another LPF that has two modulation sources for two different parameters.

In Figure 12.3 you can see a new module labeled envelope detector; it detects and follows the peak, mean-squared (MS), or root-mean-squared (RMS) value of the control signal which is the input signal here. The control signal could be taken from other sources such as another audio signal or a complex LFO signal. The effect in Figure 12.2 has a special name too: it is called an envelope follower.

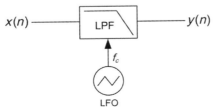

Figure 12.1: A simple LFO-modulated LPF.

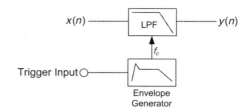

Figure 12.2: An envelope generator can also modulate a filter parameter.

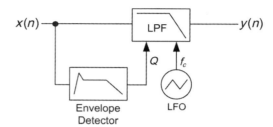

Figure 12.3: A doubly modulated LPF with both f_c and Q controls.

12.1 Design a Mod Filter Plug-In: Part I Modulated f_c

For our first effect design, we'll start with a modulated second-order LPF and modulate the cutoff frequency with an LFO. Then, we can increase the complexity by adding another LFO to control the Q and even give the option to run the two LFOs in quadrature phase. We can use the second-order digital resonant LPF you've already designed from Chapter 6 for the filter. Notice that for this initial project, we will hold the LPF Q constant at 2.0. And, we will introduce a built-in RackAFX object to handle the LFO for us. The block diagram is shown in Figure 12.4.

The parameters are as follows:

- f_c mod rate (0.2 to 10 Hz)
- f_c mod depth (0 to 100%)
- LFO type (sine, tri, saw, square)
- LPF Q: fixed at 2.0
- f_c mod range: 100 Hz to 5 kHz
- LPF is fixed as second-order bi-quad variety

By now, you should be getting really good at building RackAFX plug-ins. If you haven't just been cutting and pasting the code then you will have no problems with the next chapters. We can use the digital resonant LPF from Chapter 6 along with the built-in wave table oscillator object to quickly implement the mod filter effect. This project will use two built-in objects:

1. CBiquad for the filter
2. CWaveTable for the LFO

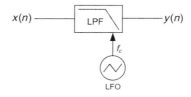

Figure 12.4: The mod filter block diagram.

You used CBiquad in Chapter 6 to begin your filtering work. The CWaveTable object was taken directly from your WTOscillator object. You can find the definition and implementation in the pluginconstants.h and pluginonbjects.cpp files. Table 12.1 shows the object's members and attributes. The flow chart for processAudioFrame() is shown in Figure 12.5.

12.1.1 Project: ModFilter

Create the project; because we are using built-in objects for the filter and LFO there are no member objects to add.

12.1.2 ModFilter GUI

For the initial design, you will need the following slider controls in Table 12.2. Note that these LFO waveform enumerations follow exactly that of the built-in CWaveTable object to make mapping the button control easy.

12.1.3 ModFilter.h File

Declare instances of the BiQuad and WaveTable objects. You don't need to #include anything since these are built-in:

```
// Add your code here: ---------------------------------------------- //
// BiQuad Objects
CBiquad m_LeftLPF;
CBiquad m_RightLPF;

// one LFO for the fc
CWaveTable m_fc_LFO;
// END OF USER CODE ------------------------------------------------- //
```

We are also going to need some variables and functions to control the effect. Specifically, we need:

- Minimum f_c variable
- Maximum f_c variable
- Function to calculate the current f_c given a LFO value
- Functions to calculate the bi-quad coefficients for left and right channels independently (we want to calculate them separately for the third part of the project where we allow for quad-phase LFOs)

Table 12.1: The CWaveTable object interface.

CWaveTable	
Member Variables	**Purpose**
float m_SinArray[1024]; float m_SawtoothArray[1024]; float m_TriangleArray[1024]; float m_SquareArray[1024];	Arrays to store the non-band limited wave tables
float m_SawtoothArray_BL5[1024]; float m_TriangleArray_BL5[1024]; float m_SquareArray_BL5[1024];	Arrays to hold the band-limited wave tables
float m_fReadIndex; float m_fQuadPhaseReadIndex;	Indices for reading the wave table
float m_f_inc;	Table increment value for current output frequency
int m_nSampleRate;	Sample rate IMPORTANT: Needed for calculation of table inc; do not forget it!
float m_fFrequency_Hz;	Oscillator frequency
bool m_bInvert;	Flag to invert output
UINT m_uOscType; enum{sine,saw,tri,square};	Enumerated unsigned integer type (UINT) for osc waveform type
UINT m_uTableMode; enum{normal,bandlimit};	Enumerated UINT for table mode (we will use normal mode in this plug-in since the oscillator is a LFO)
UINT m_uPolarity; enum{bipolar,unipolar};	Enumerated UINT for osc polarity (we will use unipolar for this plug-in)
Member Functions	
void doOscillate(float* pYn, float* pYqn) Parameters: • float* pYn • float* pYqn	The oscillate function Output: pYn is the normal output Output: pYqn is the quadrature phase output
bool prepareForPlay()	The prepareForPlay() function for the oscillator; same use as a normal plug-in
void reset ()	Reset the pointers to top
void setSampleRate(int nSampleRate) Parameters: • int nSampleRate	Called by parent to set the sample rate in Hz Input: the current sample rate
void cookFrequency()	Calculates the new *inc* value for a changed oscillation frequency

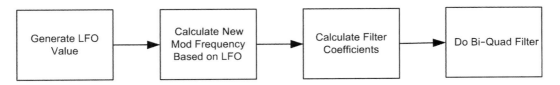

Figure 12.5: The flowchart for the mod filter process function.

Table 12.2: GUI controls for the ModFilter.

Slider Property	Value
Control Name	Mod Depth fc
Units	%
Variable Type	float
Variable Name	m_fMod_Depth_fc
Low Limit	0
High Limit	100
Initial Value	50

Slider Property	Value
Control Name	Mod Rate fc
Units	Hz
Variable Type	float
Variable Name	m_fModRate_fc
Low Limit	0.2
High Limit	10
Initial Value	1.0

Slider Property	Value
Control Name	LFO
Units	
Variable Type	enum
Variable Name	m_uLFO_waveform
Enum String	sine,saw,tri,square

Add the following to the .h file:

```
// Add your code here: ------------------------------------------------- //
// BiQuad Objects
CBiquad m_LeftLPF;
CBiquad m_RightLPF;

// one LFO for the fc
CWaveTable m_fc_LFO;

// min and max values (to make it easy to change later)
float m_fMinCutoffFreq;
float m_fMaxCutoffFreq;

// function to calculate the fc given the LFO sample
float calculateCutoffFreq(float fLFOSample);

// calculate the coeffs for a given filter
void calculateLPFCoeffs(float fCutoffFreq, float fQ, CBiquad* pFilter);

// END OF USER CODE ---------------------------------------------------- //
```

12.1.4 ModFilter.cpp File

Constructor

- Initialize the rate min and max values. The objects are self initializing upon creation.

```
CModFilter::CModFilter()
{
        <SNIP SNIP SNIP>

        // Finish initializations here

        // set our Min and Max Modulation points
        m_fMinCutoffFreq = 100.0;
        m_fMaxCutoffFreq = 5000.0;
}
```

Write the calculateCutoffFrequency() function. This function behaves exactly like the function in the flanger to calculate a new delay offset given the LFO sample and mod depth values. Here, we simply calculate the cutoff frequency between the min and max values according to the LFO sample and depth.

```
float CModFilter::calculateCutoffFreq(float fLFOSample)
{
        return (m_fMod_Depth_fc/100.0)*(fLFOSample*(
              m_fMaxCutoffFreq - m_fMinCutoffFreq)) + m_fMinCutoffFreq;
}
```

Next, we need to implement the cooking function for the LPFs to set the new coefficients for a given f_c and Q value. This is identical to the resonant LPF you designed in Chapter 6:

```
void CModFilter::calculateLPFCoeffs(float fCutoffFreq, float fQ, CBiquad* pFilter)
{
        // use same terms as book
        float theta_c = 2.0*pi*fCutoffFreq/(float)m_nSampleRate;
        float d = 1.0/fQ;

        float fBetaNumerator = 1.0 - ((d/2.0)*(sin(theta_c)));
        float fBetaDenominator = 1.0 + ((d/2.0)*(sin(theta_c)));

        float fBeta = 0.5*(fBetaNumerator/fBetaDenominator);

        float fGamma = (0.5 + fBeta)*(cos(theta_c));

        float fAlpha = (0.5 + fBeta - fGamma)/2.0;

        // apply to filter
        pFilter->m_f_a0 = fAlpha;
        pFilter->m_f_a1 = 2.0*fAlpha;
```

```
                pFilter->m_f_a2 = fAlpha;
                pFilter->m_f_b1 = -2.0*fGamma;
                pFilter->m_f_b2 = 2.0*fBeta;
}
```

prepareForPlay()

- Flush delays in LPFs.
- Set up the LFOs by setting the initial rate and LFO type (the LFO type will match our LFO_waveform variable).

```
bool __stdcall CModFilter::prepareForPlay()
{
        // Add your code here:
        //
        // Flush the LPFs
        m_LeftLPF.flushDelays();
        m_RightLPF.flushDelays();
        // Note we do NOT need to init the cutoff and Q;
        // these are done in processAudioFrame()

        // setup LFO
        m_fc_LFO.m_fFrequency_Hz = m_fModRate_fc;
        m_fc_LFO.m_uPolarity = 1; // 0 = bipolar, 1 = unipolar
        m_fc_LFO.m_uTableMode = 0; // normal, no band limiting
        m_fc_LFO.m_uOscType = m_uLFO_Waveform; // our own variable
        m_fc_LFO.setSampleRate(m_nSampleRate); // really important!

        // the LFO prepareForPlay() calls reset() and cookFrequency()
        m_fc_LFO.prepareForPlay();

        return true;
}
```

userInterfaceChange()

- If the user moves the rate slider, we need to change the LFO rate.
- If the user clicks on a radio button, we need to change the LFO waveform type.
- For now, simply recalculate everything when any control is moved; you can streamline this later.

```
bool __stdcall CModFilter::userInterfaceChange(int nControlIndex)
{
        // brute force update all
        m_fc_LFO.m_fFrequency_Hz = m_fModRate_fc;
        m_fc_LFO.m_uOscType = m_uLFO_Waveform;
```

```
            // cook to calculate
            m_fc_LFO.cookFrequency();

            return true;
    }
```

processAudioFrame()

- Calculate a new LFO value.
- Use the LFO value to calculate a new f_c value.
- Use the f_c value to calculate new filter coefficients.
- Do the bi-quad routines on the input samples.

```
    bool __stdcall CModFilter::processAudioFrame(float* pInputBuffer, float* pOutputBuffer,
                                        UINT uNumInputChannels, UINT
                                        uNumOutputChannels)
    {
            //
            // output = input -- change this for meaningful processing
            //
            float fYn = 0; // normal output
            float fYqn = 0; // quad phase output

            // call the LFO function; we only need first output
            m_fc_LFO.doOscillate(&fYn, &fYqn);

            // use the LFO value to calculate the updated fc value
            float fc = calculateCutoffFreq(fYn);

            // use the fc value and a preset Q = 2.0
            // to calculate the LPF coefficients for each channel
            calculateLPFCoeffs(fc, 2.0, &m_LeftLPF);
            calculateLPFCoeffs(fc, 2.0, &m_RightLPF);

            // do the BiQuad operation on the Left LPF
            pOutputBuffer[0] = m_LeftLPF.doBiQuad(pInputBuffer[0]);

            // Mono-In, Stereo-Out (AUX Effect)
            if(uNumInputChannels == 1 && uNumOutputChannels == 2)
                    pOutputBuffer[1] = pOutputBuffer[0]; // just copy

            // Stereo-In, Stereo-Out (INSERT Effect)
            if(uNumInputChannels == 2 && uNumOutputChannels == 2)
                    pOutputBuffer[1] = m_RightLPF.doBiQuad(pInputBuffer[1]);

            return true;
    }
```

Build and test the plug-in. You will hear the f_c modulated by the LFO; adjust the LFO rate and depth controls and try changing the LFO waveform type as well; triangle and sine seem to be the most common choices, and the saw produces an interesting pulsing effect.

12.2 Design a Mod Filter Plug-In: Part II, Modulated f_c, Q

In the second design, we will modify the current plug-in to include the modulation of the LPF Q value. We will use a second, independent LFO for the Q but will share the same LFO waveform type with the f_c LFO. The block diagram is shown in Figure 12.6.

The parameters are as follows:

- f_c mod rate (0.2 to 10 Hz)
- f_c mod depth (0 to 100%)
- Q mod rate (0.2 to 10 Hz)
- Q mod depth (0 to 100%)
- LFO type (sine, tri, saw, square)
- f_c mod range: 100 Hz to 5 kHz
- Q mod range: 0.5 to 10
- LPF is fixed as second-order bi-quad

12.2.1 ModFilter GUI

Add more sliders to the user interface (UI) for the new LFO as per Table 12.3.

Table 12.3: Additional controls for the second LFO.

Slider Property	Value	Slider Property	Value
Control Name	Mod Depth Q	Control Name	Mod Rate Q
Units	%	Units	Hz
Variable Type	float	Variable Type	float
Variable Name	m_fMod_Depth_Q	Variable Name	m_fModRate_Q
Low Limit	0	Low Limit	0.2
High Limit	100	High Limit	10
Initial Value	50	Initial Value	0.5

12.2.2 ModFilter.h File

We need to add the following new functions and variables, basically just duplicating the ones for the f_c modulation:

- LFO object for Q modulation
- calculateQ() to calculate the new Q value from the LFO sample value
- Min and max Q values

```
// Add your code here: ----------------------------------------- //
CBiquad m_LeftLPF;
CBiquad m_RightLPF;
```

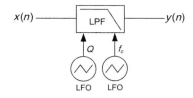

Figure 12.6: The mod filter with second LFO for Q.

```
CWaveTable m_fc_LFO;
CWaveTable m_Q_LFO;

float m_fMinCutoffFreq;
float m_fMaxCutoffFreq;

float m_fMinQ;
float m_fMaxQ;

// functions to calculate the fc or Q for a given LFO sample
float calculateCutoffFreq(float fLFOSample);
float calculateQ(float fLFOSample);

// calculate the coeffs for a given filter
void calculateLPFCoeffs(float fCutoffFreq, float fQ, CBiquad* pFilter);

// END OF USER CODE ----------------------------------------------------------
```

12.2.3 ModFilter.cpp File

Constructor

- Initialize the min and max *Q* values.

```
CModFilter::CModFilter()
{
        m_PlugInName = "ModFilter";

        // Default to Stereo Operation:
        // Change this if you want to support more/less channels
        m_uMaxInputChannels = 2;
        m_uMaxOutputChannels = 2;

        // Finish initializations here
        // set our Min and Max Modulation points
        m_fMinCutoffFreq = 100.0;
        m_fMaxCutoffFreq = 5000.0;
        m_fMinQ = 0.577;
        m_fMaxQ = 10;
}
```

Write the calculateQ() function, which behaves the same way as calculateCutoffFreq(); you only need to change the mod depth source and min and max values:

```
float CModFilter::calculateQ(float fLFOSample)
{
        return (m_fMod_Depth_Q/100.0)*(fLFOSample*(m_fMaxQ - m_fMinQ)) + m_fMinQ;
}
```

prepareForPlay()

- Set up the *Q* LFO by setting the initial rate and LFO type (the LFO type will match our m_uLFOwaveform variable).

```
bool __stdcall CModFilter::prepareForPlay()
{
        // Add your code here:
        <SNIP SNIP SNIP>

        // the LFO prepareForPlay() calls reset() and cookFrequency()
        m_fc_LFO.prepareForPlay();

        // now do the Q LFO
        m_Q_LFO.m_fFrequency_Hz = m_fModRate_Q;
        m_Q_LFO.m_uPolarity = 1; // 0 = bipolar, 1 = unipolar
        m_Q_LFO.m_uTableMode = 0; // normal, no band limiting
        m_Q_LFO.m_uOscType = m_uLFO_Waveform;
        m_Q_LFO.setSampleRate(m_nSampleRate); // really important!

        // this calls reset() and cookFrequency()
        m_Q_LFO.prepareForPlay();

        return true;
}
```

userInterfaceChange()

- Update the *Q* LFO when the user changes this value.

```
bool __stdcall CModFilter::userInterfaceChange(int nControlIndex)
{
        // add your code here
        // brute force update all
        m_fc_LFO.m_fFrequency_Hz = m_fModRate_fc;
        m_fc_LFO.m_uOscType = m_uLFO_Waveform;

        m_Q_LFO.m_fFrequency_Hz = m_fModRate_Q;
        m_Q_LFO.m_uOscType = m_uLFO_Waveform;

        // cook to calculate
        m_fc_LFO.cookFrequency();
        m_Q_LFO.cookFrequency();

        etc...
}
```

processAudioFrame()

- Calculate a new f_c LFO value.
- Use the LFO value to calculate a new f_c value.
- Calculate a new Q LFO value.
- Use the LFO value to calculate a new Q value.
- Use the f_c and Q values to calculate new filter coefficients.
- Do the bi-quad routines on the input samples.

```
bool __stdcall CModFilter::processAudioFrame(float* pInputBuffer, float* pOutputBuffer,
                                    UINT uNumInputChannels, UINT
                                    uNumOutputChannels)
{
    //
    // output = input -- change this for meaningful processing
    //

    float fYn = 0; // normal output
    float fYqn = 0; // quad phase output

    // call the fc LFO function; we only need first output
    m_fc_LFO.doOscillate(&fYn, &fYqn);

    // calculate fc
    float fc = calculateCutoffFreq(fYn);

    // call the Q LFO funciton
    m_Q_LFO.doOscillate(&fYn, &fYqn);

    // calculate the new Q
    float fQ = calculateQ(fYn);

    // use to calculate the LPF
    calculateLPFCoeffs(fc, fQ, &m_LeftLPF);
    calculateLPFCoeffs(fc, fQ, &m_RightLPF);

    // do the BiQuads
    pOutputBuffer[0] = m_LeftLPF.doBiQuad(pInputBuffer[0]);

    // Mono-In, Stereo-Out (AUX Effect)
    if(uNumInputChannels == 1 && uNumOutputChannels == 2)
        pOutputBuffer[1] = pOutputBuffer[0]; // just copy

    // Stereo-In, Stereo-Out (INSERT Effect)
    if(uNumInputChannels == 2 && uNumOutputChannels == 2)
        pOutputBuffer[1] = m_RightLPF.doBiQuad(pInputBuffer[1]);

    return true;
}
```

You can see that we only had to add a few lines of code to get the second modulation source working. Build and test the plug-in. You will hear the f_c and Q modulated by the two LFOs; adjust the LFO rate and depth controls and try changing the LFO waveform type too. The Q modulation can be more subtle to hear, but if you use RackAFX's oscillator and pump white noise through it, the modulation becomes clearly audible. We'll finish the plug-in by making one more modification: the ability to place the right and left LPF modulation sources in quadrature phase.

12.3 Design a Mod Filter Plug-In: Part III, Quad-Phase LFOs

In the third design iteration, we will modify the current plug-in to allow for quadrature phase LFOs. The block diagram is given in Figure 12.7.

The parameters are as follows:

- f_c mod rate (0.2 to 10 Hz)
- f_c mod depth (0 to 100%)
- Q mod rate (0.2 to 10 Hz)
- Q mod depth (0 to 100%)
- LFO type (sine, tri, saw, square)
- LFO phase (normal, quadrature)
- f_c mod range: 100Hz to 5kHz
- Q mod range: 0.577 to 10
- LPF is fixed as second-order bi-quad

In order to switch the LFOs into quadrature phase, we only need to add one more radio button control and make a slight modification to the processAudioFrame() function.

12.3.1 ModFilter GUI

Add the Radio Button Control in Table 12.4. Note: This is a direct control variable—there is nothing extra to add in userInterfaceChange() or prepareForPlay() since we only use it in the single processAudioFrame() function.

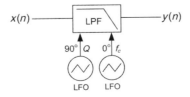

Figure 12.7: The mod filter with quad-phase LFOs.

12.3.2 ModFilter.cpp File

processAudioFrame()

- Check the enumerated variable and use the appropriate LFO output sample.

```
bool __stdcall CModFilter::processAudioFrame(float* pInputBuffer, float* pOutputBuffer,
                                              UINT uNumInputChannels, UINT
                                              uNumOutputChannels)
{
    //
    // output = input -- change this for meaningful processing
    //
    float fYn = 0; // normal output
    float fYqn = 0; // quad phase output

    // call the LFO function; we only need first output
    m_fc_LFO.doOscillate(&fYn, &fYqn);

    // calculate both fc values (can be streamlined!)
    float fc = calculateCutoffFreq(fYn);
    float fcq = calculateCutoffFreq(fYqn);

    // get the Q LFO output
    m_Q_LFO.doOscillate(&fYn, &fYqn);

    // calculate both Q values
    float fQ = calculateQ(fYn);
    float fQq = calculateQ(fYqn);

    // use the fc and Q to calculate the Left LPF coeffs
    calculateLPFCoeffs(fc, fQ, &m_LeftLPF);
    calculateLPFCoeffs(fc, fQ, &m_RightLPF);

    // test the Phase variable; if not NORM, use the
    // quad phase LFO values on the Right channel
    if(m_uLFO_Phase == NORM)
            calculateLPFCoeffs(fc, fQ, &m_RightLPF);
    else
            calculateLPFCoeffs(fcq, fQ, &m_RightLPF);
```

Table 12.4: The quad phase control.

Button Property	Value
Control Name	LFO
Units	
Variable Type	enum
Variable Name	m_uLFO_Phase
Enum String	NORM,QUAD

```
            // do the BiQuads
            pOutputBuffer[0] = m_LeftLPF.doBiQuad(pInputBuffer[0]);

            // Mono-In, Stereo-Out (AUX Effect)
            if(uNumInputChannels == 1 && uNumOutputChannels == 2)
                    pOutputBuffer[1] = pOutputBuffer[0]; // just copy

            // Stereo-In, Stereo-Out (INSERT Effect)
            if(uNumInputChannels == 2 && uNumOutputChannels == 2)
                    pOutputBuffer[1] = m_RightLPF.doBiQuad(pInputBuffer[1]);

            return true;
    }
```

Build and test the new plug-in. When you change the phase to quadrature, it will be easy to hear the two modulations chasing each other across the two speakers. Ideas for future plug-ins include:

- Change the filter type (high-pass filter, band-stop filter, band-pass filter).
- Use a double-modulator: modulate one LFO with *another* LFO, then use that to control a parameter.

12.4 Design an Envelope Follower Plug-In

While the straight-up mod filter is pretty interesting, it has a mechanical quality about it due to the LFO modulations. This might not be an issue for a synthesizer, but it does not sound musically dynamic. We can correlate the modulation to the musical input signal by implementing an envelope follower, which uses the amplitude envelope of the input signal to control the filter parameters. For example, we can code it so that when the input signal has a high amplitude, the f_c is modulated *up* to a higher value, and of course as the plug-in creators, we can also do the opposite kind of modulation. The envelope follower requires an audio envelope detector, and you can use one built into RackAFX. We will use the same second-order LPF as the filter to modulate; then you can try different filter types on your own. First, take a look at the block diagram in Figure 12.8.

The parameters are as follows:

- Pre-gain: 0 to +20 dB of pre-gain to drive the detection circuit
- Threshold (the value the signal has to cross to engage the detector)
- Envelope attack time (mSec)
- Envelope release time (mSec)
- Up/down mod control (determines whether positive amplitude excursions result in positive or negative f_c excursions)
- Digital/analog detection mode: changes detector time constant
- User-adjustable Q
- LPF is fixed as second-order bi-quad

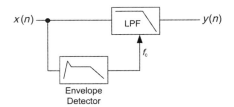

Figure 12.8: The envelope detector block diagram.

We can use the same CBiquad objects for the LFO, but we need another object to handle the envelope detection. This object is built in; the interface is in pluginconstants.h and the implementation is found in the pluginobjects.cpp file. The CEnvelopeDetector has the member variables and functions shown in Table 12.5.

This is a pretty extensive plug-in with multiple controls and interactions. Let's start with the detector itself. The CEnvelopeDetector is based on a simple detector listed in the references (musicdsp.org). The detector mimics an analog peak, mean-square, and RMS detector all in one. To use the detector, follow these steps:

1. Create an instance of the detector, for example, CEnvelopeDetector m_Detector.
2. Initialize all parameters at once with the init() function, for example, m_Detector.init((float)m_nSampleRate, 10.0, 250.0, true, DETECT_MODE_RMS, false).
3. This code initializes the detector with the following parameters:
   a. Sample rate = object's m_nSampleRate
   b. Attack time = 10 mSec
   c. Release time = 250 mSec
   d. Analog time constant = true
   e. Detect RMS signal level
   f. Linear, *not* log detection
4. To change any of the parameters on the fly, use the individual functions:
   a. m_Detector.setAttackTime()
   b. m_Detector.setReleaseTime()
   c. m_Detector.setDetectMode()
   d. m_Detector.setTCModeAnalog()

The attack and release times simply adjust these common parameters. The detection mode can be set for peak, MS, or RMS. Because audio signals must be on the range of -1.0 to $+1.0$ in RackAFX, this means that the squaring function of the MS and RMS modes will produce a value that is less than the instantaneous value. In RMS mode, the MS value is square-rooted, which makes the resulting envelope even smaller. Thus, you need to use care with the MS and RMS modes. In this envelope follower, we will use the RMS value to trigger our dynamic filter.

The time constant determines how fast the detector rises or falls to its maximum (1.0) and minimum (0.0) values. In an analog circuit, the resistor–capacitor (RC) rise and fall times are

Table 12.5: CEnvelopeDetector object definition.

CEnvelopeDetector	
Function: Implements an Envelope Detection Unit	
Member Variables	**Purpose**
int m_nSample;	A z^{-1} delay element for the detector
float m_fAttackTime; float m_fReleaseTime;	The attack and release times for the detector in mSec
float m_fSampleRate;	The sample rate, needed for detection timing
UINT m_uDetectMode;	The detection mode according to these pre-coded enumerated values: DETECT_MODE_PEAK = 0 DETECT_MODE_MS = 1 DETECT_MODE_RMS = 2
float m_fEnvelope;	The current envelope value (this is the output of the device, which it saves for the next sample to compare with)
Member Functions	
void init(arguments) Parameters:	The initialization function, called once before prepareForPlay()
• float samplerate	Input: the sample rate in Hz
• float attack_in_ms	Input: the attack time in mSec
• float release_in_ms	Input: the release time in mSec
• bool bAnalogTC	Input: If true: set the attack and release time constants using the analog definition (time to rise/fall to 36.7% of min/max) — If false: set the attack and release time constants using the 1% definition (time to rise/fall to 1% of min/max); the digital version is about 4X faster
• UINT uDetect Mode	Input: uDetect mode can be one of three constants: DETECT_MODE_PEAK DETECT_MODE_MS DETECT_MODE_RMS
• bool bLogDetector	Input: true if you want the detector to be logarithmic (usually only used with LED metering)
void setTCModeAnalog(bool bAnalogTC) Parameters: • bool bAnalogTC	Input: sets the time constant variable; see above for more info
void setAttackTime(float attack_in_ms)	Change the attack time on the fly
Parameters: • float attack_in_ms	Input: the attack time in mSec.
void setReleaseTime(float release_in_ms)	Change the release time on the fly
Parameters: • float release_in_ms	Input: the release time in mSec
void setDetectMode(UINT uDetect)	Change the detection mode on the fly
Parameters: • UINT uDetect	Input: the detect mode; see m_uDetectMode above
void prepareForPlay()	The prepareForPlay() function for this object
float detect(float fInput) Parameters: • float fInput	The detection function; it produces the current envelope output for a given input, x(n) Input: the current input sample x(n)

only 63.3% of the edge. In digital systems, the value is usually 80% to 99%. The envelope detector has a control to allow you to change the time constant from the very fast (digital) version or the slower (analog) version. We'll try both types in the envelope follower plug-in.

12.5 Envelope Detection

Software envelope detectors are based on the analog circuits that preceded them. The two basic types of detectors are half-wave and full-wave, depending on whether they track either the positive or negative portion of the input waveform (half-wave) or both portions (full-wave). Many of the classic hardware devices used the half-wave version. Either the positive or negative portion may be tracked. The circuit consists of a diode to half-wave-rectify the signal, followed by a "tank" capacitor. The capacitor charges and discharges along with the audio signal. The attack knob is a potentiometer in series with the diode; it controls the rate at which the cap charges. The release knob is a potentiometer connected to ground through which the cap discharges. The RC combinations (one for the charging side and the other for the discharging side) create the attack and release times of the detector (Figure 12.9). The positive portion of the waveform is detected and tracks the input according to the attack and release times (Figures 12.10 and 12.11).

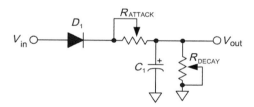

Figure 12.9: A simple positive half-wave detector.

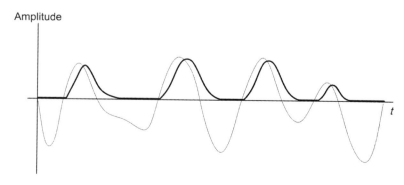

Figure 12.10: Detector output (solid dark line) with short attack and release times.

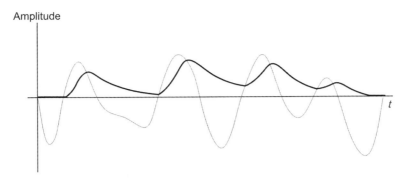

Figure 12.11: Detector output (solid dark line) with medium attack and release times.

In analog RC circuits, the cap charges and discharges exponentially, as shown in Figure 12.12. The analog attack time is calculated as the time it takes to reach 63.2% of the full charge, while the release time is the time to release from the fully charged state down to 36.8% of the full charge. Different combinations of R and C change these times. A digital envelope detector differs in the percent levels; it uses 99% and 1% as the attack and release destination values. The curves remain exponential.

It is well worth noting that there are several approaches for designing a software algorithm for the analog envelope detector. Orfanidis (1996) suggests using a full-wave rectifier followed by an LPF with separate attack and release time constants. Zöler (2011) suggests a mean-squarer followed by an LPF with only a single time constant. In either case, a running average is constantly measured and updated on each sample interval. The CEnvelopeDetector (Figure 12.13) implements a similar method using a half-wave rectifier followed by an LPF whose input is scaled by the attack or release time, depending on whether the signal is rising above the current envelope output or falling below it. You can find the original code source in the references (musicdsp.org).

The attack and release times are conveniently calculated from your controls in milliseconds. The calculations are shown in Equation 12.1:

$$\tau_a = e^{TC/(\text{attack\_in\_mSec}*\text{SampleRate}*0.001)}$$

and

$$\tau_r = e^{TC/(\text{release\_in\_mSec}*\text{SampleRate}*0.001)}$$

where

$$TC_{\text{analog}} = \log(0.368)$$
$$TC_{\text{digital}} = \log(0.01)$$

(12.1)

The code for the difference equation is:

```
if(fInput> m_fEnvelope)
     m_fEnvelope = m_fAttackTime * (m_fEnvelope - fInput) + fInput;
else
     m_fEnvelope = m_fReleaseTime * (m_fEnvelope - fInput) + fInput;
```

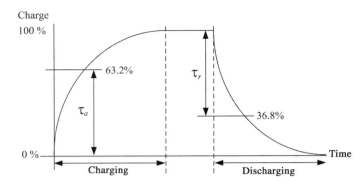

Figure 12.12: Analog attack and release times.

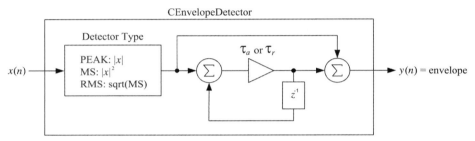

Figure 12.13: The CEnvelopeDetector block diagram.

Compare this to the block diagram in Figure 12.13 and you can see how it implements the detector.

12.5.1 Project EnvelopeFollower

Create the project; there are no member plug-ins to add.

12.5.2 EnvelopeFollower GUI

The GUI (Table 12.6) needs to have controls for the envelope detector:
- Pre-gain
- Threshold
- Attack time
- Release time
- Time constant (analog or digital)

Table 12.6: The GUI controls for the envelope follower plug-in.

Slider Property	Value	Slider Property	Value
Control Name	Pre-Gain	Control Name	Threshold
Units	dB	Units	
Variable Type	float	Variable Type	float
Variable Name	m_fPreGain_dB	Variable Name	m_fThreshold
Low Limit	0	Low Limit	0
High Limit	20	High Limit	1
Initial Value	12	Initial Value	0.2

Slider Property	Value	Slider Property	Value
Control Name	Attack Time	Control Name	Release Time
Units	mSec	Units	mSec
Variable Type	float	Variable Type	float
Variable Name	m_fAttack_mSec	Variable Name	m_fRelease_mSec
Low Limit	10	Low Limit	20
High Limit	100	High Limit	250
Initial Value	25	Initial Value	50

Slider Property	Value	Button Property	Value
Control Name	Q	Control Name	Time Constant
Units		Units	
Variable Type	float	Variable Type	enum
Variable Name	m_fQ	Variable Name	m_uTimeConstant
Low Limit	0.5	Enum String	analog, digital
High Limit	20		
Initial Value	5		

Button Property	Value
Control Name	Direction
Units	
Variable Type	enum
Variable Name	m_uDirection
Enum String	UP, DOWN

And, it needs controls for the modulated filter:

- Q
- Direction of modulation (up or down)

12.5.3 EnvelopeFollower.h File

We need to add the following new functions and variables; most can be taken directly from your ModFilter project. We will need:

- An LPF bi-quad for each channel (left, right)

- An envelope detector for each channel (left, right)
- A function to calculate a new LPF cutoff value from an envelope value
- Functions to calculate the bi-quad coefficients from the LPF cutoff and *Q* values
- Min/max LPF cutoff variables

Add the following to your .h file:

```
// Add your code here: ------------------------------------------------- //
// two LPF BiQuads for the filters
CBiquad m_LeftLPF;
CBiquad m_RightLPF;

// function to calculate the new fc from the Envelope Value
float calculateCutoffFreq(float fEnvelopeSample);

// calculate the coeffs for a given filter
void calculateLPFCoeffs(float fCutoffFreq, float fQ, CBiquad* pFilter);

// min/max variables
float m_fMinCutoffFreq;
float m_fMaxCutoffFreq;

// envelope detectors
CEnvelopeDetector m_LeftDetector;
CEnvelopeDetector m_RightDetector;

// END OF USER CODE ---------------------------------------------------- //
```

Notice that everything except the envelope detectors is straight from your last project.

12.5.4 EnvelopeFollower.cpp File

Constructor

- Initialize the rate min and max values. The member-objects are self initializing upon creation.

```
CEnvelopeFollower::CEnvelopeFollower()
{
    <SNIP SNIP SNIP>

    // Finish initializations here
    // set our Min and Max Modulation points
    m_fMinCutoffFreq = 100.0;
    m_fMaxCutoffFreq = 5000.0;
}
```

Write the calculateCutoffFrequency() function. It is similar to the previous modulation functions except that we will allow both up and down directions of modulation. Can you figure out how the code works?

```
float CEnvelopeFollower::calculateCutoffFreq(float fEnvelopeSample)
{
        // modulate from min upwards
        if(m_uDirection == UP)
                return fEnvelopeSample*(m_fMaxCutoffFreq - m_fMinCutoffFreq) +
                        m_fMinCutoffFreq;
        else // modulate from max downwards
                return m_fMaxCutoffFreq - fEnvelopeSample*(m_fMaxCutoffFreq -
                        m_fMinCutoffFreq);

        return m_fMinCutoffFreq;
}
```

Write the cooking functions for the LPFs to set the new coefficients for a given f_c and Q value. This is identical to the ModFilter project.

```
void CEnvelopeFollower::calculateLPFCoeffs(float fCutoffFreq, float fQ, CBiquad*
        pFilter)
{
        /* use same code as the Mod Filter */
}
```

prepareForPlay()
- Flush delays in LPFs.
- Initialize the detector objects.

```
bool __stdcall CEnvelopeFollower::prepareForPlay()
{
        // Add your code here:
        m_LeftLPF.flushDelays();
        m_RightLPF.flushDelays();

        // init the envelope detectors
        // set all params at once with this function;
        // false = Digital Time Constant NOT Analog one
        if(m_uTimeConstant == digital)
        {
                m_LeftDetector.init((float)m_nSampleRate, m_fAttack_mSec, m_fRelease_mSec,
                        false, DETECT_MODE_RMS, false);
                m_RightDetector.init((float)m_nSampleRate, m_fAttack_mSec,
                        m_fRelease_mSec, false, DETECT_MODE_RMS, false);
        }
        else
        {
                m_LeftDetector.init((float)m_nSampleRate, m_fAttack_mSec, m_fRelease_mSec,
                        true, DETECT_MODE_RMS, false);
                m_RightDetector.init((float)m_nSampleRate, m_fAttack_mSec,
                        m_fRelease_mSec, true, DETECT_MODE_RMS, false);
        }

        return true;
}
```

You can see that the detectors are initialized all at once, using the m_uTimeConstant to control the boolean value passed to the init function. The detection is set as MS (DETECT_MODE_MS).

userInterfaceChange()

- Decode the control ID and then call the appropriate function; we simply pass our slider values or radio button command to the functions.
- Note: Make sure your nControlIndex values match with your UI; these mappings are for my project.

```
bool __stdcall CEnvelopeFollower::userInterfaceChange(int nControlIndex)
{
    switch(nControlIndex)
    {
        case 2: // Attack
        {
            m_LeftDetector.setAttackTime(m_fAttack_mSec);
            m_RightDetector.setAttackTime(m_fAttack_mSec);
            break;
        }
        case 3: // Release
        {
            m_LeftDetector.setReleaseTime(m_fRelease_mSec);
            m_RightDetector.setReleaseTime(m_fRelease_mSec);
            break;
        }
        case 41: // Time Constant
        {
            if(m_uTimeConstant == digital)
            {
                m_LeftDetector.setTCModeAnalog(false);
                m_RightDetector.setTCModeAnalog(false);
            }
            else
            {
                m_LeftDetector.setTCModeAnalog(true);
                m_RightDetector.setTCModeAnalog(true);
            }
        }

        default:
            break;
    }

    return true;
}
```

processAudioFrame()
- The envelope follower's processAudioFrame() function will operate in the sequence shown in Figure 12.14.

```
bool __stdcall CEnvelopeFollower::processAudioFrame(float* pInputBuffer, float*
                                                    pOutputBuffer, UINT
                                                    uNumInputChannels, UINT
                                                    uNumOutputChannels)
{
        //
        // Do LEFT (MONO) Channel; there is always at least one input, one output
        float fGain = pow(10, m_fPreGain_dB/20.0);
        float fDetectLeft = m_LeftDetector.detect(fGain*pInputBuffer[0]);

        // set mod freq to minimum (un-triggered)
        float fModFreqLeft =m_fMinCutoffFreq;

        // if threshold triggered, calculate new LPF cutoff
        if(fDetectLeft >= m_fThreshold)
                fModFreqLeft = calculateCutoffFreq(fDetectLeft);

        // use the mod freq and user supplied-Q to calc the fc
        calculateLPFCoeffs(fModFreqLeft, m_fQ, &m_LeftLPF);
```

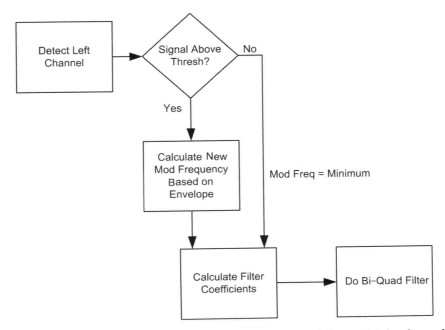

Figure 12.14: The flowchart for the left channel of the envelope follower (right channel is same).

```
            // do the BiQuads
            pOutputBuffer[0] = m_LeftLPF.doBiQuad(pInputBuffer[0]);

            // detect the other channel
            float fDetectRight = m_RightDetector.detect(fGain*pInputBuffer[1]);

            // set mod freq to minimum (un-triggered)
            float fModFreqRight = m_fMinCutoffFreq;

            // if threshold triggered, calculate new LPF cutoff
            if(fDetectLeft >= m_fThreshold)
                    fModFreqRight = calculateCutoffFreq(fDetectRight);

            // use the mod freq and user supplied-Q to calc the fc
            calculateLPFCoeffs(fModFreqRight, m_fQ, &m_RightLPF);

            // Mono-In, Stereo-Out (AUX Effect)
            if(uNumInputChannels == 1 && uNumOutputChannels == 2)
                    pOutputBuffer[1] = pOutputBuffer [0];

            // Stereo-In, Stereo-Out (INSERT Effect)
            if(uNumInputChannels == 2 && uNumOutputChannels == 2)
                            pOutputBuffer[1] = m_RightLPF.doBiQuad(pInputBuffer[1]);

            return true;
    }
```

Build and test the plug-in. Try all the controls. Note: Beware—because of the high pre-gain values possible combined with high-Q values, you can easily overload the audio. If you hear distortion, adjust the pre-gain value. Test this out with guitar, keyboard, or drum loops. Ideas for more plug-ins include:

- Change the filter type (high-pass filter, band-stop filter, band-pass filter).
- Modulate the Q in the same direction, or opposite direction, as the f_c.
- Modulate the f_c or Q with an LFO and the other parameter with the envelope.
- Allow the user to select the detection mode: peak, MS, or RMS.
- Allow the user to change between linear and log detection (see CEnvelopeDetector's init() method).

12.6 Design a Phaser Plug-In

The phaser effect will be on par with the reverb algorithms as among the more complex in the book—the theory is not that difficult, but a nice analog sounding phaser requires many nearly duplicated components, in this case APFs. The phaser effect was an attempt at creating a flanging effect before the necessary analog integrated circuits were available. A flanger uses a modulated delay line to alter the delay time. A phaser uses APFs in an attempt to do the same. APFs have a delay response that is flat across the first one-third or so of the DC-to-Nyquist band. In fact, APFs can be designed to perform fractional delay as an alternative to interpolation. The phaser didn't sound like a flanger at all. The reason is that in a flanger, the notches in the inverse-comb filtering are mathematically related as simple multiples of each other. This is not the case for the phaser. The phaser is still used today in electronic music. Like the flanger, we will get the best results by first looking at the original analog design. Our phaser topology will be based off the analog counterpart in Figures 12.15 and 12.16. Even if you can't read schematics, it's worth a moment to take a look at the overall architecture since we'll be emulating it in digital form. The complete block diagram is shown in Figure 12.17.

The phase of the signal is shifted by 90 degrees at the frequency given by Equation 12.2:

$$f_p = \frac{1}{2\pi C_1 (R_1 // r_{DS})} \qquad (12.2)$$

The positive control voltage is a triangle wave, which alters the resistance of the field effect transistor (FET) that is in parallel with R_1. This in turn changes the 90 degree phase shift frequency. The values of C_1 vary from 1 uF to 0.06 uF, and r_{DS} varies from 100 Ω (FET on) to 10 K (FET off). The input signal is run through a cascade of such modules. The output of the last module, or stage, is then mixed with the dry input signal to create the effect. You only need to decide on the number of stages and the oscillator to complete the design.

The things to note about the analog design are:

- There are six first-order APF modules.
- The APF modules are in series, one feeding the next.
- The depth control is a blend of the dry and filtered signals; a depth of 100% is actually a 50/50 mix of wet and dry.

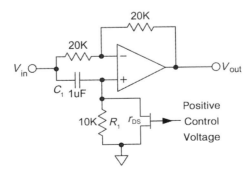

Figure 12.15: A voltage-controlled first-order all-pass filter module.

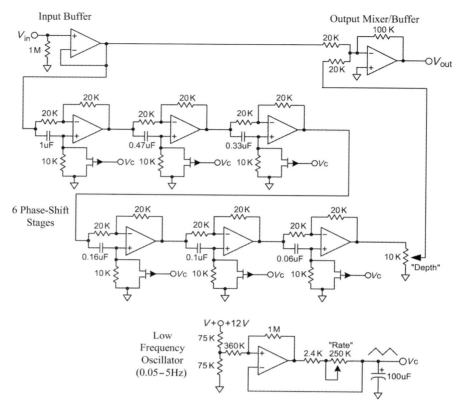

Figure 12.16: The phaser circuit (from National Semiconductor).

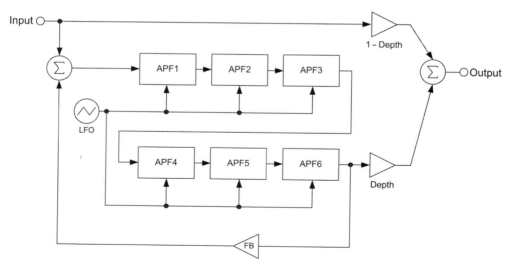

Figure 12.17: The block diagram of the analog phaser design. The feedback path has been added to the original National Semiconductor design; the feedback will be named "Intensity".

- Other phaser models also included a feedback path from the filtered output back to the filter chain input to increase intensity of the effect.
- The six APF modules have the frequency ranges in Table 12.7.

Because this is a complex design, we will break it down into two parts: first, we'll design an abbreviated version in Figure 12.18 with only three APF modules and no feedback control. Then, we'll add the other three APF stages and the feedback control.

The design equations for first-order APFs from Chapter 6 are given next.

Table 12.7: Minimum and maximum phase rotation frequencies for the phaser APFs.

APF	Minimum f_c	Maximum f_c
1	16 Hz	1.6 kHz
2	33 Hz	3.3 kHz
3	48 Hz	4.8 kHz
4	98 Hz	9.8 kHz
5	160 Hz	16 kHz
6	260 Hz	26 kHz*

*We will have to cut off at ½ Nyquist instead.

Specify:

- f_c, corner frequency

$$\begin{aligned}
\theta_c &= 2\pi f_c/f_s \\
\alpha &= \frac{\tan(\pi Q/f_s) - 1}{\tan(\pi Q/f_s) + 1} \\
\beta &= -\cos\theta_c \\
a_0 &= -\alpha \\
a_1 &= \beta(1-\alpha) \\
a_2 &= 1.0 \\
b_1 &= \beta(1-\alpha) \\
b_2 &= -1.0
\end{aligned} \quad (12.3)$$

We'll need to implement the APF using the built-in BiQuad object. The built-in oscillator will be used for the LFO; many aspects of the design and code chunks can be borrowed from your ModFilter projects. Specific to this design is that it:

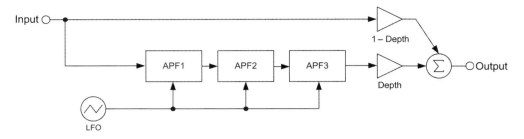

Figure 12.18: The block diagram of the simplified, first part of the design.

- Needs multiple APFs, each with its own range of frequencies
- Needs only one LFO to control all APFs
- Needs to calculate all APF cutoff frequencies and coefficients at the same time

Let's get started with the usual steps.

12.6.1 Project Phaser

Create a new project. There are no member plug-ins to add.

12.6.2 Phaser GUI

Add the controls from Table 12.8 to your GUI.

Table 12.8: The phaser GUI elements.

Slider Property	Value
Control Name	Rate
Units	Hz
Variable Type	float
Variable Name	m_fModRate
Low Limit	0.02
High Limit	10
Initial Value	0.5

Slider Property	Value
Control Name	Depth
Units	%
Variable Type	float
Variable Name	m_fMod_Depth
Low Limit	0
High Limit	100
Initial Value	50

Button Property	Value
Control Name	LFO Type
Units	
Variable Type	enum
Variable Name	m_uLFO_Waveform
Enum String	sine, saw, tri, square

12.6.3 Phaser.h File

Declare instances of the BiQuad and WaveTable objects. Also, declare the functions to calculate cutoff frequencies and filter coefficients. We will calculate all the APF coefficients in one pair of functions (one for left and one for right). Because there is so much repeated code, we will also add a helper function to calculate the APF coefficients for each bi-quad. You don't need to #include anything since these are built in. Add these to your .h files

```
// Add your code here: ----------------------------------------------- //
CBiquad m_LeftAPF_1;
CBiquad m_RightAPF_1;

CBiquad m_LeftAPF_2;
CBiquad m_RightAPF_2;

CBiquad m_LeftAPF_3;
CBiquad m_RightAPF_3;

// function to calculate the new fc from the Envelope Value
float calculateAPFCutoffFreq(float fLFOSample, float fMinFreq, float fMaxFreq);

// Two functions to calculate the BiQuad Coeffs: APF
void calculateFirstOrderLeftAPFCoeffs(float fLFOSample);
void calculateFirstOrderRightAPFCoeffs(float fLFOSample);

// helper function for APF Calculation
void calculateFirstOrderAPFCoeffs(float fCutoffFreq,
                                  CBiquad* pBiQuadFilter);

// min/max variables
float m_fMinAPF_1_Freq;
float m_fMaxAPF_1_Freq;

float m_fMinAPF_2_Freq;
float m_fMaxAPF_2_Freq;

float m_fMinAPF_3_Freq;
float m_fMaxAPF_3_Freq;

// LFO Stuff
CWaveTable m_LFO;

// END OF USER CODE ------------------------------------------------- //
```

12.6.4 Phaser.cpp File

Constructor

- Initialize the rate min and max values. The objects are self-initializing upon creation.
- Initialize the first three APF min/max pairs according to the table. We'll add the others once we have tested it and find it works properly.

```
CPhaser::CPhaser()
{
        <SNIP SNIP SNIP>

        // Finish initializations here
        // set our Min and Max Modulation points
        m_fMinAPF_1_Freq = 16.0;
        m_fMaxAPF_1_Freq = 1600.0;

        m_fMinAPF_2_Freq = 33.0;
        m_fMaxAPF_2_Freq = 3300.0;

        m_fMinAPF_3_Freq = 48.0;
        m_fMaxAPF_3_Freq = 4800.0;
}
```

Write the calculateAPFCutoffFreq() function. This function behaves exactly like the function in the mod filter except there is no mod depth. Also note that we will call this function repeatedly for each APF unit, so we need to send it the min/max values to compute with; the function itself is trivial and just like all the previous mod filters:

```
float CPhaser::calculateAPFCutoffFreq(float fLFOSample, float fMinFreq, float fMaxFreq)
{
        return fLFOSample*(fMaxFreq - fMinFreq) + fMinFreq;
}
```

Implement the helper function for calculating the APF coefficients for a given f_c.

```
void CPhaser::calculateFirstOrderAPFCoeffs(float fCutoffFreq,
                                            CBiquad* pBiQuadFilter)
{
        // coeff calculation
        float alpha_num = tan(pi*fCutoffFreq/(float)m_nSampleRate) - 1.0;
        float alpha_den = tan(pi*fCutoffFreq/(float)m_nSampleRate) + 1.0;
        float alpha = alpha_num/alpha_den;

        // set on target filter
        pBiQuadFilter->m_f_a0 = alpha;
        pBiQuadFilter->m_f_a1 = 1.0;
        pBiQuadFilter->m_f_a2 = 0.0;
        pBiQuadFilter->m_f_b1 = alpha;
        pBiQuadFilter->m_f_b2 = 0.0;
}
```

Implement the cooking functions below for the right and left banks of APFs using the helper function above:

```
void CPhaser::calculateFirstOrderLeftAPFCoeffs(float fLFOSample)
{
        // APF1 - fc -> Bi Quad
        float fCutoffFreq = calculateAPFCutoffFreq(fLFOSample, m_fMinAPF_1_Freq,
```

```
                                                    m_fMaxAPF_1_Freq);
        calculateFirstOrderAPFCoeffs(fCutoffFreq, &m_LeftAPF_1);

        // APF2 - fc -> Bi Quad
        fCutoffFreq = calculateAPFCutoffFreq(fLFOSample, m_fMinAPF_2_Freq,
                                                    m_fMaxAPF_2_Freq);
        calculateFirstOrderAPFCoeffs(fCutoffFreq, &m_LeftAPF_2);

        // APF3 - fc -> Bi Quad
        fCutoffFreq = calculateAPFCutoffFreq(fLFOSample, m_fMinAPF_3_Freq,
                                                    m_fMaxAPF_3_Freq);
        calculateFirstOrderAPFCoeffs(fCutoffFreq, &m_LeftAPF_3);
}

void CPhaser::calculateFirstOrderRightAPFCoeffs(float fLFOSample)
{
        // APF1 - fc -> Bi Quad
        float fCutoffFreq = calculateAPFCutoffFreq(fLFOSample, m_fMinAPF_1_Freq,
                                                    m_fMaxAPF_1_Freq);
        calculateFirstOrderAPFCoeffs(fCutoffFreq, &m_RightAPF_1);

        // APF2 - fc -> Bi Quad
        fCutoffFreq = calculateAPFCutoffFreq(fLFOSample, m_fMinAPF_2_Freq,
                                                    m_fMaxAPF_2_Freq);
        calculateFirstOrderAPFCoeffs(fCutoffFreq, &m_RightAPF_2);

        // APF3 - fc -> Bi Quad
        fCutoffFreq = calculateAPFCutoffFreq(fLFOSample, m_fMinAPF_3_Freq,
                                                    m_fMaxAPF_3_Freq);
        calculateFirstOrderAPFCoeffs(fCutoffFreq, &m_RightAPF_3);
}
```

prepareForPlay()

- Flush the delays in the APFs.
- Set up the LFO by setting the initial rate, LFO type (the LFO type will match our m_uLFOWaveform variable).

```
bool __stdcall CPhaser::prepareForPlay()
{
        // Add your code here:
        m_LeftAPF_1.flushDelays();
        m_RightAPF_1.flushDelays();

        m_LeftAPF_2.flushDelays();
        m_RightAPF_2.flushDelays();

        m_LeftAPF_3.flushDelays();
        m_RightAPF_3.flushDelays();
```

```
            // setup LFO
            m_LFO.m_fFrequency_Hz = m_fModRate;
            m_LFO.m_uPolarity = 1; // 0 = bipolar, 1 = unipolar
            m_LFO.m_uTableMode = 0; // normal, no band limiting
            m_LFO.m_uOscType = m_uLFO_Waveform;
            m_LFO.setSampleRate(m_nSampleRate); // really important!

            // the LFO prepareForPlay() calls reset() and cookFrequency()
            m_LFO.prepareForPlay();

            return true;
    }
```

userInterfaceChange()

- If the user moves the rate slider, we need to change the LFO rate and toggle the oscillator type if they use the buttons.
- Note: Make sure your nControlIndex values match your UI.

```
    bool __stdcall CPhaser::userInterfaceChange(int nControlIndex)
    {
            // decode the control index, or delete the switch and use brute force calls
            switch(nControlIndex)
            {
                    case 0:
                    {
                            m_LFO.m_fFrequency_Hz = m_fModRate;
                            m_LFO.cookFrequency();
                            break;
                    }
                    case 41:
                    {
                            m_LFO.m_uOscType = m_uLFO_Waveform;
                            break;
                    }
                    default:
                            break;
            }

            return true;
    }
```

processAudioFrame()

- Calculate a new LFO value.
- Cascade our three APFs one after another, feeding the output of each into the input of the next until done.

- Use the depth control to mix the output.
- Calculate the depth so that 100% depth gives a 50/50 mix ratio (done by dividing depth percent by 200.0 instead of 100).

```
bool __stdcall CPhaser::processAudioFrame(float* pInputBuffer, float* pOutputBuffer, UINT
                                          uNumInputChannels, UINT uNumOutputChannels)
{
        // Do LEFT (MONO) Channel; there is always at least one input, one output
        float fYn = 0; // normal output
        float fYqn = 0; // quad phase output

        // mod depth is at 100% when the control is at 50% !!
        float fDepth = m_fMod_Depth/200.0;

        // call the LFO function; we only need first output
        m_LFO.doOscillate(&fYn, &fYqn);

        // use the LFO to calculate all APF banks
        calculateFirstOrderLeftAPFCoeffs(fYn);

        // do the cascaded APFs
        float fAPF_1_Out = m_LeftAPF_1.doBiQuad(pInputBuffer[0]);
        float fAPF_2_Out = m_LeftAPF_2.doBiQuad(fAPF_1_Out);
        float fAPF_3_Out = m_LeftAPF_3.doBiQuad(fAPF_2_Out);

        // form the output
        pOutputBuffer[0] = fDepth*fAPF_3_Out + (1.0 - fDepth)*pInputBuffer[0];

        // calculate
        calculateFirstOrderRightAPFCoeffs(fYn);

        // do the cascade
        fAPF_1_Out = m_RightAPF_1.doBiQuad(pInputBuffer[1]);
        fAPF_2_Out = m_RightAPF_2.doBiQuad(fAPF_1_Out);
        fAPF_3_Out = m_RightAPF_3.doBiQuad(fAPF_2_Out);

        // Mono-In, Stereo-Out (AUX Effect)
        if(uNumInputChannels == 1 && uNumOutputChannels == 2)
                pOutputBuffer[1] = pOutputBuffer[0];

        // Stereo-In, Stereo-Out (INSERT Effect)
        if(uNumInputChannels == 2 && uNumOutputChannels == 2)
                pOutputBuffer[1] = fDepth*fAPF_3_Out + (1.0 - fDepth)*pInputBuffer[1];

        return true;
}
```

Build and test the plug-in. You should be able to easily hear the phaser effect, though it might not be as strong as you are used to with other devices. We're going to fix this next by added more APF stages, feedback, and the option for quadrature phase LFO operation.

12.7 Design a Stereo Phaser with Quad-Phase LFOs

Adding the extra APFs is just more of the same cut-and-paste operation; you only need to change the range of frequencies and adjust the cooking and processing functions for that. The addition of the feedback is simple—we need another slider on the UI and a feedback z^{-1} storage device. The quadrature phase is easy to implement because our LFO already produces this output for us; we only need a UI change and a branch statement in the code.

12.7.1 Phaser GUI

Update the GUI with the new controls in Table 12.9.

Table 12.9: Additional controls for the quad-phase phaser plug-in.

Button Property	Value
Control Name	LFO Type
Units	
Variable Type	enum
Variable Name	m_uLFO_Phase
Enum String	NORM, QUAD

Slider Property	Value
Control Name	Intensity
Units	%
Variable Type	float
Variable Name	m_fFeedback
Low Limit	0
High Limit	99
Initial Value	50

12.7.2 Phaser.h File

We need to add three more APF object sets (three for left, three for right) plus the feedback storage elements, one each for left and right. Add the following to your .h file:

```
// Add your code here: ----------------------------------------------

<SNIP SNIP SNIP>

CBiquad m_LeftAPF_3;
CBiquad m_RightAPF_3;

CBiquad m_LeftAPF_4;
CBiquad m_RightAPF_4;

CBiquad m_LeftAPF_5;
CBiquad m_RightAPF_5;

CBiquad m_LeftAPF_6;
CBiquad m_RightAPF_6;
```

```
// function to calculate the new fc from the Envelope Value
float calculateAPFCutoffFreq(float fLFOSample, float fMinFreq, float fMaxFreq);

<SNIP SNIP SNIP>

float m_fMinAPF_3_Freq;
float m_fMaxAPF_3_Freq;

float m_fMinAPF_4_Freq;
float m_fMaxAPF_4_Freq;

float m_fMinAPF_5_Freq;
float m_fMaxAPF_5_Freq;

float m_fMinAPF_6_Freq;
float m_fMaxAPF_6_Freq;

// LFO Stuff
CWaveTable m_LFO;

// Feedback Storage
float m_fFeedbackLeft;
float m_fFeedbackRight;

// END OF USER CODE ------------------------------------------------
```

12.7.3 Phaser.cpp File

Constructor

- Initialize the min and max APF values as per the table.

```
CPhaser::CPhaser()
{
    <SNIP SNIP SNIP>
    m_fMinAPF_3_Freq = 48.0;
    m_fMaxAPF_3_Freq = 4800.0;

    m_fMinAPF_4_Freq = 98.0;
    m_fMaxAPF_4_Freq = 9800.0;

    m_fMinAPF_5_Freq = 160.0;
    m_fMaxAPF_5_Freq = 16000.0;

    m_fMinAPF_6_Freq = 220.0;
    m_fMaxAPF_6_Freq = 22000.0;

    m_fFeedbackLeft = 0.0;
    m_fFeedbackRight = 0.0;
}
```

Modify the cooking functions to calculate the new APF coefficients:

```
void CPhaser::calculateFirstOrderLeftAPFCoeffs(float fLFOSample)
{
        <SNIP SNIP SNIP>

        // APF3 — fc -> Bi Quad
        fCutoffFreq = calculateAPFCutoffFreq(fLFOSample, m_fMinAPF_3_Freq,
                                        m_fMaxAPF_3_Freq);
        calculateFirstOrderAPFCoeffs(fCutoffFreq, &m_LeftAPF_3);

        // APF4 — fc -> Bi Quad
        fCutoffFreq = calculateAPFCutoffFreq(fLFOSample, m_fMinAPF_4_Freq,
                                        m_fMaxAPF_4_Freq);
        calculateFirstOrderAPFCoeffs(fCutoffFreq, &m_LeftAPF_4);

        // APF5 — fc -> Bi Quad
        fCutoffFreq = calculateAPFCutoffFreq(fLFOSample, m_fMinAPF_5_Freq,
                                        m_fMaxAPF_5_Freq);
        calculateFirstOrderAPFCoeffs(fCutoffFreq, &m_LeftAPF_5);

        // APF6 - fc -> Bi Quad
        fCutoffFreq = calculateAPFCutoffFreq(fLFOSample, m_fMinAPF_6_Freq,
                                        m_fMaxAPF_6_Freq);
        calculateFirstOrderAPFCoeffs(fCutoffFreq, &m_LeftAPF_6);

}

void CPhaser::calculateFirstOrderRightAPFCoeffs(float fLFOSample)
{
        <SNIP SNIP SNIP>

        // APF3 - fc -> Bi Quad
        fCutoffFreq = calculateAPFCutoffFreq(fLFOSample, m_fMinAPF_3_Freq,
                                        m_fMaxAPF_3_Freq);
        calculateFirstOrderAPFCoeffs(fCutoffFreq, &m_RightAPF_3);

        // APF4 - fc -> Bi Quad
        fCutoffFreq = calculateAPFCutoffFreq(fLFOSample, m_fMinAPF_4_Freq,
                                        m_fMaxAPF_4_Freq);
        calculateFirstOrderAPFCoeffs(fCutoffFreq, &m_RightAPF_4);

        // APF5 - fc -> Bi Quad
        fCutoffFreq = calculateAPFCutoffFreq(fLFOSample, m_fMinAPF_5_Freq,
                                        m_fMaxAPF_5_Freq);
        calculateFirstOrderAPFCoeffs(fCutoffFreq, &m_RightAPF_5);

        // APF6 - fc -> Bi Quad
        fCutoffFreq = calculateAPFCutoffFreq(fLFOSample, m_fMinAPF_6_Freq,
                                        m_fMaxAPF_6_Freq);
        calculateFirstOrderAPFCoeffs(fCutoffFreq, &m_RightAPF_6);
}
```

prepareForPlay()

- Flush the new APF buffers.
- Reset the feedback variables.

```
bool __stdcall CPhaser::prepareForPlay()
{
        // Add your code here:
        <SNIP SNIP SNIP>

        m_LeftAPF_3.flushDelays();
        m_RightAPF_3.flushDelays();

        m_LeftAPF_4.flushDelays();
        m_RightAPF_4.flushDelays();

        m_LeftAPF_5.flushDelays();
        m_RightAPF_5.flushDelays();

        m_LeftAPF_6.flushDelays();
        m_RightAPF_6.flushDelays();

        m_LFO.m_fFrequency_Hz = m_fModRate;
        m_LFO.m_uPolarity = 1; // 0 = bipolar, 1 = unipolar
        m_LFO.m_uTableMode = 0; // normal, no band limiting
        m_LFO.m_uOscType = m_uLFO_Waveform;
        m_LFO.setSampleRate(m_nSampleRate); // really important!

        // the LFO prepareForPlay() calls reset() and cookFrequency()
        m_LFO.prepareForPlay();

        m_fFeedbackLeft = 0.0;
        m_fFeedbackRight = 0.0;

        return true;
}
```

processAudioFrame()

- Add more cascaded APFs to each channel.
- Store and use the feedback values.
- Check for quad-phase LFO and use the quad LFO output on right channel if needed.

```
bool __stdcall CPhaser::processAudioFrame(float* pInputBuffer, float* pOutputBuffer, UINT
                                          uNumInputChannels, UINT uNumOutputChannels)
{
        //
        // output = input -- change this for meaningful processing
        //
        // Do LEFT (MONO) Channel; there is always at least one input, one output
```

```
float fYn = 0; // normal output
float fYqn = 0; // quad phase output

// mod depth is at 100% when the control is at 50% !!
float fDepth = m_fMod_Depth/200.0;
float fFeedback = m_fFeedback/100.0;

// call the LFO function; we only need first output
m_LFO.doOscillate(&fYn, &fYqn);

// use the LFO to calculate all APF banks
calculateFirstOrderLeftAPFCoeffs(fYn);

// do the cascaded APFs
float fAPF_1_Out = m_LeftAPF_1.doBiQuad(pInputBuffer[0] +
                                        m_fFeedbackLeft*fFeedback);

float fAPF_2_Out = m_LeftAPF_2.doBiQuad(fAPF_1_Out);
float fAPF_3_Out = m_LeftAPF_3.doBiQuad(fAPF_2_Out);
float fAPF_4_Out = m_LeftAPF_4.doBiQuad(fAPF_3_Out);
float fAPF_5_Out = m_LeftAPF_5.doBiQuad(fAPF_4_Out);
float fAPF_6_Out = m_LeftAPF_6.doBiQuad(fAPF_5_Out);

// for next sample period
m_fFeedbackLeft = fAPF_6_Out;

pOutputBuffer[0] = fDepth*fAPF_6_Out + (1.0 - fDepth)*pInputBuffer[0];

// use the fc and Q to calculate the Left LPF coeffs
if(m_uLFO_Phase == QUAD)
        calculateFirstOrderRightAPFCoeffs(fYqn);
else
        calculateFirstOrderRightAPFCoeffs(fYn);

// do the cascaded APFs
fAPF_1_Out = m_RightAPF_1.doBiQuad(pInputBuffer[1] +
                                   m_fFeedbackRight*fFeedback);
fAPF_2_Out = m_RightAPF_2.doBiQuad(fAPF_1_Out);
fAPF_3_Out = m_RightAPF_3.doBiQuad(fAPF_2_Out);
fAPF_4_Out = m_RightAPF_4.doBiQuad(fAPF_3_Out);
fAPF_5_Out = m_RightAPF_5.doBiQuad(fAPF_4_Out);
fAPF_6_Out = m_RightAPF_6.doBiQuad(fAPF_5_Out);

// for next sample period
m_fFeedbackRight = fAPF_6_Out;

// Mono-In, Stereo-Out (AUX Effect)
if(uNumInputChannels == 1 && uNumOutputChannels == 2)
        pOutputBuffer[1] = pOutputBuffer[0];
```

```
        // Stereo-In, Stereo-Out (INSERT Effect)
        if(uNumInputChannels == 2 && uNumOutputChannels == 2)
                pOutputBuffer[1] = fDepth*fAPF_6_Out + (1.0 - fDepth)
                *pInputBuffer[1];
        return true;
}
```

Build and test the plug-in. Try out the different modes; turn up the intensity and depth controls for very intense effects. This phaser should be one of the finest you have heard as it is not like some digital phasers that only use one giant APF. The cascaded APFs are the way to go with this effect. The saw LFO with copious feedback sounds very much like a classic pulsating "vibe" effect.

Bibliography

Anderton, C. 1981. *Electronic Projects for Musicians*. London: Wise Publications.
Dodge, C. and Jerse, T. 1997. *Computer Music Synthesis, Composition and Performance*. Chapter 2. New York: Schirmer.
Giles, M., ed. 1980. *The Audio/Radio Handbook*. Santa Clara, CA: National Semiconductor Corp.
Roads, C. 1996. *The Computer Music Tutorial*. Chapter 2. Cambridge, MA: The MIT Press.

References

musicdsp.org. "Envelope follower with different attack and release." Accessed 2008, http://www.musicdsp.org/archive.php?classid=0#205.
Orfanidis, S. 1996. *Introduction to Signal Processing*, Chapter 8. Englewood Cliffs, NJ: Prentice-Hall.
Zöler, U. 2011. *DAFX—Digital Audio Effects*, Chapter 4. West Sussex, U.K.: John Wiley & Sons.

CHAPTER 13
Dynamics Processing

Dynamics processors are designed to automatically control the amplitude, or gain, of an audio signal and consist of two families: compressors and expanders. Technically speaking, compressors and expanders both change the gain of a signal after its level rises *above* a predetermined threshold value. A compressor reduces the gain of the signal once it goes over the threshold. An expander raises the gain of a signal after it crosses above the threshold. With the exception of noise reduction systems, true expanders are rare since they can easily cause instabilities, runaway gain, and distortion. What is normally called an "expander" today is technically a *downward expander*. A downward expander reduces the gain of a signal after it drops *below* the threshold. We will be designing a downward expander but will use the common lingo and refer to it simply as an expander. Both compressors and expanders require the user to decide on the threshold of operation as well as a *ratio* value that tells the device how much gain reduction to implement.

Figure 13.1 shows the input/output transfer functions for the compressor family. The line marked 1:1 is unity gain. A given input x(dB) results in the same output y(dB). Above the threshold, the output level is compressed according to the ratio. For example, above the threshold on the 2:1 line, every increase in 1 dB at the input results in an increase of only 0.5 dB at the output. On the 4:1 line, each 1 dB input increase results in a 0.25 dB output increase. On the ∞:1 line, increasing the input amplitude beyond the threshold results in no output increase at all, thus y(dB) is limited to the threshold value. This version of the device is called a *limiter* and represents the most extreme form of compression.

Figure 13.2 shows the input/output transfer functions for the expander family. The ratios are reversed. You can see that as the signal falls below the threshold, it is attenuated by the ratio amount. For example, below the threshold on the 1:2 line, every decrease in 1dB at the input results in a decrease of 2 dB at the output. On the 1:∞ line, when the input falls below the threshold it receives infinite attenuation, that is, it is muted. This version of the device is called a *gate* (or noise gate) and represents the most extreme form of downward expansion. Perhaps the most important aspect of both families is that their gain reduction curves are linear for logarithmic axes. These devices operate in the *log domain*. The two families of dynamics processors yield four combinations: compression and limiting, and expansion and gating. We will implement all four of them.

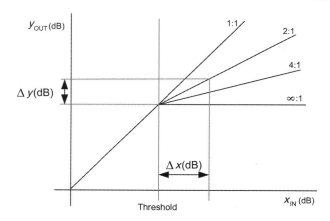

Figure 13.1: The generalized input/output transfer curves for a compressor at various reduction ratios. The ratio is in the form $\Delta x : \Delta y$.

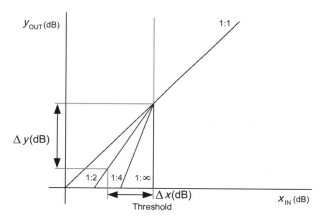

Figure 13.2: The generalized input/output transfer curves for a downward expander at various reduction ratios. The ratio is in the form $\Delta x : \Delta y$.

Figure 13.3 shows the feed-forward and feedback topologies for dynamics processors. There is some debate as to which is best; we'll stay out of that argument and design the feed-forward version first. Then, you can design a feedback version for yourself. Both designs include a detector/computer and a *digitally controlled amplifier/attenuator* (DCA). The detector analyzes the input signal and computes a new gain value to be applied to the signal. The detector/computer block lies in the *side-chain* of the circuit.

Because the compressor and downward expander are both gain reduction devices, an additional stage of make-up gain is required. In some designs, the make-up gain is calculated in the side-chain while in others it sits outside of the DCA. Both accomplish the same task. We are going to add another gain stage at the input (similar to the mod filter in Chapter 12) to drive the detector; this pre-gain block is useful when the source material is very low in amplitude, making it difficult to trigger the detector. Figure 13.4 shows the block diagram of the first dynamic processor plug-in.

Figure 13.4 labels the input and control signals as follows:

- $x(n)$, $y(n)$ = input/output
- $d(n)$ = detector output
- $G(n)$ = gain value for current sample
- d_o = threshold in dB
- ρ = ratio

The CEnvelopeDetector object you used in the last chapter has a log output mode so we can combine the detector and log conversion blocks together, as shown in the dotted box in Figure 13.4. These feed the gain calculation block, which uses the threshold (in dB) and the ratio to set the gain for a given detector output. Orfanidis (1996), Zöler (2011), and Reiss (2011) have given equations that describe the gain reduction calculation. The Zöler equations (Equations 13.1 and 13.2) seem to be the most concise. These equations require additional slope variables (CS and ES) that are calculated from the ratio. In the compressor, CS varies

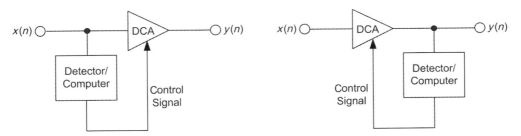

Figure 13.3: The feed-forward and feedback topologies for a dynamics processor. A detector calculates the peak or RMS value of the signal to create a control signal. The DCA attenuates the input signal according to its control value.

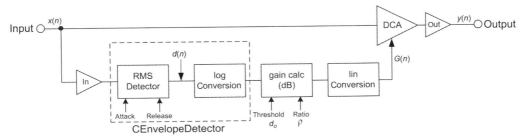

Figure 13.4: A generalized feed-forward topology for the compressor/downward expander families.

between 0 and 1 as the ratio ρ varies from 1 to ∞. In the downward expander, *ES* varies between 0 and −∞ as the ratio ρ varies from ∞ to 1.

Compressor gain (dB):

$$G(n) = f(d(n))$$

where

$$CS = \begin{cases} 1 - (1/\rho) & \text{for compression} \\ 1.0 & \text{for limiting} \end{cases}$$

$$f = \begin{cases} CS \cdot (d_0 - d(n)) & \text{if } d(n) \geq d_0 \\ 0.0 & \text{if } d(n) < d_0 \end{cases}$$

(13.1)

Downward expander gain (dB):

$$G(n) = f(d(n))$$

where

$$ES = \begin{cases} (1/\rho) - 1 & \text{for downward expansion} \\ 0.0 & \text{for gating} \end{cases}$$

$$f = \begin{cases} ES \cdot (d_0 - d(n)) & \text{if } d(n) < d_0 \\ 0.0 & \text{if } d(n) \geq d_0 \end{cases}$$

(13.2)

The point in the input/output plot where the signal hits the threshold, thereby engaging the device, is called the *knee* of the compression curve. With a hard knee the signal is either above the threshold or below it and the device kicks in and out accordingly. The soft-knee compressor allows the gain control to change more slowly over a width of dB around the threshold called the *knee width* (*W*). In this way, the device moves more smoothly into and out of compression or downward expansion. Figure 13.5 compares hard- and soft-knee curves.

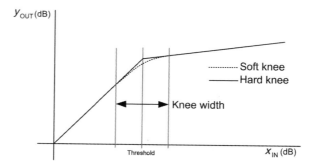

Figure 13.5: Hard and soft-knee compression curves.

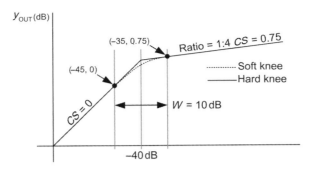

Figure 13.6: Setting up the endpoints for interpolating the CS value for a soft-knee compressor.

There are several approaches to generating the soft-knee portion of the curve when calculating the gain reduction value. Curve fitting by interpolating across the knee width is popular. There are a variety of different ways to implement the curve fitting. We already have a Lagrange interpolator in the software, so we can use this to fit a second-order polynomial curve to the knee. This is done by altering the *CS* or *ES* values. The two values above and below the threshold are the x values, while the two values of the slope variable at the end points are the y components. When the detector signal is in the knee zone around the threshold, we use its value to calculate a new slope. Figure 13.6 shows an example. The knee width (W) is 10 dB with a −40 dB threshold. For a compression ratio of 1:4, *CS* is 0.75. Therefore, the soft-knee *CS* will vary between 0 at −45 dB to 0.75 at −35 dB. The Lagrange interpolation will find the *CS* value for a given input detection value across this range.

13.1 Design a Compressor/Limiter Plug-In

We are going to build a complete dynamics processor with the following features and functions:

- Compressor
- Limiter
- Expander
- Gate
- Analog or digital time constant
- Attack, release controls for detector
- Threshold, ratio controls for gain computer
- Knee-width control: 0 = hard knee processing
- Gain reduction meter to view the reduction

You'll notice a new feature in this plug-in: the ability to monitor a signal and display it on a meter. We will start with the compressor/limiter function and build up from there.

13.1.1 Project: DynamicsProcessor

Create a new project called "Dynamics Processor." You don't need any stock objects or other advanced features.

13.1.2 DynamicsProcessor: GUI

You need to add sliders and radio button controls to implement the eight control blocks required for the plug-in. Table 13.1 shows the graphical user interface (GUI) slider properties for the design.

Table 13.1: The GUI controls for the dynamics processor plug-in.

Slider Property	Value	Slider Property	Value
Control Name	Det Gain	Control Name	Threshold
Units	dB	Units	dB
Variable Type	float	Variable Type	float
Variable Name	m_fInputGain_dB	Variable Name	m_fThreshold
Low Limit	−12	Low Limit	−60
High Limit	20	High Limit	0
Initial Value	0	Initial Value	0

Slider Property	Value	Slider Property	Value
Control Name	Attack Time	Control Name	Release Time
Units	mSec	Units	mSec
Variable Type	float	Variable Type	float
Variable Name	m_fAttack_mSec	Variable Name	m_fRelease_mSec
Low Limit	1	Low Limit	20
High Limit	300	High Limit	5000
Initial Value	20	Initial Value	1000

Slider Property	Value	Slider Property	Value
Control Name	Ratio	Control Name	Output Gain
Units		Units	dB
Variable Type	float	Variable Type	float
Variable Name	m_fRatio	Variable Name	m_fOutputGain_dB
Low Limit	1	Low Limit	0
High Limit	20	High Limit	20
Initial Value	1	Initial Value	0

Slider Property	Value
Control Name	Knee Width
Units	dB
Variable Type	float
Variable Name	m_fKneeWidth
Low Limit	0
High Limit	20
Initial Value	0

Button Property	Value
Control Name	Processor
Units	
Variable Type	enum
Variable Name	m_uProcessorType
Enum String	COMP, LIMIT, EXPAND, GATE

Button Property	Value
Control Name	Time Constant
Units	
Variable Type	enum
Variable Name	m_uTimeConstant
Enum String	Digital, Analog

13.1.3 DynamicsProcessor.h File

We need to add the following new functions and variables; most can be taken directly from your ModFilter project. There's not much to the basic device. You will need the following:

- An envelope detector for each channel (left, right).
- A function to calculate a compression gain value, given the detector input; this function will also implement the limiting function.

```
// Add your code here: ------------------------------------------------ //
// envelope detectors
CEnvelopeDetector m_LeftDetector;
CEnvelopeDetector m_RightDetector;

// calculate the compressor G(n) value from the Detector output
float calcCompressorGain(float fDetectorValue, float fThreshold, float fRatio,
                         float fKneeWidth, bool bLimit);
// END OF USER CODE --------------------------------------------------- //
```

13.1.4 DynamicsProcessor.cpp File

We need to implement Equation 13.1 and also need to add the code to provide the Lagrange interpolation when the soft-knee function is enabled in the calcCompressorGain() function. The gain calculations simply follow Equation 13.1. See the Lagrange interpolation function in the pluginconstants.h file for documentation. You provide it with *x* and *y* arrays for the end points along with the order (two, since we have two endpoints) and the *x* value to calculate the interpolated *y* value.

```
// calculate the compressor G(n) value from the Detector output
float CDynamicsProcessor::calcCompressorGain(float fDetectorValue, float fThreshold,
                                             float fRatio, float fKneeWidth, bool bLimit)
{
    // slope variable
    float CS = 1.0 - 1.0/fRatio; // [Eq. 13.1]

    // limiting is infinite ratio thus CS->1.0
    if(bLimit)
        CS = 1;

    // soft-knee with detection value in range?
    if(fKneeWidth > 0 && fDetectorValue > (fThreshold - fKneeWidth/2.0) &&
       fDetectorValue < fThreshold + fKneeWidth/2.0)
    {
        // setup for Lagrange
        double x[2];
        double y[2];
        x[0] = fThreshold - fKneeWidth/2.0;
        x[1] = fThreshold + fKneeWidth/2.0;
        x[1] = min(0, x[1]); // top limit is 0dBFS
        y[0] = 0;       // CS = 0 for 1:1 ratio
        y[1] = CS;      // current CS

        // interpolate & overwrite CS
        CS = lagrpol(&x[0], &y[0], 2, fDetectorValue);
    }

    // compute gain; threshold and detection values are in dB
    float yG = CS*(fThreshold - fDetectorValue); // [Eq. 13.1]

    // clamp; this allows ratios of 1:1 to still operate
    yG = min(0, yG);

    // convert back to linear
    return pow(10.0, yG/20.0);
}
```

prepareForPlay()

- Initialize the detector objects; note the boolean flags that set the analog/digital and log or linear calculation; we use log.

```
bool __stdcall CDynamicsProcessor::prepareForPlay()
{
        // Add your code here:
        // init the envelope detectors
        // set all params at once with this function; see function definition
        if(m_uTimeConstant == Digital)
        {
                m_LeftDetector.init((float)m_nSampleRate, m_fAttack_mSec, m_fRelease_mSec,
                                false, DETECT_MODE_RMS, true);
                m_RightDetector.init((float)m_nSampleRate, m_fAttack_mSec,
                                m_fRelease_mSec, false, DETECT_MODE_RMS, true);
        }
        else
        {
                m_LeftDetector.init((float)m_nSampleRate, m_fAttack_mSec, m_fRelease_mSec,
                                true, DETECT_MODE_RMS, true);
                m_RightDetector.init((float)m_nSampleRate, m_fAttack_mSec,
                                m_fRelease_mSec, true, DETECT_MODE_RMS, true);
        }

        return true;
}
```

userInterfaceChange()

- Set the new attack time on the detectors.
- Set the new release time on the detectors.
- Set the time constant mode on the detectors.
- Note: Make sure you check your variable control ID values to match your UI.

```
bool __stdcall CDynamicsProcessor::userInterfaceChange(int nControlIndex)
{
        // decode the control index, or delete the switch and use brute force calls
        switch(nControlIndex)
        {
                case 2:
                {
                        m_LeftDetector.setAttackTime(m_fAttack_mSec);
                        m_RightDetector.setAttackTime(m_fAttack_mSec);
                        break;
                }
```

```
                    case 3:
                    {
                        m_LeftDetector.setReleaseTime(m_fRelease_mSec);
                        m_RightDetector.setReleaseTime(m_fRelease_mSec);
                        break;
                    }

                    case 42:
                    {
                        if(m_uTimeConstant == Digital)
                        {
                            m_LeftDetector.setTCModeAnalog(false);
                            m_RightDetector.setTCModeAnalog(false);
                        }
                        else
                        {
                            m_LeftDetector.setTCModeAnalog(true);
                            m_RightDetector.setTCModeAnalog(true);
                        }
                    }
            }
            default:
                break;
    }

    return true;
}
```

processAudioFrame()

- Follow the block diagram to complete the operations.
- Apply input gain to detector.
- Detect input sample.
- Calculate gain.
- Apply dynamic gain reduction and static make-up gain.

```
        bool __stdcall CDynamicsProcessor::processAudioFrame(float* pInputBuffer, float*
                                                    pOutputBuffer, UINT
                                                    uNumInputChannels, UINT
                                                    uNumOutputChannels)
{
        //
        // output = input -- change this for meaningful processing
        //
        // Do LEFT (MONO) Channel; there is always at least one input, one output
        // calculate gains
        float fInputGain = pow(10.0, m_fInputGain_dB/20.0);
        float fOutputGain = pow(10.0, m_fOutputGain_dB/20.0);

        // detect left channel
        float fLeftDetector = m_LeftDetector.detect(pInputBuffer[0]);
```

```
            // gain calc
            float fGn = 1.0;

            // branch
            if(m_uProcessorType == COMP)
                    fGn = calcCompressorGain(fLeftDetector, m_fThreshold, m_fRatio,
                                            m_fKneeWidth, false);
            else if(m_uProcessorType == LIMIT)
                    fGn = calcCompressorGain(fLeftDetector, m_fThreshold, m_fRatio,
                                            m_fKneeWidth, true);

            // form left output and apply make up gain
            pOutputBuffer[0] = fGn*pInputBuffer[0]*fOutputGain;

            // Mono-In, Stereo-Out (AUX Effect)
            if(uNumInputChannels == 1 && uNumOutputChannels == 2)
                    pOutputBuffer[1] = pOutputBuffer[0];

            // Stereo-In, Stereo-Out (INSERT Effect)
            if(uNumInputChannels == 2 && uNumOutputChannels == 2)
            {
                    // detect right channel
                    float fRightDetector = m_RightDetector.detect(pInputBuffer[1]);

                    // gain calc
                    float fGn = 1.0;

                    // branch
                    if(m_uProcessorType == COMP)
                            fGn = calcCompressorGain(fRightDetector, m_fThreshold, m_fRatio,
                                                    m_fKneeWidth, false);
                    else if(m_uProcessorType == LIMIT)
                            fGn = calcCompressorGain(fRightDetector, m_fThreshold, m_fRatio,
                                                    m_fKneeWidth, true);

                    // form right output and apply make up gain
                    pOutputBuffer[1] = fGn*pInputBuffer[1]*fOutputGain;
            }

        return true;
}
```

Build and test the plug-in. Try all the controls. Test the analog and digital time constants. Try extreme settings and listen for artifacts called "pumping and breathing." The soft-knee function can sometimes be subtle, so you may have to try a variety of different audio files to hear it. It would be nice to have a visual way of seeing the gain reduction, so we will use an advanced RackAFX feature that will enable a metering device. To make it even better, we can invert the meter so that gain reduction results in a downward meter view—common in dynamics processor devices, both analog and digital.

In RackAFX, right click on one of the LED meters in the meter section of the main UI, shown in Figure 13.7. The meters accept a value between 0.0 and 1.0 and display accordingly. This is perfect for us since our gain values are also fluctuating on that same range. You only need to declare a floating point variable to set the meter; RackAFX does everything else. A dialog box will pop up that allows you to name the metering variable you are going to attach to the meter as well as customize its look and behavior. You will be adding two meters, one for the left and one for the right channel gain reduction.

GRL stands for gain reduction for the left and GRR stands for gain reduction for the right channel. Set up the two meters according to Table 13.2.

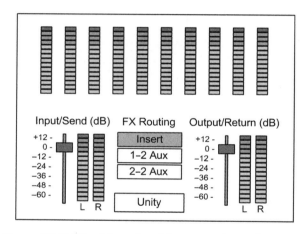

Figure 13.7: The 10 assignable meters in RackAFX.

Table 13.2: Meter settings for the dynamics processor.

Meter Property	Value	Meter Property	Value
Meter Blurb	GRL	Meter Blurb	GRR
Invert Meter	true	Invert Meter	true
Enable Meter	true	Enable Meter	true
Meter Cal	linear	Meter Cal	linear
Detect Mode	PEAK	Detect Mode	PEAK
Meter Variable	m_fGRMeterValue_L	Meter Variable	m_fGRMeterValue_R
Attack Time	0	Attack Time	0
Release Time	0	Release Time	0

The gain reduction value we will be metering is linear, from 0 to 1.0. The gain reduction value will have its own attack and release times according to our settings, so we want to monitor the PEAK variable type with no attack or release time.

All you need to do is make sure the variables m_fGRMeterValue_L and m_fGRMeterValue_R are reset to the current gain reduction value on each sample interval, and the meter will track it. The meter value needs to go up as the gain reduction goes down to visualize it properly. To set the meter, add one line of code to your processAudioFrame() function for each of the left and right channels.

13.1.5 DynamicsProcessor.cpp File

processAudioFrame()

- Add the metering variables to track the gain reduction values.

```
bool __stdcall CDynamicsProcessor::processAudioFrame(float* pInputBuffer, float*
                              pOutputBuffer, UINT uNumInputChannels,
                              UINT uNumOutputChannels)
{
    // Do LEFT (MONO) Channel; there is always at least one input/one output

    <SNIP SNIP SNIP>

    // form left output and apply make up gain
    pOutputBuffer[0] = fGn*pInputBuffer[0]*fOutputGain;

    // set the meter to track 1-gain value
    m_fGRMeterValue_L = 1.0 - fGn;

    <SNIP SNIP SNIP>

            // form right output and apply make up gain
            pOutputBuffer[1] = fGn*pInputBuffer[1]*fOutputGain;

            // set the meter to track 1-gain value
            m_fGRMeterValue_R = 1.0 - fGn;
    }
    etc...
```

Build and test the plug-in now and you'll get a visual cue about how much gain reduction is taking place. Try the analog limiter setting and adjust the attack, release, and threshold values to give smooth limiting (note the ratio slider doesn't work in limiter mode). Once you have that running properly, we'll finish off the dynamics processor by adding the downward expansion/gating capability.

13.2 Design a downward expander/gate plug-in

We'll add on to the current project. Here is what is needed:

- A function to calculate downward expander gain
- Branching in the processAudioFrame() function to implement all four dynamics processor operations

13.2.1 DynamicsProcessor.h File

Add the following function declaration for the downward expander calculation:

```
// Add your code here: ---------------------------------------------- //
// envelope detectors
CEnvelopeDetector m_LeftDetector;
CEnvelopeDetector m_RightDetector;

// calcualte the compressor G(n) value from the Detector output
float calcCompressorGain(float fDetectorValue, float fThreshold, float fRatio,
                        float fKneeWidth, bool bLimit);

// calculate the downward expander G(n) value from the Detector output
float calcDownwardExpanderGain(float fDetectorValue, float fThreshold, float
                        fRatio, float fKneeWidth, bool bGate);

// END OF USER CODE -------------------------------------------------- //
```

13.2.2 DynamicsProcessor.cpp File

Implement the downward expander function including the soft-knee operation.

```
// calculate the downward expander G(n) value from the Detector output
float CDynamicsProcessor::calcDownwardExpanderGain(float fDetectorValue, float
                                    fThreshold, float fRatio, float
                                    fKneeWidth, bool bGate)
{
    // slope variable
    float ES = 1.0/fRatio - 1; // [Eq. 13.2]

    // gating is infinite ratio; ES -> -1.0
    if(bGate)
        ES = -1;

    // soft-knee with detection value in range?
    if(fKneeWidth > 0 && fDetectorValue > (fThreshold - fKneeWidth/2.0) &&
       fDetectorValue < fThreshold + fKneeWidth/2.0)
```

```
    {
            // setup for Lagrange
            double x[2];
            double y[2];
            x[0] = fThreshold - fKneeWidth/2.0;
            x[1] = fThreshold + fKneeWidth/2.0;
            x[1] = min(0, x[1]); // top limit is 0dBFS
            y[0] = ES;  // current ES
            y[1] = 0;   // 1:1 ratio

            // interpolate the value
            ES = lagrpol(&x[0], &y[0], 2, fDetectorValue);
    }

    // compute gain; threshold and detection values are in dB
    float yG = ES*(fThreshold - fDetectorValue); // [Eq. 13.2]

    // clamp; this allows ratios of 1:1 to still operate
    yG = min(0, yG);

    // convert back to linear
    return pow(10.0, yG/20.0);
}
```

processAudioFrame()

- Similar to the compressor, add the branching code to calculate the downward expansion gain.

```
bool __stdcall CDynamicsProcessor::processAudioFrame(float* pInputBuffer, float*
                                                     pOutputBuffer,
                                                     UINT uNumInputChannels,
                                                     UINT uNumOutputChannels)
{
    // Do LEFT (MONO) Channel; there is always at least one input/one output

    <SNIP SNIP SNIP>

    // gain calc
    float fGn = 1.0;

    // branch
    if(m_uProcessorType == COMP)
         fGn = calcCompressorGain(fLeftDetector, m_fThreshold, m_fRatio,
                                  m_fKneeWidth, false);
    else if(m_uProcessorType == LIMIT)
         fGn = calcCompressorGain(fLeftDetector, m_fThreshold, m_fRatio,
                                  m_fKneeWidth, true);
    else if(m_uProcessorType == EXPAND)
         fGn = calcDownwardExpanderGain(fLeftDetector, m_fThreshold, m_fRatio,
                                        m_fKneeWidth, false);
```

```
            else if(m_uProcessorType == GATE)
                    fGn = calcDownwardExpanderGain(fLeftDetector, m_fThreshold, m_fRatio,
                                                    m_fKneeWidth, true);

            // form left output and apply make up gain
            pOutputBuffer[0] = fGn*pInputBuffer[0]*fOutputGain;

            <SNIP SNIP SNIP>

            // Stereo-In, Stereo-Out (INSERT Effect)
            if(uNumInputChannels == 2 && uNumOutputChannels == 2)
            {
                    // detect right channel
                    float fRightDetector = m_RightDetector.detect(pInputBuffer[1]);

                    // gain calc
                    float fGn = 1.0;

                    // branch
                    if(m_uProcessorType == COMP)
                            fGn = calcCompressorGain(fRightDetector, m_fThreshold, m_fRatio,
                                                    m_fKneeWidth, false);
                    else if(m_uProcessorType == LIMIT)
                            fGn = calcCompressorGain(fRightDetector, m_fThreshold, m_fRatio,
                                                    m_fKneeWidth, true);
                    else if(m_uProcessorType == EXPAND)
                            fGn = calcDownwardExpanderGain(fLeftDetector, m_fThreshold,
                                                    m_fRatio, m_fKneeWidth, false);
                    else if(m_uProcessorType == GATE)
                            fGn = calcDownwardExpanderGain(fLeftDetector, m_fThreshold,
                                                    m_fRatio, m_fKneeWidth, true);

                    // form right output and apply make up gain
                    pOutputBuffer[1] = fGn*pInputBuffer[1]*fOutputGain;
                    etc...
```

Build and test the plug-in now and try all four modes with several different input files; drums and guitars make good test files. The analog time constant along with the soft-knee settings generally provide a smoother, less obtrusive gain reduction function.

13.3 Design a Look-Ahead Compressor Plug-In

A problem with the detector circuit is that for any non-zero attack and release time settings, the detector output lags behind the input in time. Remember from Chapter 12 how the output of the detector tracks the input signal, as shown in Figure 13.8. The dark line is the detector output, which is clearly lagging behind the input.

Since the detector output lags, the gain reduction actually misses the event that needs to be attenuated. In the digital world, we can accommodate for this detector lag by using a

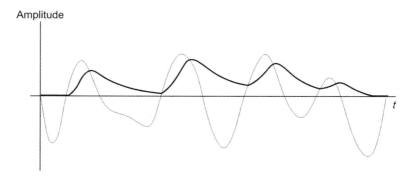

Figure 13.8: The detector output with the medium attack and medium decay times.

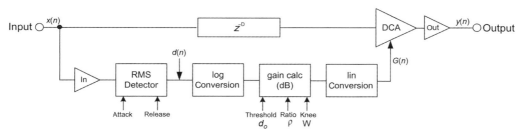

Figure 13.9: The look-ahead compressor block diagram.

look-ahead technique. We can't look ahead into the future of a signal, but we can delay the present. If we insert a delay line in the forward signal path (not the side-chain) we can make up for the detector delay so that the detector begins charging before the signal actually gets to the DCA, as shown in Figure 13.9.

You can use the CDelay object that comes as a "stock reverb object" and add it to your project now for our look-ahead pre-delay. Do this by using the Edit button and edit the project. Check the box marked "Stock Reverb Objects" and then hit OK. Your compiler will ask you if you want to reload the project, so answer yes. You will then have the reverb objects included in your project. We'll add a single new slider to control the look-ahead time. Remember when using the CDelay-based objects that you need to make sure you initialize the object to have the same maximum delay time as your slider. The amount of look-ahead delay will depend on the compressor settings. We'll make the maximum value match our maximum attack time of 300 mSec.

13.3.1 DynamicsProcessor: GUI

Add a new slider for the look-ahead delay time in mSec using Table 13.3.

Table 13.3: The look-ahead slider properties.

Slider Property	Value
Control Name	Look-Ahead
Units	mSec
Variable Type	float
Variable Name	m_fLookAheadDelay_mSec
Low Limit	0
High Limit	300
Initial Value	0

13.3.2 DynamicsProcessor.h File

Declare the delays for the look-ahead:

```
// Add your code here: ------------------------------------------------ //
// envelope detectors
CEnvelopeDetector m_LeftDetector;
CEnvelopeDetector m_RightDetector;

<SNIP SNIP SNIP>

// Delay lines for a look-ahead compressor
CDelay m_LeftDelay;
CDelay m_RightDelay;
// END OF USER CODE -------------------------------------------------- //
```

13.3.3 DynamicsProcessor.cpp File

prepareForPlay()

- Initialize the delays now that we know the sample rate.
- Set the current delay time in mSec.
- Flush the delays.

```
bool __stdcall CDynamicsProcessor::prepareForPlay()
{
    // init lookahead delays
    m_LeftDelay.init(0.3*(float)m_nSampleRate);
    m_RightDelay.init(0.3*(float)m_nSampleRate);

    // set the current value
    m_LeftDelay.setDelay_mSec(m_fLookAheadDelay_mSec);
    m_RightDelay.setDelay_mSec(m_fLookAheadDelay_mSec);

    // flush delays
    m_LeftDelay.resetDelay();
    m_RightDelay.resetDelay();

    etc…
```

userInterfaceChange()

- Set the look-ahead delay time on the delay elements.
- Note: Make sure your control IDs match your GUI.

```
bool __stdcall CDynamicsProcessor::userInterfaceChange(int nControlIndex)
{
        <SNIP SNIP SNIP>

            case 7:
            {
                    m_LeftDelay.setDelay_mSec(m_fLookAheadDelay_mSec);
                    m_RightDelay.setDelay_mSec(m_fLookAheadDelay_mSec);
                    break;
            }
        etc...
```

processAudioFrame()

- Delay the signal going to the DCA, not the detector side-chain.

```
bool __stdcall CDynamicsProcessor::processAudioFrame(float* pInputBuffer, float*
                                                      pOutputBuffer,
                                                      UINT uNumInputChannels,
                                                      UINT uNumOutputChannels)
{
        <SNIP SNIP SNIP>

        // delay the left DCA after gain calculation
        float fLookAheadOut = 0;
        m_LeftDelay.processAudio(&pInputBuffer[0], &fLookAheadOut);

        // form left output and apply make up gain
        pOutputBuffer[0] = fGn*fLookAheadOut*fOutputGain;

            <SNIP SNIP SNIP>

            // delay the right DCA after gain calculation
            float fLookAheadOut = 0;
            m_RightDelay.processAudio(&pInputBuffer[1], &fLookAheadOut);

            // form right output and apply make up gain
            pOutputBuffer[1] = fGn*fLookAheadOut*fOutputGain;

            // set the meter to track 1-gain value
            m_fGRMeterValue_R = 1.0 - fGn;
        }
        return true;
}
```

Build and test the plug-in now and try the look-ahead feature. It will take some critical listening to adjust the look-ahead value; start by matching the attack time, and then tweak it from there.

13.4 Stereo-Linking the Dynamics Processor

You can see from the block diagram and your own code that our stereo dynamics processor shares all of the controls: gains, attack, release, threshold, and ratio. However, the ultimate value of the dynamic gain factor $G(n)$ really depends on the audio content of each channel, which is fed into the independent left and right detectors. If the left and right channels have very different signals, then the resulting very different $G(n)$ values for left and right can create a confusing output. Stereo-linking a dynamics processor means that you tie together the two detector outputs, sum and average them, and then apply that one signal to the log converter. The resulting $G(n)$ value is applied to both left and right channels. Figure 13.10 shows how the right channel's gain computation has been removed since it now shares the left channel's side-chain path. In order to modify the existing processor, we need to do two things: modify the UI to add a stereo link switch and add more branching in processAudioFrame() to do the linking.

13.4.1 DynamicsProcessor: GUI

For a change, I'll use a slider control in enum mode to make a switch for the stereo link. Remember that you can use sliders to make enumerated-list variables in addition to the buttons. Use Table 13.4 to set up the slider.

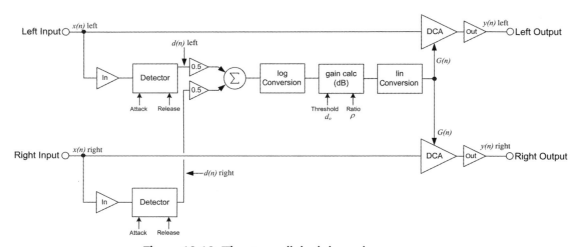

Figure 13.10: The stereo-linked dynamics processor.

NOTE: The detectors are running in log output mode. This means that we need to convert their values back to linear before summing them. After they are summed, they must be converted back to the log domain prior to being used in the gain calculation function.

Table 13.4: The stereo link slider properties.

Slider Property	Value
Control Name	Stereo Link
Variable Type	enum
Variable Name	m_uStereoLink
Enum String	ON, OFF

13.4.2 DynamicsProcessor.cpp File

There are many different ways to do this branching and you should definitely try to figure out your own way. My code is for education purposes and not very streamlined either. For the branching code, I chose to set up the stereo link and right channel detection early on.

```
bool __stdcall CDynamicsProcessor::processAudioFrame(float* pInputBuffer, float*
                                                     pOutputBuffer,
                                                     UINT uNumInputChannels,
                                                     UINT uNumOutputChannels)
{
    // calculate gains
    float fInputGain = pow(10.0, m_fInputGain_dB/20.0);
    float fOutputGain = pow(10.0, m_fOutputGain_dB/20.0);

    // setup stereo link
    float xn_L = fInputGain*pInputBuffer[0];
    float xn_R = 0.0; // for later

    // do left
    float fLeftDetector = m_LeftDetector.detect(xn_L);
    float fRightDetector = fLeftDetector; // use in case of mono file

    // check for right side; can use again in right side code
    if(uNumOutputChannels == 2)
    {
        // get the right side
        xn_R = fInputGain*pInputBuffer[1];

        // detect it
        fRightDetector = m_RightDetector.detect(xn_R);
    }

    // start with Left
    float fLinkDetector = fLeftDetector;
    float fGn = 1.0;

    if(m_uStereoLink == ON) // this works even with mono files
    {
```

```
            // detectors output log values; convert to linear to sum them
            fLinkDetector = 0.5*(pow(10.0, fLeftDetector/20.0) + pow(10.0,
                                    fRightDetector/20.0));

            // convert back to log
            fLinkDetector = 20.0*log10(fLinkDetector);
}
// branch
if(m_uProcessorType == COMP)
        fGn = calcCompressorGain(fLinkDetector, m_fThreshold, m_fRatio,
                                    m_fKneeWidth, false);
else if(m_uProcessorType == LIMIT)
        fGn = calcCompressorGain(fLinkDetector, m_fThreshold, m_fRatio,
                                    m_fKneeWidth, true);

etc… same for other branches

// delay the left DCA
float fLookAheadOut = 0;
m_LeftDelay.processAudio(&pInputBuffer[0], &fLookAheadOut);

<SNIP SNIP SNIP>

// Stereo-In, Stereo-Out (INSERT Effect)
if(uNumInputChannels == 2 && uNumOutputChannels == 2)
{
        float fGn = 1.0;

        // if not linked, overwrite the variable
        if(m_uStereoLink == OFF)
            fLinkDetector = fRightDetector;

        // branch
        if(m_uProcessorType == COMP)
            fGn = calcCompressorGain(fLinkDetector, m_fThreshold, m_fRatio,
                                        m_fKneeWidth, false);
        else if(m_uProcessorType == LIMIT)
            fGn = calcCompressorGain(fLinkDetector, m_fThreshold, m_fRatio,
                                        m_fKneeWidth, true);
        else if(m_uProcessorType == EXPAND)
            fGn = calcDownwardExpanderGain(fLinkDetector, m_fThreshold,

        etc… same for other branches

        // delay the righr DCA
        float fLookAheadOut = 0;
        m_RightDelay.processAudio(&pInputBuffer[1], &fLookAheadOut);

        etc…
```

13.5 Design a Spectral Compressor/Expander Plug-In

In this project, you will design a spectral dynamics processor. A spectral processor splits up the input signal into two or more frequency bands, then applies signal processing independently to each band. Finally, the processed filtered outputs are summed to produce the final output. In this way, you can apply signal processing to only one band of frequencies, or apply different types of processing to different bands. In this design, we will create a two-band spectral compressor/expander. We will use complementary low-pass filter (LPF) and high-pass filter (HPF) units to split the incoming audio into two bands: low-frequency (LF) and high-frequency (HF). Then, we will process each band independently and recombine the outputs as shown in Figure 13.11. You can compress, say, only the HF content to squash cymbals and other sibilance. Or you can smooth out the bass by compressing it slightly, or any combination of both that you like.

Take a look at Figure 13.8 and check out the features. The input LPF and HPF are adjusted with a single f_c control. This means their cut-off points are always overlapping. The outputs are processed through two independent dynamics processors with independent make-up gains, then the results are summed back together. This means the UI is going to contain a lot of sliders. But, RackAFX will be taking care of most of the GUI details. We need to focus on the input filter-bank first.

For the input LPF and HPF we will be using a special variant on the Butterworth filter called the Linkwitz–Riley filter (Equation 13.3). It has the same second-order roll-off as the Butterworth, but the cutoff point is −6 dB rather than −3 dB so that recombining the filtered signals results in a flat response. Remember that filters alter the phase response as well, so recombining them in parallel offers a special problem with how the phases recombine. *For the two signals to recombine properly, you must invert one of the filter outputs*. It doesn't matter whether you invert the LPF or HPF output as long as you invert one of them. This is specific to the Linkwitz-Riley filters. The filter design equations are shown from Chapter 6.

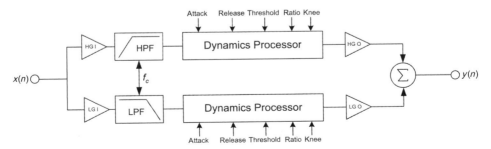

Figure 13.11: A two-band spectral compressor.

$$\begin{array}{ll}
\text{LPF} & \text{HPF} \\
\theta_c = \pi f_c / f_s & \theta_c = \pi f_c / f_s \\
\Omega_c = \pi f_c & \Omega_c = \pi f_c \\
\kappa = \dfrac{\Omega_c}{\tan(\theta_c)} & \kappa = \dfrac{\Omega_c}{\tan(\theta_c)} \\
\delta = \kappa^2 + \Omega_c^2 + 2\kappa\Omega_c & \delta = \kappa^2 + \Omega_c^2 + 2\kappa\Omega_c \\
a_0 = \dfrac{\Omega_c^2}{\delta} & a_0 = \dfrac{\kappa^2}{\delta} \\
a_1 = 2\dfrac{\Omega_c^2}{\delta} & a_1 = \dfrac{-2\kappa^2}{\delta} \\
a_2 = \dfrac{\Omega_c^2}{\delta} & a_2 = \dfrac{\kappa^2}{\delta} \\
b_1 = \dfrac{-2\kappa^2 + 2\Omega_c^2}{\delta} & b_1 = \dfrac{-2\kappa^2 + 2\Omega_c^2}{\delta} \\
b_2 = \dfrac{-2\kappa\Omega_c + \kappa^2 + \Omega_c^2}{\delta} & b_2 = \dfrac{-2\kappa\Omega_c + \kappa^2 + \Omega_c^2}{\delta}
\end{array} \quad (13.3)$$

Because these are second-order filters, we can use RackAFX's built-in bi-quad to do the filtering; all we need to do is supply the calculation of the coefficients. Thus, we're going to need to design a stereo dynamics processor with independent attack, release, threshold, ratio, and gain make-up controls. We will need a shared f_c control for the two filters. Although this plug-in shares most of the features of the previous project, for simplicity there are a few modifications:

- This design will not include the look-ahead function.
- The knee width will be a common shared control that both bands will use.

13.5.1 Project: SpectralDynamics

Create a new project named "SpectralDynamics." Since we are not implementing the look-ahead feature you don't have to add the stock objects.

13.5.2 SpectralDynamics: GUI

The GUI is basically a double version of the previous design without the look-ahead function. However, we also need to add a cutoff frequency slider to control the point where the bands

> NOTE: Unlike the last plug-in, the processor will be hard-coded in stereo link mode. We can then run the detectors in *linear* operation mode, perform the summation, and then convert the linked detector output to dB prior to calling the gain calculation functions. This only serves to simplify the processAudioFrame() method a bit.

split and a knee-width control. Tables 13.5 and 13.6 show all the controls you will need. In Table 13.5, all of the settings are identical to the previous project so only the variable names are shown.

Table 13.5: Duplicated sets of controls.

Control	Variable Name	Control	Variable Name
LF Detector gain	m_LF_DetectorGain_dB	HF Detector gain	m_HF_DetectorGain_dB
LF Threshold	m_LF_Threshold	HF Threshold	m_HF_Threshold
LF Attack Time	m_LF_Attack_mSec	HF Attack Time	m_HF_Attack_mSec
LF Release Time	m_LF_Release_mSec	HF Release Time	m_HF_Release_mSec
LF Ratio	m_LF_Ratio	HF Ratio	m_HF_Ratio
LF Make Up Gain	m_LF_MakeUpGain_dB	HF Make Up Gain	m_HF_MakeUpGain_dB

Table 13.6: The two additional sliders for cutoff frequency and knee-width.

Slider Property	Value	Slider Property	Value
Control Name	fc	Control Name	Knee Width
Units	Hz	Units	dB
Variable Type	float	Variable Type	float
Variable Name	m_fFilterBankCutoff	Variable Name	m_fKneeWidth
Low Limit	100	Low Limit	0
High Limit	8000	High Limit	20
Initial Value	1000	Initial Value	0

13.5.3 Additional Slider Controls

Note: These are duplicated sets of controls that are set up exactly like the previous project, with the exception of the variable names.

13.5.4 Spectral Dynamics Buttons

You need the same two button banks as the previous project, one for the processing type and the other for the time-constant mode.

13.5.5 Spectral Dynamics Metering

This plug-in will feature extensive monitoring using the LED meter bank to indicate the LF and HF input, gain reduction, and output levels. The gain reduction meters are inverted. You can also choose different color schemes for the meters. Other than color and inversion, the meters are set up identically to the previous project. You will need to set up six meters to connect to the following variables:

- m_fMeterLFIn
- m_fMeterHFIn

- m_fMeterLFGr
- m_fMeterHFGr
- m_fMeterLFOut
- m_fMeterHFOut

The metering is especially important for the spectral processing because of the differences in energy in the bands; you will get a clearer picture of the equalization nature of this effect. The middle two meters show the gain reduction and are colored green and red and inverted. The input and output use the Volume Unit or "VU meter" color scheme. Figure 13.12 shows the GUI after all the controls have been added.

13.5.6 SpectralDynamics.h File

Because the bulk of the work is in the UI code, you only need a few more components to add to the original dynamics processor. You need the left and right detectors and the gain calculation functions as before. You also need four bi-quads: an LPF/HPF pair for each of the left and right channels. Finally, you need one function to calculate all the bi-quad coefficients at once for a given f_c value.

```
// Add your code here: ---------------------------------------- //
// envelope detectors X2
CEnvelopeDetector m_Left_LF_Detector;
```

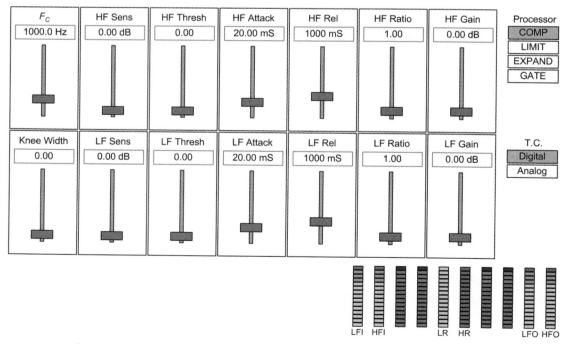

Figure 13.12: The final SpectralDynamics GUI sliders, buttons, and meters.

```
CEnvelopeDetector m_Right_LF_Detector;
CEnvelopeDetector m_Left_HF_Detector;
CEnvelopeDetector m_Right_HF_Detector;

// calculate the compressor G(n) value from the Detector output
float calcCompressorGain(float fDetectorValue, float fThreshold, float fRatio);

// calculate the downward expander G(n) value from the Detector output
float calcDownwardExpanderGain(float fDetectorValue, float fThreshold, float
                                fRatio);
// our input filter banks X2
CBiquad m_LeftLPF;
CBiquad m_LeftHPF;
CBiquad m_RightLPF;
CBiquad m_RightHPF;

// function to set all the filter cutoffs
void calculateFilterBankCoeffs(float fCutoffFreq);

// END OF USER CODE -------------------------------------------------- //
```

The detector variables are the same as your previous dynamics processor but are split into LF and HF sections or bands.

13.5.7 SpectralDynamics.cpp File

The calcCompressorGain() and calcDownwardExpanderGain() functions are *identical* to the last project you did, so you can cut and paste them. The calculateFilterBankCoeffs() function is pretty straightforward and follows the design equations for the Linkwitz–Riley filters. Since all the filters share the same cutoff, they are pretty easy to deal with.

```
// all filters have same cutoff frequency; only need LPF and HPF calcs
void CSpectralDynamics::calculateFilterBankCoeffs(float fCutoffFreq)
{
        // Shared Factors:
        float omega_c = pi*fCutoffFreq;
        float theta_c = pi*fCutoffFreq/(float)m_nSampleRate;

        float k = omega_c/tan(theta_c);
        float k_squared = k*k;

        float omega_c_squared = omega_c*omega_c;

        float fDenominator = k_squared + omega_c_squared + 2.0*k*omega_c;
```

```
        float fb1_Num = -2.0*k_squared + 2.0*omega_c_squared;
        float fb2_Num = -2.0*k*omega_c + k_squared + omega_c_squared;

        // the LPF coeffs
        float a0 = omega_c_squared/fDenominator;
        float a1 = 2.0*omega_c_squared/fDenominator;
        float a2 = a0;
        float b1 = fb1_Num/fDenominator;
        float b2 = fb2_Num/fDenominator;

        // set the LPFs
        m_LeftLPF.m_f_a0 = a0;
        m_LeftLPF.m_f_a1 = a1;
        m_LeftLPF.m_f_a2 = a2;
        m_LeftLPF.m_f_b1 = b1;
        m_LeftLPF.m_f_b2 = b2;

        // right
        m_RightLPF.m_f_a0 = a0;
        m_RightLPF.m_f_a1 = a1;
        m_RightLPF.m_f_a2 = a2;
        m_RightLPF.m_f_b1 = b1;
        m_RightLPF.m_f_b2 = b2;

        // the HPF coeffs
        a0 = k_squared/fDenominator;
        a1 = -2.0*k_squared/fDenominator;
        a2 = a0;
        b1 = fb1_Num/fDenominator;
        b2 = fb2_Num/fDenominator;

        // set the HPFs
        m_LeftHPF.m_f_a0 = a0;
        m_LeftHPF.m_f_a1 = a1;
        m_LeftHPF.m_f_a2 = a2;
        m_LeftHPF.m_f_b1 = b1;
        m_LeftHPF.m_f_b2 = b2;

        // right
        m_RightHPF.m_f_a0 = a0;
        m_RightHPF.m_f_a1 = a1;
        m_RightHPF.m_f_a2 = a2;
        m_RightHPF.m_f_b1 = b1;
        m_RightHPF.m_f_b2 = b2;
}
```

prepareForPlay()

- Flush delays in all filters.
- Calculate the initial setting of the filters.
- Init the detectors—identical to the previous project except that there are more of them.

```cpp
bool __stdcall CSpectralDynamics::prepareForPlay()
{
    // Add your code here:
    // Flush the filters
    m_LeftLPF.flushDelays();
    m_LeftHPF.flushDelays();
    m_RightLPF.flushDelays();
    m_RightHPF.flushDelays();

    // calculate the Coeffs all at once!
    calculateFilterBankCoeffs(m_fFilterBankCutoff);

    // init detectors
    // set all params at once with this function; false = Digital Time Constant NOT
    // NOTE: Setting detector for linear operaton so we can sum the results;
    //       we have to convert back to log manually
    if(m_uTimeConstant == Digital)
    {
        m_Left_LF_Detector.init((float)m_nSampleRate, m_LF_Attack_mSec,
                                m_LF_Release_mSec, false, DETECT_MODE_RMS,
                                false);

        m_Right_LF_Detector.init((float)m_nSampleRate, m_LF_Attack_mSec,
                                m_LF_Release_mSec, false, DETECT_MODE_RMS,
                                false);

        m_Left_HF_Detector.init((float)m_nSampleRate, m_HF_Attack_mSec,
                                m_HF_Release_mSec, false, DETECT_MODE_RMS,
                                false);

        m_Right_HF_Detector.init((float)m_nSampleRate, m_HF_Attack_mSec,
                                m_HF_Release_mSec, false, DETECT_MODE_RMS,
                                false);
    }
    else
    {
        m_Left_LF_Detector.init((float)m_nSampleRate, m_LF_Attack_mSec,
                                m_LF_Release_mSec, true, DETECT_MODE_RMS,
                                false);

        m_Right_LF_Detector.init((float)m_nSampleRate, m_LF_Attack_mSec,
                                m_LF_Release_mSec, true, DETECT_MODE_RMS,
                                false);

        m_Left_HF_Detector.init((float)m_nSampleRate, m_HF_Attack_mSec,
                                m_HF_Release_mSec, true, DETECT_MODE_RMS,
                                false);

        m_Right_HF_Detector.init((float)m_nSampleRate, m_HF_Attack_mSec,
                                m_HF_Release_mSec, true, DETECT_MODE_RMS,
                                false);
    }
    return true;
}
```

userInterfaceChange()

- Decode the nControlIndex and apply changes to filter and detectors.
- Basically the same as the last project except the addition of the filter and the fact that there are four times as many detectors.

```
bool __stdcall CSpectralDynamics::userInterfaceChange(int nControlIndex)
{
        // decode the control index, or delete the switch and use brute force calls
        switch(nControlIndex)
        {
                // filter cutoff
                case 0:
                {
                        calculateFilterBankCoeffs(m_fFilterBankCutoff);
                        break;
                }

                // HF Attack
                case 3:
                {
                        m_Left_HF_Detector.setAttackTime(m_HF_Attack_mSec);
                        m_Right_HF_Detector.setAttackTime(m_HF_Attack_mSec);
                        break;
                }

                // LF Attack
                case 13:
                {
                        m_Left_LF_Detector.setAttackTime(m_LF_Attack_mSec);
                        m_Right_LF_Detector.setAttackTime(m_LF_Attack_mSec);
                        break;
                }

                // HF Release
                case 4:
                {
                        m_Left_HF_Detector.setReleaseTime(m_HF_Release_mSec);
                        m_Right_HF_Detector.setReleaseTime(m_HF_Release_mSec);
                        break;
                }

                // LF Release
                case 14:
                {
                        m_Left_LF_Detector.setReleaseTime(m_LF_Release_mSec);
                        m_Right_LF_Detector.setReleaseTime(m_LF_Release_mSec);
                        break;
                }
```

```
            // Time Constant
            case 42:
            {
                if(m_uTimeConstant == Digital)
                {
                    m_Left_HF_Detector.setTCModeAnalog(false);
                    m_Right_LF_Detector.setTCModeAnalog(false);
                    m_Right_HF_Detector.setTCModeAnalog(false);
                    m_Right_LF_Detector.setTCModeAnalog(false);
                }
                else
                {
                    m_Left_HF_Detector.setTCModeAnalog(true);
                    m_Right_LF_Detector.setTCModeAnalog(true);
                    m_Right_HF_Detector.setTCModeAnalog(true);
                    m_Right_LF_Detector.setTCModeAnalog(true);
                }
            }
            default:
                break;
    }

    return true;
}
```

processAudioFrame()

- Calculate input/output gain values from their dB UI versions.
- Filter the input into LF and HF components; *invert one filter's output*.
- Detect the LF and HF components separately.
- Calculate the dynamic gain factors $G(n)$ for LF and HF components.
- Implement the gain reduction for each band.
- Recombine the outputs of each band.

```
bool __stdcall CSpectralDynamics::processAudioFrame(float* pInputBuffer, float*
                                                    pOutputBuffer, UINT
                                                    uNumInputChannels, UINT
                                                    uNumOutputChannels)
{
    // Do LEFT (MONO) Channel: there is always at least one input/one output
    float fLeftInput = pInputBuffer[0];
    float fRightInput = uNumInputChannels == 1 ? fLeftInput : pInputBuffer[1];

    // setup the input and output gains
    float fLFGain = pow(10.0, m_LF_DetectorGain_dB/20.0);
    float fHFGain = pow(10.0, m_HF_DetectorGain_dB/20.0);

    float fLFOutputGain = pow(10.0, m_LF_MakeUpGain_dB/20.0);
    float fHFOutputGain = pow(10.0, m_HF_MakeUpGain_dB/20.0);
```

```
// split the signal into m_Left LF and HF parts
float fLeft_LF_Out = m_LeftLPF.doBiQuad(fLeftInput*fLFGain);
float fLeft_HF_Out = m_LeftHPF.doBiQuad(fLeftInput*fHFGain);

// invert ONE of the outputs for proper recombination
fLeft_HF_Out *= -1.0;

// send these to the detectors: NOTE OUTPUTS ARE LINEAR
float fLeft_LF_Detector = m_Left_LF_Detector.detect(fLeft_LF_Out);
float fLeft_HF_Detector = m_Left_HF_Detector.detect(fLeft_HF_Out);

// split the signal into m_Left LF and HF parts
float fRight_LF_Out = m_RightLPF.doBiQuad(fRightInput*fLFGain);
float fRight_HF_Out = m_RightHPF.doBiQuad(fRightInput*fHFGain);

// invert ONE of the outputs for proper recombination
fRight_HF_Out *= -1.0;

// send these to the detectors: NOTE OUTPUTS ARE LINEAR
float fRight_LF_Detector = m_Right_LF_Detector.detect(fRight_LF_Out);
float fRight_HF_Detector = m_Right_HF_Detector.detect(fRight_HF_Out);

// This is the stereo linking of the detector paths;
// The detectors were set for linear operation so we could sum them
// they must be converted back to dB before use in the gain calculation
float LFDetectorSum = 0.5*(fRight_LF_Detector, fLeft_LF_Detector);
float HFDetectorSum = 0.5*(fRight_HF_Detector, fLeft_HF_Detector);

// convert back to dB
LFDetectorSum = 20.0*log10(LFDetectorSum);
HFDetectorSum = 20.0*log10(HFDetectorSum);

// sum for input metering
m_fMeterLFIn = 0.5*(fLeft_LF_Out + fRight_LF_Out);
m_fMeterHFIn = 0.5*(fLeft_HF_Out + fRight_HF_Out);

// calculate the gain factors
float fGn = 1.0;

// --- LF BAND ---------------
// branch: all are LF: detect, thresh, ratio
if(m_uProcessorType == COMP)
        fGn = calcCompressorGain(LFDetectorSum, m_LF_Threshold, m_LF_Ratio,
                                m_fKneeWidth, false);
else if(m_uProcessorType == LIMIT)
        fGn = calcCompressorGain(LFDetectorSum, m_LF_Threshold, m_LF_Ratio,
                                m_fKneeWidth, true);
else if(m_uProcessorType == EXPAND)
        fGn = calcDownwardExpanderGain(LFDetectorSum, m_LF_Threshold,
                                m_LF_Ratio, m_fKneeWidth, false);
```

```cpp
        else if(m_uProcessorType == GATE)
                fGn = calcDownwardExpanderGain(LFDetectorSum, m_LF_Threshold,
                                        m_LF_Ratio, m_fKneeWidth, true);

        // create left and right LF outputs
        float fLFOutputL = fGn*fLeft_LF_Out*fLFOutputGain;
        float fLFOutputR = fGn*fRight_LF_Out*fLFOutputGain;

        // gain reduction meter
        m_fMeterLFGr = 1.0 - fGn;

        // --- HF BAND ---------------
        // branch: all are HF: detect, thresh, ratio
        if(m_uProcessorType == COMP)
                fGn = calcCompressorGain(HFDetectorSum, m_HF_Threshold, m_HF_Ratio,
                                        m_fKneeWidth, false);
        else if(m_uProcessorType == LIMIT)
                fGn = calcCompressorGain(HFDetectorSum, m_HF_Threshold, m_HF_Ratio,
                                        m_fKneeWidth, true);
        else if(m_uProcessorType == EXPAND)
                fGn = calcDownwardExpanderGain(HFDetectorSum, m_HF_Threshold,
                                        m_HF_Ratio, m_fKneeWidth, false);
        else if(m_uProcessorType == GATE)
                fGn = calcDownwardExpanderGain(HFDetectorSum, m_HF_Threshold,
                                        m_HF_Ratio, m_fKneeWidth, true);

        // create left and right HF outputs
        float fHFOutputL = fGn*fLeft_HF_Out*fHFOutputGain;
        float fHFOutputR = fGn*fRight_HF_Out*fHFOutputGain;

        // meter output
        m_fMeterLFOut = 0.5*(fLFOutputL + fLFOutputR);
        m_fMeterHFOut = 0.5*(fHFOutputL + fHFOutputR);

        // meter GR
        m_fMeterHFGr = 1.0 - fGn;

        // sum outputs
        float fLeftOutput  = fLFOutputL + fHFOutputL;
        float fRightOutput = fLFOutputR + fHFOutputR;

        // write the outputs
        pOutputBuffer[0] = fLeftOutput;

        // Mono-In, Stereo-Out (AUX Effect)
        if(uNumInputChannels == 1 && uNumOutputChannels == 2)
                pOutputBuffer[1] = fRightOutput;

        // Stereo-In, Stereo-Out (INSERT Effect)
        if(uNumInputChannels == 2 && uNumOutputChannels == 2)
                pOutputBuffer[1] = fRightOutput;

}
```

Build and test the plug-in. This is a complex device with many interactions. It can sound a lot like an equalizer because it is a very special case of one. Be careful with gain and threshold settings.

Ideas for future projects might include:

- Add separate left, right, HF, and LF meters instead of summing them.
- Add the look-ahead function.
- Split the input into three or more bands (tricky—make a band-pass filter by cascading two Linkwitz–Riley filters, and be careful how you sum the band edges) and run dynamics processing on each band.
- Make a feed-back topology compressor.
- Add parallel compression—a dry path that mixes with the effected signal in a user-chosen ratio.
- Try alternate side-chain configurations (next).

13.6 Alternate Side-Chain Configurations

As Reiss points out, there are several different configurations for the contents of the side-chain. The version we've used so far is a standard linear timing placement configuration. One alternate configuration is called the biased configuration, in which the attack and release levels start at the threshold, not zero. Figure 13.13 shows this configuration. Reiss notes that this has the effect of allowing the release part of the envelope to fade out when the input drops below the threshold. If you want to try to implement this one, you will need to make a modification to the built-in CEnvelopeDetector object to set the low limit levels of the attack and release times—which depend on the linear magnitude of the signal—to match the threshold (which is in dB).

The other two configurations involve the location of the attack and release controls. In the post-gain configuration shown in Figure 13.14 the attack and release controls follow the output of the gain computer and are not intermixed with the RMS detector. The post-gain configuration can also be implemented in RackAFX. You will need to use two CEnvelopeDetectors. The first one runs in RMS mode with zero attack and zero release times. It feeds the gain calculator. The second detector follows the gain computer, runs in PEAK mode and has the user-controlled attack and release controls connected to it. Example code is available at the www.willpirkle.com.

A second variation called log-domain, shown in Figure 13.15, places the timing controls after the gain computer but before the conversion to the linear domain. It could also be implemented using an extra detector placed in the proper location in the side-chain.

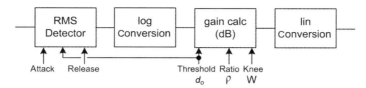

Figure 13.13: In the biased configuration, the threshold is used to set the attack and release starting levels.

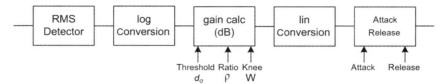

Figure 13.14: The post-gain configuration with timing controls placed after the gain computer.

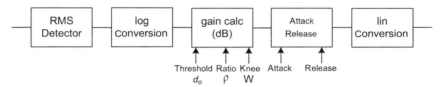

Figure 13.15: The log-domain timing configuration places the timing controls after the gain computer but before linear conversion.

Bibliography

Ballou, G. 1987. *Handbook for Sound Engineers*, pp. 850–860. Carmel, IN: Howard W. Sams & Co.
Floru, F. 1998. "Attack and Release Time Constants in RMS-Based Feedback Compressors." *Journal of the Audio Engineering Society* preprint 4703: 1–4.
Pirkle, W. C. 1995. "Applications of a New High-Performance Digitally-Controlled Audio Attenuator in Dynamics Processing." *Journal of the Audio Engineering Society* preprint 3970: 1–8.

References

Orfanidis, S. 1996. *Introduction to Signal Processing*, Chapters 10–11. Englewood Cliffs, NJ: Prentice-Hall.
Reiss, J. 2011 (October). Under the hood of a dynamic range compressor. Paper presented at the 131st Audio Engineering Society Convention, New York (Tutorials).
Zöler, U. 2011. *DAFX—Digital Audio Effects*, Chapter 4. West Sussex, U.K.: John Wiley & Sons.

CHAPTER 14
Miscellaneous Plug-Ins

There are a few effects left over that didn't have an exact chapter to reside in, so they are presented here. Interestingly, they are some of the simplest to implement but can have a massive sonic impact on the audio they process. These effects include

- Tremolo
- Auto-panning
- Ring modulation
- Wave shaping

14.1 Design a Tremolo/Panning Plug-In

The tremolo is a modulated amplitude effect that uses a low-frequency oscillator (LFO) to directly modulate the output. The LFO waveform is usually triangular or sinusoidal. If the LFO is a square wave, it produces a gapping effect, where intermittent chunks of audio are alternatively muted then restored to unity gain. An auto-panning algorithm pans the signal from left to right following an LFO. Since they are both amplitude-based effects we can combine them into one plug-in. The block diagrams are shown in Figure 14.1.

We will allow the user to switch between tremolo and auto-panning modes. For the tremolo, the LFO will be in unipolar mode, swinging between 0 and 1.0, and the LFO depth will

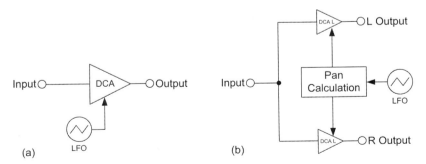

Figure 14.1: (a) A tremolo effect uses an LFO to modulate the amplitude of the input. (b) The auto-panning plug-in uses the LFO to calculate scaling values for the two digitally controlled amplifiers DCA L and DCA R to pan the signal from the left to the right.

modulate the amplitude of the input signal. The only tricky part is to decide how far below 1.0 the effect will reduce gain. For the tremolo, we define the depth as follows:

- Depth = 100%, the gain will swing from 0.0 to 1.0
- Depth = 75%, the gain will swing from 0.25 to 1.0
- Depth = 50%, the gain will swing from 0.5 to 1.0

For auto-panning, a depth of 0 yields no effect while 100% pans the signal through the full left–right stereo field. The panning is calculated using the constant power pan rule (Equation 14.1). In fact, any plug-in you design that has a pan control on it needs to use the same calculation. Instead of linearly adjusting the left and right gains, they follow the curves of the first quarter cycle of a sin/cos pair. This way, they overlap at the 0.707 points. It should be noted there are other panning schemes, but this one seems to be the most common. The LFO must be in bipolar mode for this to work easily.

$$p = \frac{\pi}{4}(\text{LFO} + 1)$$
$$\text{Gain}_L = \cos(p)$$
$$\text{Gain}_R = \sin(p)$$
$$\text{LFO is bipolar}$$

(14.1)

14.1.1 Project: TremoloPanner

Create the project and name it "TremoloPanner." There are no other objects or options to add.

14.1.2 TremoloPanner: GUI

We need rate and depth controls for the LFO and buttons to change the LFO type and the operational mode of the plug-in shown in Table 14.1.

TremoloPanner.h File

We need to add a wave table object and two gain calculation functions, one for the tremolo and one for the panner; notice the pass-by-pointer method used to return the left and right gain values for the panner.

```
// Add your code here: ----------------------------------------- //
// a single LFO
CWaveTable m_LFO;

// a function to calculate the amplitude based on LFO
float calculateGainFactor(float fLFOSample);

void calculatePannerGainFactor(float fLFOSample, float* pLeftVolume,
                               float* pRightVolume);
// END OF USER CODE ------------------------------------------- //
```

Table 14.1: TremoloPanner graphical user interface (GUI) controls.

Slider Property	Value	Slider Property	Value
Control Name	Mod Rate	Control Name	Mod Depth
Units	Hz	Units	%
Variable Type	float	Variable Type	float
Variable Name	m_ModRate	Variable Name	m_fModDepth
Low Limit	0.02	Low Limit	0
High Limit	25	High Limit	100
Initial Value	5	Initial Value	50

Button Property	Value	Button Property	Value
Control Name	LFO	Control Name	Mode
Units		Units	
Variable Type	enum	Variable Type	enum
Variable Name	m_uLFO_Waveform	Variable Name	m_uMode
Enum String	sine, saw, tri, square	Enum String	Tremolo, Panner

TremoloPanner.cpp File

The tremolo gain calculation will:

- Multiply the LFO sample by depth/100.0.
- Add the value 1.0-depth/100.0 to the result.

That provides the mapping we need for the effect.

```
float CTremoloPanner::calculateGainFactor(float fLFOSample)
{
        // first multiply the value by depth/100
        float fOutput = fLFOSample*(m_fModDepth/100.0);

        // then add the value (1 - m_fModDepth/100.0)
        fOutput += 1 - m_fModDepth/100.0;

        return fOutput;
}

void CTremoloPanner::calculatePannerGainFactor(float fLFOSample,
                                               float* pLeftVolume, float* pRightVolume)
{
        // calc sin/cos quadrant location
        float fPan = ((m_fModDepth/100.0)*fLFOSample + 1)*pi/4.0;

        // equal power calculation
        *pLeftVolume = cos(fPan);
        *pRightVolume = sin(fPan);
}
```

prepareForPlay()

- Initialize the LFO.
- Forward the prepareForPlay() method to the LFO.

```
bool __stdcall CTremoloPanner::prepareForPlay()
{
    // Add your code here:
    m_LFO.m_fFrequency_Hz = m_fModRate;
    m_LFO.m_uPolarity = m_uMode == Panner ? 0 : 1; // 0 = bipolar, 1 = unipolar
    m_LFO.m_uTableMode = 0; // normal, no band limiting
    m_LFO.m_uOscType = m_uLFO_Waveform;
    m_LFO.setSampleRate(m_nSampleRate); // really important!

    // the LFO prepareForPlay() calls reset() and cookFrequency()
    m_LFO.prepareForPlay();

    return true;
}
```

userInterfaceChange()

- Decode the control index and update variables as needed.
- Note: Make sure you check your nControlIndex values to match your GUI.

```
bool __stdcall CTremoloPanner::userInterfaceChange(int nControlIndex)
{
    // decode the control index, or delete the switch and use brute force calls
    switch(nControlIndex)
    {
        case 0:
        {
            m_LFO.m_fFrequency_Hz = m_fModRate;
            m_LFO.cookFrequency();
            break;
        }

        // 41: LFO Waveform change
        case 41:
        {
            m_LFO.m_uOscType = m_uLFO_Waveform;
            break;
        }

        case 42:
        {
            if(m_uMode == Panner)
                m_LFO.m_uPolarity = 0; // 0 = bipolar, 1 = unipolar
            else // is tremolo
                m_LFO.m_uPolarity = 1; // 0 = bipolar, 1 = unipolar
            break;
        }
    }
}
```

processAudioFrame()

- Generate the LFO value.
- Calculate the channel gain values according to the mode.

In processAudioFrame() we need to generate a new LFO value, calculate the new gain factor, then implement the gain function.

```
bool __stdcall CTremoloPanner::processAudioFrame(float* pInputBuffer,
                                                 float* pOutputBuffer,
                                                 UINT uNumInputChannels,
                                                 UINT uNumOutputChannels)
{
        // Do LEFT (MONO) Channel; there is always at least one input/one output
        float fYn = 0; // normal output
        float fYqn = 0; // quad phase output

        // call the LFO function; we only need first output
        m_LFO.doOscillate(&fYn, &fYqn);

        // setup necessary variables
        float fGnL = 1.0;
        float fGnR = 1.0;
        float fMonoIn = 0.0;

        // branch
        if(m_uMode == Tremolo)
                fGnL = calculateTremoloGainFactor(fYn);
        else // panner sums inputs
        {
                if(uNumInputChannels == 1)
                        fMonoIn = pInputBuffer[0];
                else
                        fMonoIn = 0.5*(pInputBuffer[0] + pInputBuffer[1]);

                calculatePannerGainFactor(fYn, &fGnL, &fGnR);
        }

        // do MONO (Left) channel
        pOutputBuffer[0] = pInputBuffer[0]*fGnL;

        // Mono-In, Stereo-Out (AUX Effect)
        if(uNumInputChannels == 1 && uNumOutputChannels == 2)
                pOutputBuffer[1] = pOutputBuffer[0]; // just copy

        // Stereo-In, Stereo-Out (INSERT Effect)
        if(uNumInputChannels == 2 && uNumOutputChannels == 2)
        {
                // branch
                if(m_uMode == Tremolo)
                {
                        // do right channel, use same gain as left
                        pOutputBuffer[1] = pInputBuffer[1]*fGnL;
                }
```

```
                        else
                        {
                                // do right channel, its value
                                pOutputBuffer[1] = pInputBuffer[1]*fGnR;
                        }
                }

                return true;
        }
```

Build and test the plug-in. Try raising the LFO rate maximum from 20 Hz to something like 200 or 2 kHz; the results are quite interesting.

14.2 Design a Ring Modulator Plug-In

Ring modulation is accomplished by multiplying the input signal against a sinusoid called the *carrier*, as shown in Figure 14.2. This is technically amplitude modulation or frequency mixing. The resulting signal contains sum and difference pairs of each frequency present in the signal. For example, if the carrier is 250 Hz and the input signal is 1 kHz, the output will consist of two frequencies, 1 kHz − 250 Hz = 750 Hz and 1 kHz + 250 Hz = 1.25 kHz. The original input signal at 1 kHz is gone. For complex audio signals, this process happens for every frequency component that the signal contains. The resulting output can sound nothing like the input. Drums change their pitches and full spectrum signals become obliterated. Because of the sum and difference pair, aliasing will occur when the sum signal goes outside Nyquist. One way to handle the aliasing is to band-split the signal just like the spectral compressor and only process the lower frequencies. Though not a perfect solution, it will at least alleviate some of the aliasing. The ring modulator project here only implements the frequency modulation without addressing aliasing problems.

14.2.1 Project: RingModulator

Create a project and name it "RingModulator." There are no other objects or options to add.

14.2.2 RingModulator: GUI

We need frequency and depth controls for the carrier oscillator in Table 14.2. The depth control will increase or decrease the amount of modulation.

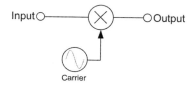

Figure 14.2: The ring modulator block diagram.

Table 14.2: RingModulator plug-in GUI controls.

Slider Property	Value	Slider Property	Value
Control Name	Mod Freq	Control Name	Mod Depth
Units	Hz	Units	%
Variable Type	float	Variable Type	float
Variable Name	m_fModFrequency	Variable Name	m_fModDepth
Low Limit	100	Low Limit	0
High Limit	5000	High Limit	100
Initial Value	1000	Initial Value	50

14.2.3 RingModulator.h File

We only need to add a single wave table object to implement the carrier oscillator.

```
// Add your code here: ----------------------------------------- //
// the Carrier Oscillator
CWaveTable m_Oscillator;

// END OF USER CODE -------------------------------------------- //
```

14.2.4 RingModulator.cpp File

prepareForPlay()

- Initialize the oscillator.
- Forward the preareForPlay() method to the oscillator.

```
bool __stdcall CRingModulator::prepareForPlay()
{
        // Add your code here:
        m_Oscillator.m_fFrequency_Hz = m_fModFrequency;
        m_Oscillator.m_uPolarity = 0; // 0 = bipolar, 1 = unipolar
        m_Oscillator.m_uTableMode = 0; // normal, no band limiting
        m_Oscillator.m_uOscType = 0; // 0 = sin()
        m_Oscillator.setSampleRate(m_nSampleRate); // really important!

        // the prepareForPlay() calls reset() and cookFrequency()
        m_Oscillator.prepareForPlay();

        return true;
}
```

userInterfaceChange()

- If the frequency slider moves, change the oscillator frequency and call the cooking function.
- The depth control is a direct-control variable, so it does not need any attention here.

```
bool __stdcall CRingModulator::userInterfaceChange(int nControlIndex)
{
    // decode the control index, or delete the switch and use brute force calls
    switch(nControlIndex)
    {
        case 0:
        {
            m_Oscillator.m_fFrequency_Hz = m_fModFrequency;
            m_Oscillator.cookFrequency();
            break;
        }

        default:
            break;
    }

    return true;
}
```

processAudioFrame()

- Generate a new carrier oscillator value.
- Multiply the input signal against it.

```
bool __stdcall CRingModulator::processAudioFrame(float* pInputBuffer, float*
                                                 pOutputBuffer, UINT uNumInputChannels,
                                                 UINT uNumOutputChannels)
{
    // Do LEFT (MONO) Channel; there is always at least one input/one output

    // generate carrier
    float yn, yqn;
    m_Oscillator.doOscillate(&yn, &yqn);

    // multiply
    pOutputBuffer[0] = pInputBuffer[0]*yn*(m_fModDepth/100.0);

    // Mono-In, Stereo-Out (AUX Effect)
    if(uNumInputChannels == 1 && uNumOutputChannels == 2)
        pOutputBuffer[1] = pOutputBuffer[0];

    // Stereo-In, Stereo-Out (INSERT Effect)
    if(uNumInputChannels == 2 && uNumOutputChannels == 2)
        pOutputBuffer[1] = pInputBuffer[1]*yn*(m_fModDepth/100.0);

    return true;
}
```

Build and test the plug-in. You may find it to be a remarkable effect or you may think it sounds awful. Some things you could do to modify this project include:

- Use multiple carrier frequencies.
- Use series or parallel multiple modulation.
- Use an LFO to modulate the carrier frequency.

14.3 Design a Wave Shaper Plug-In

Wave shaping is a method of nonlinear processing that adds harmonics to the input signal as a result. It is used in synthesizers and distortion algorithms. Because harmonics are added to the signal, wave shapers suffer from aliasing problems, like the ring modulator. The process of wave shaping is simple because no memory elements are required. You treat the input signal x as the argument to a function. The function is often a trigonometric or polynomial function. An interesting function to use is the arctangent. It is part of a family of functions called *sigmoid functions*. Sigmoid functions produce S-shaped curves. The amplitude transfer function of a vacuum tube generally resembles a sigmoid, so this function can be used to try to mimic a tube sound. Figure 14.3 shows the arctangent function atan(kx), where k controls the amplitude of the input value and thus the amount of nonlinear processing applied. The exact equation is shown in Equation 14.2, where a normalization factor has been added to restrict the output to the range of -1 to $+1$. You can see in Figure 14.3 that with $k = 1$, the input/output relationship is nearly linear. As k increases, the S-shaped curve emerges and adds gain. For example, with $k = 5$ and $x(n) = 0.25$, $y(n)$ is about 0.6.

$$y(n) = \frac{1}{\arctan(k)} \arctan(kx(n)) \qquad (14.2)$$

Asymmetrical distortion can be implemented easily by using two different k-values, one for positive input samples and the other for negative ones. This asymmetrical distortion is found in Class-A tube circuits. Cascading multiple stages will result in more harmonic distortion. In some amplifiers, many stages are cascaded in series. Class A tube circuits also invert the signal. This means that the asymmetrical distortion curves are also inverted in between each stage. The resulting sound is quite different than simply cascading the modules without

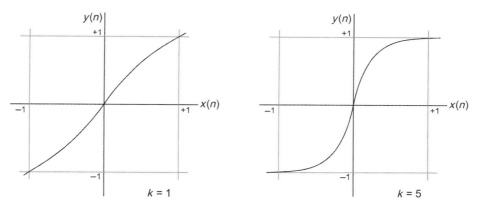

Figure 14.3: The input/output relationship for Equation 14.1 with $k = 1$ and $k = 5$.

inversion. We will design a plug-in that will implement the arctangent wave shaping and allow the user to adjust the following:

- The *k*-value for both the positive and negative halves of the input signal
- The number of stages in series, up to 10
- Inverting or not inverting every other stage when cascaded

14.3.1 Project: WaveShaper

Create a project and name it "WaveShaper." There are no other objects or options to add.

14.3.2 WaveShaper: GUI

Add the controls in Table 14.3 to implement the GUI we will need for the WaveShaper project.

WaveShaper.h File

There's nothing to do in the .h file as we require no additional variables, memory storage spaces, methods, or objects!

WaveShaper.cpp File

The only method that needs to be altered is processAudioFrame().

Table 14.3: WaveShaper plug-in GUI controls.

Slider Property	Value	Slider Property	Value
Control Name	+Atan Exp	Control Name	−Atan Exp
Units		Units	
Variable Type	float	Variable Type	float
Variable Name	m_fArcTanKPos	Variable Name	m_fArcTanKNeg
Low Limit	0.10	Low Limit	0.10
High Limit	20	High Limit	20
Initial Value	1	Initial Value	1

Slider Property	Value	Slider Property	Value
Control Name	Stages	Control Name	Invert Stages
Units		Units	
Variable Type	float	Variable Type	enum
Variable Name	m_nStages	Variable Name	m_uInvertStages
Low Limit	1		
High Limit	10		
Initial Value	1	Enum String	OFF,ON

```cpp
bool __stdcall CWaveShaper::processAudioFrame(float* pInputBuffer, float* pOutputBuffer,
                                              UINT uNumInputChannels,
                                              UINT uNumOutputChannels)
{
    // Do LEFT (MONO) Channel; there is always at least one input/one output
    // (INSERT Effect)
    float f_xn = pInputBuffer[0];

    // cascaded stages
    for(int i=0; i<m_nStages; i++)
    {
        if(f_xn >= 0)
            f_xn = (1.0/atan(m_fArcTanKPos))*atan(m_fArcTanKPos*f_xn);
        else
            f_xn = (1.0/atan(m_fArcTanKNeg))*atan(m_fArcTanKNeg*f_xn);

        // invet every other stage
        if(m_uInvertStages == ON && i%2 == 0)
            f_xn *= -1.0;
    }

    pOutputBuffer[0] = f_xn;

    // Mono-In, Stereo-Out (AUX Effect)
    if(uNumInputChannels == 1 && uNumOutputChannels == 2)
        pOutputBuffer[1] = pOutputBuffer[0];

    // Stereo-In, Stereo-Out (INSERT Effect)
    if(uNumInputChannels == 2 && uNumOutputChannels == 2)
    {
        float f_xn = pInputBuffer[0];

        // cascaded stages
        for(int i=0; i<m_nStages; i++)
        {
            if(f_xn >= 0)
                f_xn = (1.0/atan(m_fArcTanKPos))*atan(m_
                    fArcTanKPos*f_xn);
            else
                f_xn = (1.0/atan(m_fArcTanKNeg))*atan(m_
                    fArcTanKNeg*f_xn);

            // invet every other stage
            if(m_uInvertStages == ON && i%2 == 0)
                f_xn *= -1.0;
        }

        pOutputBuffer[1] = f_xn;
    }

    return true;
}
```

Build and test the plug-in. Note that in order to fully replicate a tube circuit you will need much more than just the distortion component, as tube circuits are amplitude sensitive and band limited. However, this plug-in is a good place to start for experimenting with nonlinear processing.

Bibliography

Black, H. S. 1953. *Modulation Theory*. New York: Van Nostrand-Reinhold.
Dodge, C. and T. Jerse. 1997. *Computer Music Synthesis, Composition and Performance*, Chapter 4. New York: Schirmer.
Roads, C. 1996. *The Computer Music Tutorial*, Chapter 2. Cambridge, MA: The MIT Press.
U.S. Navy. 1973. *Basic Electronics*, *Rate Training Manual NAVPES, 10087-C*. New York: Dover Publications.

APPENDIX A
The VST® and AU® Plug-In APIs

Once you learn how to design plug-ins with RackAFX you will find that learning other plug-in application programming interfaces (APIs) is fairly straightforward. The main reason is that most of the plug-in APIs share very similar structuring and function calls as RackAFX. An analogy might be learning to fly: once you have the fundamentals of solo flight internalized, learning to fly different *kinds* of airplanes is a matter of understanding the peculiarities of each one. You don't need to worry about the principles of lift, drag, pitch, and yaw after you've mastered them. The current popular plug-in APIs include Steinberg's Virtual Studio Technology (VST), Apple Computer's Audio Units (AU), and Avid's Avid Audio eXtension (AAX)® formats. If you want to write for these APIs, the first step is to set up a developer account with the manufacturers via their websites. Once you have an account you can download their API documentation and SDKs (software development kits), which include sample code templates. The VST and AU APIs are publicly documented, while AAX is not. Like RackAFX, VST plug-ins are written in straight C++ without the need for external frameworks or foundation classes. AU plug-ins are not and require a good working knowledge of Mac programming to develop. Once you have a RackAFX plug-in, how can you make it work in other plug-in APIs?

A.1 Compiling as a VST Plug-In in Windows

The simplest way to use your RackAFX plug-in in a third-party client is to take advantage of the fact that you can set your project to compile as a dynamic link library (DLL) that is compatible with both RackAFX and VST on the Windows operating system (OS). You can enable this functionality when you create your project or later in development. If you already have a project in development, use File → Edit Project or the Edit button on the user interface (UI). On the edit project dialog, check the "Make VST Compatible" box, as shown in Figure A.1. The resulting DLL will work with both RackAFX and any Windows VST client. Just copy the DLL from your /PlugIns folder to the designated VST folder for the client.

The GUI options are as follows:

- Use VST client's GUI: VST clients will provide you with a *default interface* whose look and feel are determined by the client's manufacturer. These range from a simple

```
                    .h File:  | Volume.h   |
                   .cpp File: | Volume.cpp |

          ☐ Output Only Synthesizer Plug-In
          ☑ Make VST Compatible
               ⦿ Use VST Client's GUI
               ○ Use RackAFX Custom GUI
               ○ I will write my own GUI
```

Figure A.1: The VST Compatible switch reveals three options for the GUI.

bank of sliders to more elaborate GUIs. The client will query your plug-in for control information and build the GUI at run time. You do not have to provide any extra code for this option.

- Use RackAFX custom GUI: If you have the proper version of RackAFX, this button will be enabled. Choosing this option will allow the GUI you create with GUI designer (Appendix B) to be used by the VST client. It will look and work identically to the one you create in RackAFX. You do not have to provide any extra code for this option.
- I will write my own GUI: If you have the skills to do this, then you can check this box and write your own GUI. The client will call a method on your plug-in to launch your GUI. All details of the GUI are your responsibility and you must provide the code and have the necessary resource editor.

For the last option, first look in the PlugIn.h file and find the flag and three methods you need to override.

```
// set this to TRUE if you have your own GUI to use; not for RackAFX GUIs
bool m_bUserCustomGUI;
```

First, set the m_bUserCustomGUI flag to "true" in your constructor. This will tell the client that you have your own GUI. Next, override the following three methods:

```
// for user (NOT RackAFX) generated resources/GUIs
virtual void __stdcall showGUI(HWND hParentWnd);
virtual void __stdcall hideGUI(HWND hParentWnd);
virtual void __stdcall refreshGUI(HWND hParentWnd);
```

- The client calls showGUI() to show it.
- The client calls hideGUI() to hide it.
- The client calls refreshGUI() to update it. See Appendix B for more information about updating.

In the three methods, the argument is the client's window handle. You will need to create your child window using the client's handle as your parent. If you don't know what that means, you probably shouldn't be using this option.

A.2 Wrapping Your RackAFX Plug-In

Your RackAFX plug-in will compile on Windows (Visual Studio) or Mac (Xcode) because it is pure C++ with some #define statements added to split out the Windows- and Mac-specific components. The audio signal processing and GUI component objects are all done in C++ so your plug-in object can be wrapped with an outer C++ object that translates function calls from the VST or AU client into RackAFX function calls. This is exactly what happens when you choose the "Make VST Compatible" option; when the client requests a VST object, it receives one with your plug-in embedded inside as a member object. The wrapper object then forwards calls to your plug-in and returns information or audio data to the client, as shown in Figure A.2. Your plug-in exposes an interface of public methods (gray boxes). The wrapper exposes its API specific methods (white boxes) which the client calls. The translation usually involves a simple forwarding method call. Sometimes more than one wrapper method connects to a single plug-in method. Other times, the wrapper method only needs to return a success code or special string so there is nothing to call on the plug-in object.

If you want to write a wrapper object, it is fairly straightforward, as you will see in Section A.3 when we compare the different API function calls. However, you need to know a bit more about how your GUI objects are designed so you can use the GUI controls you set up in RackAFX. Alternatively, you can implement your own scheme. When you set up a control in RackAFX,

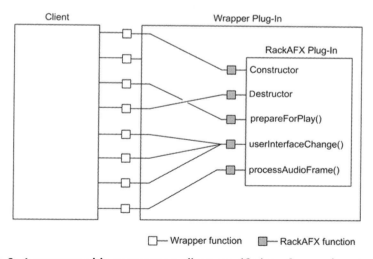

Figure A.2: A wrapper object exposes a client specific interface and routes calls to and from the plug-in.

you are really creating a new C++ object that is added to a linked list. The linked list is a member of your CPlugIn class named "m_UIControlList." It contains a list of C++ CUICtrl objects. A CUICtrl object can represent one of the following:

- Slider
- Radio button bank
- Meter

The simplest way to explain is by example. Look in your initUI() method for any plug-in and you will see the instantiation and initialization of the GUI objects. Here's the first part of the volume slider code from the very first project. The highlights are in bold. Remember, *do not ever edit this code manually.*

```
m_fVolume = 0.750000;
CUICtrl ui0;
ui0.uControlType = FILTER_CONTROL_CONTINUOUSLY_VARIABLE;
ui0.uControlId = 0;
ui0.fUserDisplayDataLoLimit = 0.000000;
ui0.fUserDisplayDataHiLimit = 1.000000;
ui0.uUserDataType = floatData;
ui0.fInitUserIntValue = 0;
ui0.fInitUserFloatValue = 0.750000;
ui0.fInitUserDoubleValue = 0;
ui0.fInitUserUINTValue = 0;
ui0.m_pUserCookedIntData = NULL;
ui0.m_pUserCookedFloatData = &m_fVolume;
ui0.m_pUserCookedDoubleData = NULL;
ui0.m_pUserCookedUINTData = NULL;
ui0.cControlUnits = "                    ";
ui0.cVariableName = "m_fVolume";
ui0.cEnumeratedList = "SEL1,SEL2,SEL3";
ui0.dPresetData[0] = 0.000000;ui0.dPresetData[1] <SNIP SNIP SNIP>
ui0.cControlName = "Volume";
```

The very first line initializes the underlying variable, m_fVolume, as you set it up when you created the slider. The control type is FILTER_CONTROL_CONTINUOUSLY_VARIABLE, which means a continuous controller. The uControlID is 0—this is the value that is passed to userInterfaceChange() when a control is manipulated. This "unique ID" paradigm for identifying the control is universal among the other plug-in APIs. The connection to the variable itself is via a pointer. Since this is a float variable, m_pUserCookedFloatData is set to the address of the underlying variable. This is how RackAFX manipulates it for you. When you wrap your plug-in you can use this to control the variable as well. Of course, you can always use other schemes. The units variable codes several items and will need to be trimmed of white spaces before you use it (do not alter its length in the object, however). Thus, you can see the fundamentals about each variable: the ID, minimum, maximum, and initial value, and the name and units. These are all used in the other plug-in APIs as well.

> RackAFX automatically groups all the legal VST controls at the top of the control list. It provides you with these methods so you can iterate through your GUI objects when you write your wrapper translation code.

RackAFX automatically sorts the GUI component objects by type. The sliders and radio button banks are supported by the default GUIs for the VST and AU formats but the meters are not. The m_UIControlList is defined in pluginconstants.h and exposes several methods for you to use:

```
int count();              // returns the total number of GUI objects
int countLegalVSTIF();    // returns the number of legal VST Default GUI objects
CUICtrl* getAt(int nIndex); // returns a pointer to a GUI object at a list index
```

The CPlugIn object also has some helper functions for you to use. Once you know the ID values for the controls, you can use the following CPlugIn methods to access the objects:

```
void setParameter (UINT index, float value);   // set a parameter 0 -> 1
float getParameter (UINT index);               // get a parameter 0 -> 1
```

These two methods are very important because they mimic the way a VST client will be getting or setting values through the default interface. *The index value is the index in the linked list and not the ID value.* Finally, for Windows plug-ins, you will often need to know the name of the directory where your DLL exists for dealing with presets or loading/saving custom information. CPlugIn exposes the method for you to use to get that folder name.

```
char* getMyDLLDirectory();   // return the DLL's owning directory
```

A.3 Comparison of Objects/Methods/GUIs

RackAFX, VST, and AU share many similarities. First, they all require that you derive your plug-in from a base class (or classes) that the manufacturer specifies. They also implement the same basic set of functionality that you've been coding all along. The base classes are shown in Table A.1. The comparison of methods is shown in Table A.2.

Table A.1: Comparison of base classes in the APIs.

API	Base Class(es)
RackAFX	CPlugIn
VST	AudioEffect, AudioEffectX
AU	AUBase, AUEffectBase, AUKernelBase

Table A.2: Comparison of the functions and the methods that implement them.

Function	RackAFX	VST	AU
Instantiation	CPlugIn()	AudioEffectX()	AUEffectBase() AUKernelBase()
Destruction	~CPlugIn()	~AudioEffectX()	~AUEffectBase() ~AUKernelBase()
Initialization	prepareForPlay()	resume()	Initialize()
Prepare For Play	prepareForPlay()	setSampleRate() updateSampleRate() resume()	N/A
Process Audio	processAudioFrame()	processReplacing()	Process() ProcessBufferList()
Get UI Control Value	getParameter() (optional)	getParameter()	GetParameter()
Set UI Control Value	userInterfaceChange() (optional)	setParameter()	SetParameter()
Update GUI (see Appendix B)	sendUpdateGUI()	updateDisplay()	N/A
Get the DLL's Directory	getMyDLLDirectory()	getDirectory()	N/A
Show a Custom GUI	showGUI()	dispatcher()	AudioUnitGetProperty()

RackAFX, VST, and AU all provide for a default standard interface—a GUI that the user does not have to code. For RackAFX, you can optionally design your own custom GUI with the GUI designer (Appendix B) that requires no coding; alternatively, you can design your own (see Section A.1). VST and AU also allow you to design your own GUI. VST provides a GUI class of objects that you can optionally use in your coding that allows for cross-platform compatibility. AU only runs on Mac and does not provide cross-platform compatibility. The default interface setup for RackAFX involves you simply filling in a properties dialog for the components you want; the GUI object code is written for you. In VST and AU, you must set up a way to describe your controls. In all three formats, when you load your plug-in the client queries it for its default interface controls and then builds the GUI at run time. In RackAFX, the client simply iterates through your linked list of objects and sets up the GUI.

A.4 VST Plug-in without Wrapping

Let's start by pretending to write a VST plug-in from scratch *without* wrapping any RackAFX code—of course, you can always cut and paste processing or cooking functions directly from your source, since it is pure C++. The pretend plug-in will be a delay with feedback called CDelay.

A.4.1 Default GUI

First, you define an enumerated list of unsigned integer-type (UINT) constant values in your Plugin.h file where "Plugin" is the name and you have derived it from the AudioEfffectX base class. These constants are the control ID values the host will use when querying about, getting, or setting a control variable. A delay plug-in might have delay time, feedback, and wet/dry mix controls, so you might define your enum as follows:

```
enum
{
        uDelayTime,      /* ID = 0 */
        uFeedBack,       /* ID = 1 */
        uWetDry,         /* ID = 2 */

        uNumParams       /* Always last = 3 = number of controls */
};
```

Next, you declare a *parallel set of member variables* that will be used in your plug-in. These are for cooked data values.

```
float m_fDelayTime;
float m_fFeedBack;
float m_fWetDry;
```

Finally, you declare the minimum and maximum values for each control.

```
float m_fMinDelayTime, m_fMaxDelayTime;
float m_fMinFeedBack, m_fMaxFeedback;
float m_fMinWetDry, m_fMaxWetDry;
```

The last item in the enumeration keeps track of the control count by virtue of being last in the list. During operation, the VST client will use these ID values passed as the variable "index" in the following methods. First, there are three methods to fill out that describe the name, display, and units for each control.

1. getParameterName: You create the name of the control as a string.

    ```
    void CDelay::getParameterName (long index, char *label)
    {
            switch (index)
            {
                    case uDelayTime:strcpy (label, "Delay"); break;
                    case uFeedBack: strcpy (label, "FeedBack"); break;
                    case uWetDry:   strcpy (label, "Wet/Dry"); break;
            }
    }
    ```
2. getParameterDisplay: You create a string that represents the value to display on the UI; the float2String() method is a helper function that is provided in the VST object file.

> All default VST controls use values from 0.0 to 1.0 regardless of how they map in your actual plug-in. You must write the cooking and uncooking functions that will convert your internal variable values to and from values from 0.0 to 1.0.

```
void CDelay::getParameterDisplay (long index, char *text)
{
    switch (index)
    {
        case uDelayTime:float2string (m_fDelayTime, text);   break;
        case uFeedBack: float2string (m_fFeedBack, text);    break;
        case uWetDry:   float2string (m_fWetDry, text);      break;
    }
}
```

3. getParameterLabel: You create a string that represents the units for the control.

```
void CDelay::getParameterLabel (long index, char *label)
{
    switch (index)
    {
        case uDelayTime:strcpy (label, "mSec");  break;
        case uFeedBack: strcpy (label, "%");     break;
        case uWetDry:   strcpy (label, "%");     break;
    }
}
```

Next, there are two methods you need to implement to allow the client to get and set your control variables. However, these methods deal with *raw VST client data*.

1. getParameter: You return the raw data (0–1) for the control.

```
float CDelay::getParameter (long index)
{
    float fRawData = 0;

    switch (index)
    {
        case uDelayTime:
        {
            float fDiff = m_fMaxDelayTime - m_fMinDelayTime;
            fRawData = (m_fDelayTime - m_fMinDelayTime)/fDiff;
            break;
        }

        // same thing for the other variables
    }

    return fRawData;
}
```

2. setParameter: You cook the raw data (0–1) for use in your plug-in.
```
void CDelay::setParameter (long index, float value)
{
        switch (index)
        {
               case uDelayTime:
               {
                      m_fDelayTime = (m_fMaxDelayTime - m_fMinDelayTime)* value +
                                              fMinDelayTime;
               }

               // same thing for the other variables
        }
}
```

Thus, every time you add or remove a variable, you have to update your enumerated list, change the variables and min/max pairs, and update the five functions above.

A.4.2 Signal Processing

You will notice that you've already had to write a lot of code just to deal with the UI; you have not written any code to do any meaningful processing. In RackAFX, you alter three functions for signal processing:

1. Constructor
2. prepareForPlay()
3. processAudioFrame()

Constructor

- Init all variables.
- Set our input and output channel count (this is also done in the CPlugIn constructor).
- Set a unique four-character identifier (required by VST).
- Call the base class's constructor to pass it the number of presets (0 for now) and number of controls.
- The audioMasterCallback you see is a mechanism for client ↔ plug-in communication; its details are hidden from developers.

```
CDelay::CDelay (audioMasterCallback audioMaster)
         : AudioEffectX (audioMaster, 0, uNumParams)  // 0 presets, uNumParams controls
{
        m_fDelayTime = 1000;
        m_fFeedback = 0;
        m_fWetDry = 50;

        // VST Specific setup
        setNumInputs (2);
```

```
            setNumOutputs (2);
            setUniqueID ('WDly');
}
```

prepareForPlay() → setSampleRate() and resume()

prepareForPlay() is special for RackAFX because it is called after the sample rate is set. In VST, this is done in setSampleRate(). We also use prepareForPlay() to flush out our delay line buffers and do any other preparations for the next run of the plug-in. The resume() function is called when the VST plug-in is turned on so we can flush buffers there.

```
void CDelay::setSampleRate(float sampleRate)
{
        // Left
        if(m_pLeftBuffer)
                delete m_pLeftBuffer;

        // create a 2 second buffer
        m_pLeftBuffer = new float[2*(int) sampleRate];

        // Right
        if(m_pRightBuffer)
                delete m_pRightBuffer;

        // create a 2 second buffer
        m_pRightBuffer = new float[2*(int) sampleRate];

        // flush buffers
        memset (m_pLeftBuffer, 0, 2*(int) sampleRate * sizeof (float));
        memset (m_pRightBuffer, 0, 2*(int) sampleRate * sizeof (float));
}

void CDelay::resume()
{
        // flush buffer
        memset (m_pLeftBuffer, 0, 2*(int) sampleRate * sizeof (float));
        memset (m_pRightBuffer, 0, 2*(int) sampleRate * sizeof (float));

        // reset read/write indices, etc…
        m_nReadIndex = 0;
        m_nWriteIndex = 0;

        etc…
}
```

processAudioFrame() → processReplacing()

The fundamental difference in the processing functions is in how the data arrives. In RackAFX, the default mechanism is that it arrives in frames, where each frame is a sample

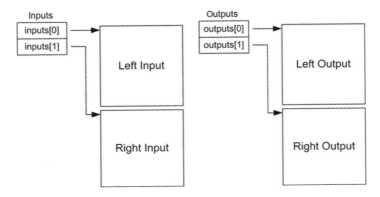

Figure A.3: VST audio data is delivered in buffers of frames.

from each channel. The frames arrive as pointers to buffers. In VST, the audio is processed in *buffers* of frames, rather than single frames. Figure A.3 shows a two-channel in and two-channel out VST setup.

Instead of passing a pointer to a buffer, you pass a pointer to a buffer of pointers, or float** variables. The processReplacing() method also passes you a count of the number of frames. You use these variables to iterate through and access your audio data. VST was designed to allow you to have any number of input and output channels, which you specify in the constructor. The parameters that are passed to the processReplacing() method are as follows:

- float** inputs: A pointer to a buffer of pointers to input buffers
- float** outputs: A pointer to a buffer of pointers to output buffers
- Long sampleframes: The count of frames in each buffer

```
void CDelay::processReplacing(float **inputs, float **outputs, long sampleframes)
{
        // pick up pointers to buffers
        float *inL  = inputs[0];
        float *inR  = inputs[1];
        float *outL = outputs[0];
        float *outR = outputs[1];

        UINT uFrameCount = 0;

        while (--sampleFrames >= 0)
        {
                // pick up samples from buffers
                float fxnL = inL[uFrameCount];
                float fxnR = inR[uFrameCount];

                // calculate and write out
                outL[uFrameCount] = doLeftDelayFunction(fxnL);
                outR[uFrameCount] = doRightDelayFunction(fxnR);
```

```
                // inc frame counter
                uFrameCount++;
        }
}
```

You can now see the similarities and differences in RackAFX and VST:

- There's more overhead for dealing with the GUI—even if you write your own GUI, you must provide the default implementation functions because not all VST clients have custom GUI capability.
- VST uses setSampleRate() and resume() to change the sample rate or restart the plug-in.
- VST processes in buffers of frames rather than frames.

A.5 VST Plug-In with RackAFX Wrapping

To wrap the RackAFX plug-in you first need to create a new AudioEffectX derived class. Then, you make one of the member variables a RackAFX plug-in object. Suppose you have a delay plug-in named "CStereoDelay" and you want to embed it into the VST plug-in. In the .h file you simply declare an instance of it: CStereoDelay m_Delay.

A.5.1 Default GUI

You still need to fill in those five VST functions, but you will be using your plug-in to supply the information.

getParameterName

- Copy the name of object; you use the getAt() method to find the UI object in the linked list.

```
void CDelay::getParameterName (long index, char *label)
{
        // get the GUI object
        CUICtrl* pUICtrl = m_Delay.m_UIControlList.getAt(index);
        if(!pUICtrl) return;

        // copy the ControlName
        strcpy (label, pUICtrl->cControlName);
}
```

getParameterDisplay

- Access the variable via the pointer; this example assumes all parameters are float variables, but you can always check the value of the uUserDataType variable to find out what kind of variable the object stores and branch accordingly.

```
void CDelay::getParameterDisplay (long index, char *text)
{
        // get the GUI object
        CUICtrl* pUICtrl = m_Delay.m_UIControlList.getAt(index);
        if(!pUICtrl) return;

        // use the helper function to copy the data
        float2string (*pUICtrl->m_pUserCookedFloatData, text);
}
```

getParameterLabel

- You create a *string* that represents the units for the control.

```
void CDelay::getParameterLabel (long index, char *label)
{
        // get the GUI object
        CUICtrl* pUICtrl = m_Delay.m_UIControlList.getAt(index);
        if(!pUICtrl) return;

        // copy the ControlUnits
        strcpy (label, pUICtrl->cControlUnits);
}
```

setParameter/getParameter

- Here is where the CPlugIn helper functions work for you; all you need to do is forward the call.

```
float CDelay::getParameter(long index)
{
        return m_Delay.getParameter(index);
}
void CDelay::setParameter(long index, float value)
{
        m_Delay.setParameter(index, value);
}
```

Constructor

For the constructor you will need to handle one minor detail: you need to know how many legal VST controls you have. Unfortunately, you need to pass these to the AudioEffect base class constructor before your member object plug-in is constructed. You will need to count your slider controls (including any embedded in the LCD control in Appendix B) and radio button bank controls. You can also look through the initUI() method and count all control objects with either the FILTER_CONTROL_CONTINUOUSLY_VARIABLE or FILTER_CONTROL_RADIO_SWITCH_VARIABLE as the uControlType. In this case, we have three legal VST controls.

```
CDelay::CDelay (audioMasterCallback audioMaster)
    : AudioEffectX (audioMaster, 0, 3) // 0 presets, 3 controls
{
    // child object is now instantiated
    //
    // VST Specific setup; get info from embedded plug-in
    setNumInputs (m_Delay.m_uMaxInputChannels);
    setNumOutputs (m_Delay.m_uMaxOutputChannels);
    setUniqueID ('WDly');
}
```

processReplacing

- In order to use processReplacing() you need to combine samples from the input/output (I/O) buffers into frames to send the processAudioFrame() function. This version is for a stereo insert effect.

```
void CDelay::processReplacing(float **inputs, float **outputs, long sampleframes)
{
    // pick up pointers to buffers
    float *inL  = inputs[0];
    float *inR  = inputs[1];
    float *outL = outputs[0];
    float *outR = outputs[1];

    UINT uFrameCount = 0;

    // mini frames
    float fInputFrame[2];
    float fOutputFrame[2];

    while (--sampleFrames >= 0)
    {
        // pick up samples from buffers
        fInputFrame[0] = inL[uFrameCount];
        fInputFrame[1] = inR[uFrameCount];

        // calculate and write out (...2, 2) = 2 in/2 out
        m_Delay.processAudioFrame(&fInputFrame[0], &fOutputFrame[0], 2, 2);

        // inc frame counter
        uFrameCount++;
    }
}
```

A.6 AU Overview

VST was a bit more complex in that you had to do some of the GUI work yourself. Whether you write one from scratch or wrap the RackAFX plug-in, you are still stuck with the default

GUI. AU plug-ins are another order of magnitude in complexity. The actual audio processing function is fairly simple. The large majority of the work goes into setting up the default GUI and maintaining your presets. Unlike both RackAFX and VST, AU plug-ins are not ANSI C/C++; they require the Audio Units and Core Foundation components, so you need to be familiar with Mac programming if you are going to get into AU programming. AU uses two base class objects together to implement a plug-in. When you create a new Audio Unit plug-in in XCode, it sets up the derived classes for you, one from each base class. You then override methods on both of them.

- AUEffectBase handles the GUI controls and any other non-audio-processing details.
- AUKernelBase handles the processing and reset controls.

A.6.1 Default GUI

Suppose we have a simple volume plug-in with just one volume control. AU calls a control a "parameter." Similar to VST, you first define an enumerated list of UINT constant values in your Plugin.h file where "Plugin" is the name and you have derived it from the AUEffectBase base class. These constants are the control ID values, just as before.

```
enum {
        kParam_One =0,
        //Add your parameters here... increment kNumberOfParameters

        kNumberOfParameters=1
};
```

In the same file, you create constant definitions for the initial value and name.

```
// set the initial value
static const float kDefaultValue_ParamOne = 0.5;

// set the name; CFStringRef is a Core Foundation String object
static CFStringRef kParameterOneName = CFSTR("Volume");
```

You declare your controls by filling in the GetParameterInfo() method. It is beyond the scope of this book to analyze the code line by line, but you can get a feel for how it works by looking at the way the objects are set up.

```
OSStatus AUVolume::GetParameterInfo(AudioUnitScope       inScope,
                                    AudioUnitParameterID inParameterID,
                                    AudioUnitParameterInfo &outParameterInfo)
{
        <SNIP SNIP SNIP>

        switch(inParameterID)
        {
           case kParam_One:
               AUBase::FillInParameterName (outParameterInfo, kParameterOneName,
               false);
```

```
                outParameterInfo.unit = kAudioUnitParameterUnit_LinearGain;
                outParameterInfo.minValue = 0.0;
                outParameterInfo.maxValue = 1;
                outParameterInfo.defaultValue = kDefaultValue_ParamOne;
                break;
        default:
                result = kAudioUnitErr_InvalidParameter;
                break;
        }

    etc…
```

Here, when the Client queries the plug-in on its first parameter (kParam_One) it fills out a structure about the control. You can see the minimum, maximum, and default values being set. The "unit" variable of the structure is set to kAudioUnitParameterUnit_ LinearGain. A linear gain control moves between 0 and 1.0. Another example is kAudioUnitParameterUnit_Boolean, which specifies a boolean control variable. What is interesting here is that you don't declare the actual variable that stores the control data. The variable is a float variable that the framework stores for you. You tell the framework what kind of variable it represents. When you need to get or set the variable, you use GetParameter() and SetParameter(), which get and set a float data type.

A.6.2 Signal Processing

The AU processing function is simply named Process(). The audio data moves in buffers. The buffers can contain any arbitrary number of channels with the simplest systems being $N \times N$ or equal input/outputs. Although the inNumChannels variable is passed to the function, it is always set to 1.0 for the current version of AU. This is OK for mono or when you don't need to process channels independently. A boolean flag is passed to the method, indicating the mute condition and the function simply returns without processing. The parameters that are passed to the Process() method are as follows:

- const Float32: *inSourceP (a pointer to an input buffer)
- Float32: *inDestP (a pointer to the output buffer)
- UInt32: InSamplesToProcess (number of samples in the input buffer)
- UInt32: inNumChannels (number of channels to process)
- bool: &ioSilence (indicates no processing, if true)

Here is the AU process function for a volume plug-in.

```
    void AUVolume::AUVolumeKernel::Process(const Float32    *inSourceP,
                                                 Float32    *inDestP,
                                                 UInt32     inFramesToProcess,
                                                 UInt32     inNumChannels,
                                                 bool       &ioSilence)
```

```
{
        // setup counter
        UInt32 nSampleFrames = inFramesToProcess;

        // setup pointers to I/O
        const Float32 *sourceP = inSourceP;
        Float32 *destP = inDestP;

        // use our own Get method
        Float32 gain = GetParameter( kParam_One );

        while (nSampleFrames-- > 0) {
              Float32 inputSample = *sourceP;

              sourceP += inNumChannels;

              // here's where you do your DSP work
              Float32 outputSample = inputSample * gain;

              *destP = outputSample;
              destP += inNumChannels;
        }
}
```

If you need to process multi-channel audio data independently, then you need to remove the Process() method override and replace it by overriding ProcessBufferLists().

```
ProcessBufferLists(AudioUnitRenderActionFlags &ioActionFlags,
                   const AudioBufferList &inBuffer,
                   AudioBufferList &outBuffer,
                   UInt32 inFramesToProcess)
```

This method passes data back and forth by using AudioBufferList objects. These objects contain pointers so buffers of interleaved data along with a variable that tells you how many channels the buffer holds. For example, for the input buffer:

```
inBuffer.mNumberBuffers = the number of channels in this Buffer List
inBuffer.mBuffers[0].mData = a float* to the buffer for channel 0
inBuffer.mBuffers[1].mData = a float* to the buffer for channel 1, etc...
```

You would typically write something like this to extract them (in this case, we know the channel count is 2):

```
// create the mini-frame arrays
float **inputs = new float*[2 * sizeof(float*)];
inputs[0] = (float*)inBuffer.mBuffers[0].mData;
inputs[1] = (float*)inBuffer.mBuffers[1].mData;

float **outputs = new float*[2 * sizeof(float*)];
outputs[0] = (float*)outBuffer.mBuffers[0].mData;
outputs[1] = (float*)outBuffer.mBuffers[1].mData;
```

```
// call your processing function
doProcess(inputs, outputs, inFramesToProcess);
```

In this case, your processing function is just like the VST processReplacing()—it is taking pointers to buffer pointers as arguments, then it would split out the data as in the VST version above. Lastly, the AUKernelBase method Reset() is implemented to reset your plug-in to its just-instantiated state.

… APPENDIX B

More RackAFX Controls and GUI Designer

RackAFX has several optional graphical user interface (GUI) controls that were not covered in the book chapters. It also has a GUI designer that lets you convert your slider-based default GUI into a fully customized version. The website www.willpirkle.com is always the best way to get the latest information on the GUI designer and any additional controls that may be added at a later date. The two additional GUI controls are the alpha wheel and liquid crystal display (LCD) and the vector joystick control.

B.1 The Alpha Wheel and LCD Control

In the upper right of the main RackAFX UI is a control that represents an alpha wheel, LCD matrix view, and the control knob. During the course of designing your project, you've noticed that the LCD is used to give you feedback about the state of your project and compiler. However, it can also be used as a GUI control just like the other sliders. The LCD control allows you to store up to 1024 continuous slider controls inside it. Anything you can set up with a normal slider control will work as one of the LCD embedded controls. And, the GUI designer lets you place one of these controls on your own GUI. There are several uses for the control:

- It's a different way to store the continuous controls.
- If you run out of sliders, you can always put 1024 more of them in the LCD.
- You can use the enumerated list variable to create a set of presets or other global functions or configuration controls.

To set up the LCD control, right-click on it when you are in development mode (not with a plug-in loaded). A box pops up as shown in Figure B.1. Click on New to open the very familiar slider properties dialog. Fill it out just as you would any other slider control. After adding a few more controls, you might have something resembling Figure B.2, with three controls embedded. You can edit the controls, remove them, or move them up and down the list (which changes their sequence in the LCD control).

You treat these embedded slider controls just as any other slider, so there's nothing to add to your normal operational code. When you run your plug-in, use the alpha wheel to select one

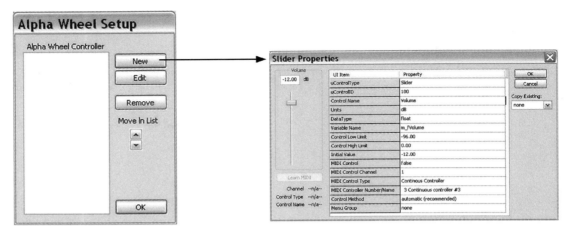

Figure B.1: The New button pops up the slider properties to add a control to the LCD.

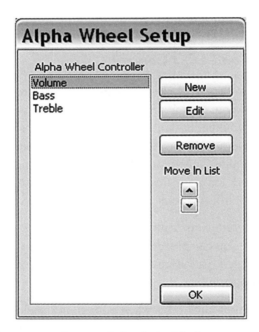

Figure B.2: The LCD is loaded with three controls.

of the controls and the value knob to adjust its value. Note there is currently no edit control connected to the LCD for manual entry of control values. Figure B.3 shows the LCD control with the alpha wheel adjusted for the bass control.

An alternative way to use the LCD control is for global operations (like effects configurations) or to store presets, either because you need more than RackAFX provides or

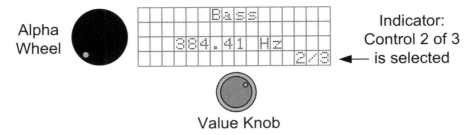

Figure B.3: The alpha wheel and LCD control is reminiscent of vintage synthesizer and rack effects.

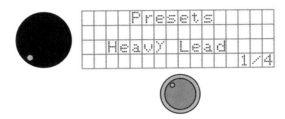

Figure B.4: The underscores are automatically removed from enumerated list variables in the LCD control.

you need to write your own methods for loading and storing data. You can use the enumerated list variable to create up to 256 individual string names. You can also take advantage of a feature of the LCD control: You can use the underscore (_) to separate strings in your list (you can't have spaces in an enumerated name). When the LCD control uses them, it replaces the underscores with single white spaces. This allows you much more flexibility than the radio buttons or normal sliders provide. For example, you could set up a preset list with names like "Heavy_Lead," "Liquid_Pad," "Lush_Space," and so on. Then, you can use the up/down arrows to place your "presets" control at the top of the control list. Now, as shown in Figure B.4, the user can select one of your custom presets (remember, it's up to you to code your own preset format) and then scroll to adjust the other controls. The underscores have been replaced with spaces.

B.2 The Vector Joystick Control

RackAFX also has a dedicated vector joystick control that communicates with your plug-in. Dave Smith designed the first vector synthesizers for Sequential Circuits, Inc. and then later for Yamaha and Korg. The vector joystick is a combination of a traditional *x/y* track-pad and an exponential joystick. The calculation documentation is available on the website. In the original synthesizers, the vector joystick was used to program or alter in real time the mixture of two or four different waveforms. The position of the joystick determines the mix

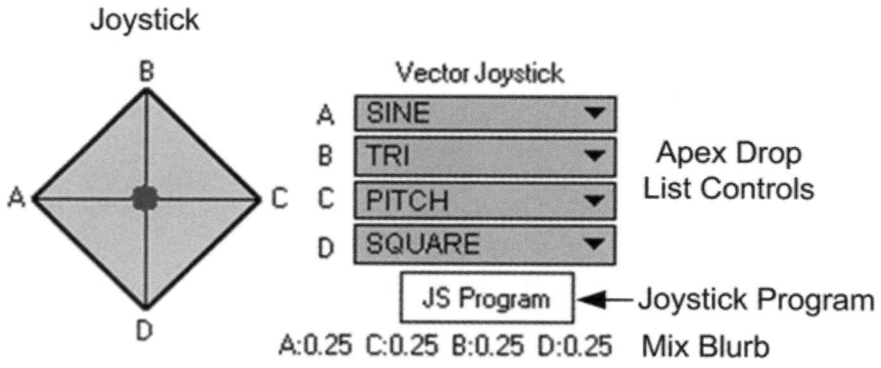

Figure B.5: The RackAFX vector joystick.

ratios, which are not linear but rather exponential. In RackAFX, you can use the joystick to alter up to four different parameters at once, exponentially. In addition, a linear *x/y* value of the joystick location is available for your plug-in. Figure B.5 shows the vector joystick and components.

The joystick is arranged in a diagonal with four corners labeled A, B, C, D, clockwise; this is in keeping with Dave Smith's original lingo. As the joystick moves around, a new set of mix ratios is generated and appears in the mix blurb. The mix ratios will always add up to 1.0. In the center, the ratios are all 0.25. A second set of values also exists, the AC-mix and the BD-mix, which are linear. On the AC axis, 0 is at the A apex and 1.0 is at the C corner. On the BD axis, 0 is at D and 1.0 is at B. In the center position, the AC-mix is 0.5 and the BD-mix is also 0.5. Figure B.6 shows the joystick roughly three-quarters of the way from the origin to the A corner. However, the mix blurb shows the value of A to be 0.79, but the other three have dropped off substantially. With the joystick all the way in the A corner, the ratios are {A, B, C, D} = {1, 0, 0, 0} and interestingly, as Figure B.7 demonstrates, when the joystick is halfway between A and B on the outer edge of the diamond, the A/B ratio is 50/50 while the rest are 0. In Figure B.7, the AC-mix is 0.25 and the BD-mix is 0.75.

By sweeping around the field, you can generate many different combinations. You could use these combinations as mix ratios for something in your plug-in. Or, you could use them to control any parameter or other aspect of the plug-in that you like. You are responsible for mapping the 0.0 to 1.0 values to your plug-in's parameters. The vector joystick does not connect to any variable in your plug-in; like the assignable buttons, these passive controls simply tell you when they've moved. When the joystick moves, a function called joystickControlChange() is called.

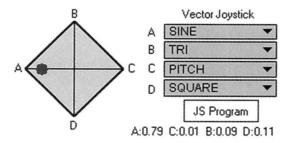

Figure B.6: The exponential mix ratios are evident in the vector joystick.

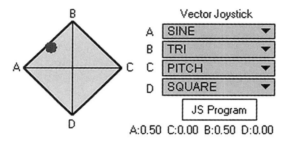

Figure B.7: This position results in the 50/50 mix of A/B only. AC-mix = 0.25, BD-mix = 0.75.

```
/* joystickControlChange

        Indicates the user moved the joystick point; the variables are the relative
        mixes of each axis; the values will add up to 1.0
                        B
                        |
                A       x       C
                        |
                        D
        The point in the very center (x) would be:
        fControlA = 0.25
        fControlB = 0.25
        fControlC = 0.25
        fControlD = 0.25

        AC Mix = projection on X Axis (0 -> 1)
        BD Mix = projection on Y Axis (0 -> 1)

*/
```

```
bool __stdcall CVolume::joystickControlChange(float fControlA,
                                              float fControlB,
                                              float fControlC,
                                              float fControlD,
                                              float fACMix,
                                              float fBDMix)
{
    // add your code here

    return true;
}
```

You can see that RackAFX leaves you a comment on the coding of the joystick location. The six variables arrive as function parameters and you use them however you wish. In Figure B.5 you can also see four drop list controls to the right of the joystick. I have already set them up to have something meaningful in them for the purpose of demonstration. These controls are generally used for two purposes:

1. They describe the current "meaning" of the apex lettering system; in Figure B.5 the A-corner is the "SINE" corner.
2. They allow the user to change the meaning of the apex in the plug-in by selecting different values.

You set up the drop list boxes when you are in development mode by clicking on them and choosing the item "Select To Setup," which allows you to connect these drop list boxes to different enumerated lists. They follow the same rules as the other enumerated lists, however, like the LCD control, you can use the underscore (_) to create a blank space in the string at run time. When the user selects an item from the box, the client calls your userInterfaceChange() function, passing it the parameters as described in the comment section for that method. You can then handle those messages as you wish.

```
Joystick Drop List Boxes            Index
-----------------------------------------
        Drop List A                 60
        Drop List B                 61
        Drop List C                 62
        Drop List D                 63
```

The joystick program is an advanced user's control for automating movement of the joystick in reaction to some event. Please see the website for tutorials and sample code for using the program.

B.3 Using the sendUpdateGUI() Method

This is a perfect place to explain how to use the sendUpdateGUI() method. You can force the user interface (UI) (RackAFX or a custom graphical user interface [GUI]) to change its settings based on your internal variable values. This is opposite the normal flow where the GUI calculates and alters your variables. In this case, you alter your own variable and tell the GUI to update accordingly. This can be used to move sliders in real time or toggle radio buttons programmatically. The joystick drop lists present an interesting use of this method. Suppose you wanted to have four identical lists that represent the joystick corners, such as {A, B, C, D} = {SINE, SAW, TRIANGLE, NOISE} and the user will move the joystick to produce a mixture of low-frequency oscillators (LFOs) in your plug-in. You would like the lists to contain the same values, but you would like to initialize them with other values than the first one. For example, you might want the initial state of the boxes to show SINE for A, SAW for B, NOISE for C, and TRIANGLE for D (notice this is different than the order of the list too). When you set up the enumerated list, you give it a list variable to connect to (e.g., m_uApexA) and a list that goes along with that. By clicking "Use Existing String List," then double-clicking on any of the items in the list, you can legally share your string lists. This is something we did not cover in the book chapters because it is a more advanced feature. Then, in your code (prepareForPlay() or wherever you need to) you can alter your own unsigned integer-type (UINT) variables then call sendUpdateGUI() to show the changes. For example, to initialize the drop lists as described above, you would add this code to your prepareForPlay() method:

```
// set my variables
m_uApexA = SINE;
m_uApexB = SAW;
m_uApexC = NOISE;
m_uApexD = TRIANGLE;

// tell the GUI to change accordingly
sendUpdateGUI();
```

Theoretically, you can do this from any point in the code *except the constructor* (because the messaging pointer isn't assigned until after construction), even from the joystick handler if needed. However, you should never initiate client communication like this in the processAudioFrame() method as it relies on a slow window-messaging system. This is true for other plug-in formats as well. Remember that you can use the sendUpdateGUI() method to alter the GUI for any of the RackAFX ↔ Plug-In connected variables.

B.4 Using GUI Designer

RackAFX has a powerful GUI designer that lets you arrange visually appealing GUIs in a matter of just a few moments. The GUI designer is always being updated with more features, so always check the latest additions at the website. Depending on which version you have, your GUI designer will look more or less like that shown in Figure B.8. The flow of operations is as follows:

1. In prototype view (the main RackAFX view), you assemble your plug-in. You create controls, give them min, max, and initial values, connect them to variables, and so on. Because this is a development mode, you will probably change some controls, add or remove them, and so on. Also, because we are prototyping, we can set up a bunch of controls we really would not want the user to be able to adjust in the final plug-in. An excellent example is the reverb plug-in; once we've got those parallel comb filters tuned, we don't want the user messing with them.
2. After your plug-in is complete, debugged, and ready to go, you click on the GUI Designer tab to reveal the blank GUI surface, shown in Figure B.8.
3. You drag and drop controls from the left side and arrange them however you like on the surface. Because they have transparent backgrounds, they can be overlapped.
4. For the slider, radio buttons, and knob controls, you must connect the control with the variable in your plug-in that it will control.
5. The LCD and vector joystick controls don't need any additional setup.

Suppose after testing we decide that the final GUI for the reverb project needs to have the following controls, while hiding all the others:

- Pre-delay time
- Pre-delay attenuation
- Bandwidth

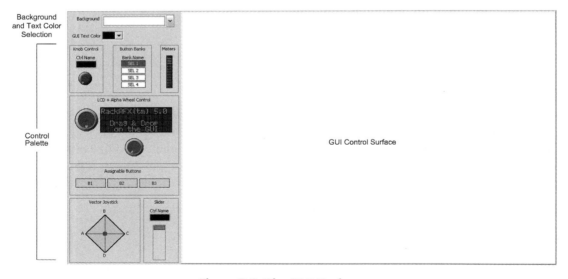

Figure B.8: The GUI Designer.

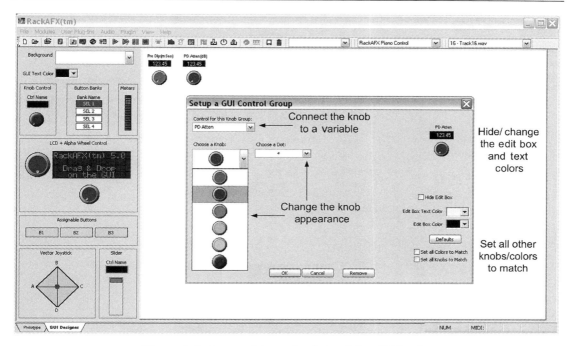

Figure B.9: Customizing a knob by right-clicking it.

- Damping
- RT60
- Wet/dry

Any of the prototype's sliders can be dragged and dropped either as knobs or sliders using the palette on the left. Figure B.9 shows the GUI designer after one knob control has been placed and mapped to the pre-delay time. A second knob has just been dropped next to it. Right-clicking on the knob itself pops up another customization box. Here you can do the following:

- Connect the knob to a variable via the drop list.
- Change the appearance of the knob and dot combination.
- Change the edit box behavior, hide it, or change its colors.
- Set all the other knobs to match this one.

You can do similar things for the radio buttons, meter, or the slider control, as shown in the pop-up boxes in Figures B.10 and B.11. The current GUI designer rules are as follows:

- For sliders, you link to a control variable, just like the knobs. You can customize many aspects of the slider control's appearance.
- When you run out of continuous controls (sliders) you can't drag any more knobs/slider controls.

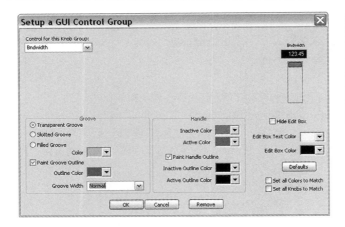
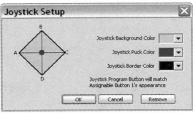
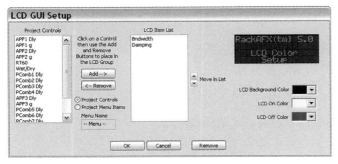

Figure B.10: The slider, joystick, LCD, and meter GUI customization dialog boxes.

- Each time you drag a knob or slider, you must set it up before dragging any other knobs or sliders.
- The joystick can have its colors altered and you can set the program button to look like the assignable buttons. There can be only one joystick.
- You can change the IN and OUT button bitmaps for the assignable buttons. There can be only one assignable button bank.
- You can add any continuous control (slider) to the GUI LCD group; it is a separate entity from the one on your main RackAFX GUI. Here I added bandwidth and damping. There can be only one LCD control.
- When you drag meter controls, RackAFX starts with the leftmost meter on the main GUI and automatically links your control variable for you. It then moves from left to right as you drag and drop meters. If you need to skip meters, add all the meters first, then remove the ones you don't want.
- When you add radio button controls, RackAFX starts with the topmost button bank on the main GUI and automatically links your control variable for you. It then moves from top to bottom as you drag the radio buttons. If you need to skip button banks, add all of them first, then remove the ones you don't want. The buttons banks are highly customizable.

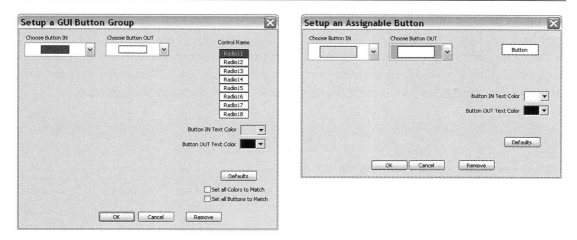

Figure B.11: The Radio and Assignable button GUI customization dialog boxes.

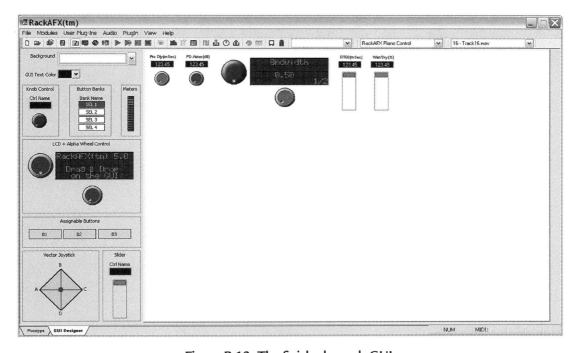

Figure B.12: The finished reverb GUI.

Figure B.12 shows my finished reverb GUI. I chose the white background just to make it easier to see in print. After you are done, go back to the prototype view and rebuild your plug-in. You need this rebuild to embed the new GUI information into your code. The GUI styles are coded into UINT arrays in your control variables as well as the plug-in object.

the website for more details on the encoding. Now, load your plug-in. You will notice that the blue knob button on the toolbar is now active. You can open your GUI with that button or the menu item View → Custom Plug-In GUI. Figure B.13 shows the finished GUI. The designer automatically crops it to fit the controls you've placed on it. When you alter the new GUI controls, the RackAFX controls change as well (but not the other way around).

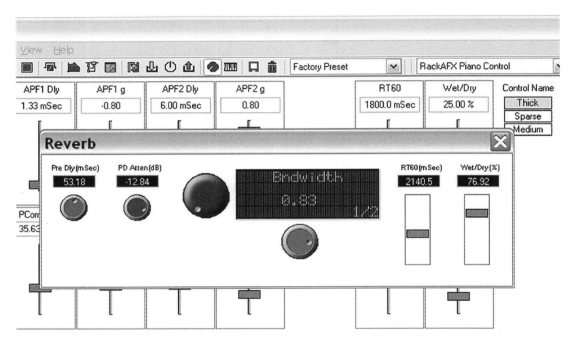

Figure B.13: The reverb GUI in action!

If you have enabled your plug-in to be compiled to produce a Virtual Studio Technology (VST) plug-in (Appendix A) then your RackAFX custom GUI will be available in any Windows VST client that supports custom GUIs (which is just about all of them). You launch it in whatever way the client digital audio workstation (DAW) requires.

Index

A

absorbent all-pass filters (AAPFs) 390
addition operation 17
address generation unit (AGU) 207
all-pass filter (APF) reverberators 368–71
alpha wheel GUI controls 519
analog-to-digital conversion 1, 170–8
anti-aliasing filters 3
applications programming interfaces (APIs) 27–9, 501
 functions typically required of 29–30
 and RackAFX philosophy 31–3
asynchronous operation 6
attack times of envelope detectors 428–9
attenuation function and attenuators 18–19
AU programming and processing 515–18
audio filter designs 253–87
audio specific filters 188–96

B

beats-per-minute (BPM) synchronization 235
bilinear transform 172–5, 179–80
bipolar/unipolar functionality 324–6
bi-quadratic filters (bi-quads) 157–62
 design of 181–8
 modified 189
Bohn, D. 196

book-keeping and book-keeping records 13–15, 114–15
Browne, S. 357
Butterworth filters 184–7, 475

C

C programming language 24
C++ programming language 24, 29, 31
Cartesian coordinate system 100–1
cascaded graphic filters 194–6
Chaigne, A. 385
Chemistruck, M. 390
child plug-in objects 239
chorus effect 331–4
circular buffers 209–10, 273
comb filter reverberator 364–8
comb filters 221
comparison of objects, GUIs and functions 505–6
complementary filter design 266–70
complex conjugates 133–4, 144
complex numbers 100–1, 112
complex sinusoid function 97–100
compressor/limiter plug-in design 457–65
compressors 453–6
conjugate poles of resonators 167–70
constant-Q filters 192–4
convolution plug-ins 273–81
convolution process 254–8, 274
cooking functions 57, 227, 230–1

D

Dahl, L. 390
Dattorro, J. 381
Dattorro plate reverb 382–5

Decca chorus 354
delay effects 207–9
 modulated 327–54
delay operators 16
delaying all-pass filter reverberators 368–71
difference equations 18
 z transform of 117–18
digital audio technology, history of 1
digital delay line (DDL) 209–15, 327
 advanced form of module 248
 module plug-in 224–33
digital signal processing (DSP):
 address generation unit 207
 basic theory of 97–122
 chips 4
 filters 71–95
 practical application of theory 123–62
 test signals 10–12
digital-to-analog converters (DACs) 3
digitally-controlled amplifiers/attenuators (DCAs) 454
dimension-style choruses 351–3
direct z-plane design techniques 163–4
Doppler effect 328
downward expander/gate plug-in design 466–8
dynamic link library (DLL): access to 22
 building of 35
dynamic linking 21–2
dynamic processing 453–87

Index

E

early reflections 358
echoes 360
eigenfrequencies 361
energy decay relief (EDR)
    plots 362–4
enumerated slider variables 245–8
envelope detection 426–36
envelope follower plug-in
    design 425–8
Euler's equation 98–9, 104–7,
    112, 138–9, 148, 152, 155,
    215, 221
expanders 453–6

F

feed-back filters: first-order 88–95,
    123–4
    second-order 142–9
feed-forward filters 119
    first-order 74–88, 94–5,
    103–6
    second-order 132–41
feedback 221–2
    effect on impulse
    response 218–19, 222–3
feedback delay network (FDN)
    reverbs 385–91
    generalized 385–9
finite impulse response (FIR)
    filters 161–2, 253, 262,
    266–7, 281–7
flanger/vibrato effect 328–31
flanger/vibrato/chorus plug-in
    334–42
floating-point data 9–10
fractional delay 235–9
Frenette, J. 381–2
frequency response: direct
    evaluation 127–30
    estimation 121, 133–8, 143–7,
    151–2, 158–60
frequency response plots 71–2,
    83–5, 104, 111
frequency sampling method 263–6
full-wave envelope detectors 428

G

gain control 122–3
gain function 18

Gardner, W. G. 357, 366–8, 377–81
Gardner's nested APF
    reverberators 377–81
gates 453
genetic algorithm (GA) 390–1
Gerzon, M.A. 377
Gordon–Smith oscillator 299–301
graphical user interfaces
    (GUIs) 25–8, 35
    for DDL Module 225–6
Griesinger, D. 358, 360

H

half-wave envelope detectors 428
hard-knee compression curves 456

I

impulse response: effect of
    feedback on 218–19,
    222–3; *see also* room impulse
    response
infinite impulse response (IIR)
    filters 161–204, 207, 253
interpolation functions 236–7
interrupt-based designs 6–7
inverse comb filtering 217–18

J

Jot, J.-M. 385–90

K

knee of the compression curve 456
Korg Triton 250–1, 345, 350
Kuttruff, H. 361–2

L

Lagrange interpolator 284
late reverberations 358
limiters 453
linear interpolation 236–8
linear phase filters 83, 266–7
Linkwitz–Riley filters 186–7,
    475, 479
liquid crystal display (LCD) GUI
    controls 519
look-ahead compressor plug-in
    design 468–71
low-frequency oscillation
    (LFO) 328, 331
low-pass filter–comb
    reverberator 373–5

M

Mac programming 501, 503,
    506, 515
McGrath, D. S. 357
Marcolini, K. 390
Massberg, Michael 201
Massberg analog-matched low-pass
    filter 201–4
median filter algorithm 284–7
member objects 224
mixing algorithms 19
mod filter plug-in design 412–25
modal density 361–2
modes (room resonances) 360–1
modulated APF reverberator 381–2
modulated delay effects 327–54
modulated delay line (MDL) 327
modulated filter effects 411–51
Moorer, J. A. 375, 377
Moorer's reverberator 365–71
moving average (MA)
    interpolator 282–3
multi-tap delay 249–50
multi-unit LCR chorus plug-
    in 345–50
multiplication operation 16–17

N

National Semiconductor 438
negative frequencies 104–5, 113
nested filter structures 379–81
normalized frequency 104
numerical coding of audio data
    7–9
numerical method FIR filters
    281–7
Nyquist frequency 1–2, 11–12,
    76–7, 95

O

one-sample delay 15–16
optimal method for designing
    filters 270–3
order of a filter 114
Orfanidis, S. 429, 455
oscillators 289–326
    bipolar/unipolar
    functionality 324–6
    direct-form 289–99
    wave-table type 301–8

Index

P

panning *see* tremolo/panning plug-in design
parametric filters 191–4
parent plug-ins 233, 239
Parks–McClellan algorithm 262, 270
peaking filters *see* parametric filters
phase inversion function 18
phase response plots 71–2, 83–5, 101, 104, 111
phaser plug-in design 436–45
ping-pong delay 250
plate reverb algorithm 382–5
plug-in, definition of 21
polar notation 100–1
pole-zero filters 149–57
poles of the transfer function 124–32, 145
poles or zeros at infinity, effect of 178–80
polynomials, factoring of 119–20
pre-delay 358
Puckette, M. 385, 388
'pumping and breathing' artifacts 463

R

RackAFX software 29–70
    design of tone control plug-ins 58–69
    design of volume control plug-ins 40–1, 54–8
    destruction of plug-in objects 38
    dynamic link library 35–6
    FIR filter designer 262
    frequency sampling method tool 267–9
    graphical user interfaces 36–7, 525–30
    impulse convolver 258–62
    optimal method tool 271–3
    processing audio 37–8
    setting up plug-ins for use and setup preferences 41–54
    stock objects 394–8
    testing 38–40
    user menu 69–70
reconstruction filters 3
Reilly, A. 357
Reiss, J. 455, 486
release times of envelope detectors 428–9
Remez exchange algorithm 271–3
resonant LPF plug-in 196–201
resonators 165–70
reverb algorithms 357–409
reverb modules 363
reverb time 359–60
ring modulator plug-in design 494–6
Roland Dimension D chorus 351–3
room impulse response 358–60
room resonances *see* modes
room reverb: design of 398–408
    example of 391–3
RT_{60} reverb time 359–60

S

Sabine, W. 359–60
samples of audio signal, acquisition of 1–3
sampling theorem 2, 7, 106
saw-tooth waveform 317–18
scalar multiplication operation 16–17
scattering matrices 389–90
Schroeder, M. 357, 360–4, 368, 370, 376–7, 385
Schroeder's reverberator 372–3
'sendUpdate GUI()' method 525
shelving filters 156–7, 189–90, 201
sigmoid functions 497
signal-processing algorithms 13
signal-processing flow 6–7
signal-processing systems 4–5
single pole filters 164–5
slider controls 245
Smith, Dave 522
Smith, J. O. 389
Smith-Angell improved resonator 168–70
soft-knee compression curves 456–7
software development kits (SDKs) 501
Sony DPS–M7 354
spectral compressor/expander plug-in design 475–86
square wave form 317, 319
static linking 21–2
Stautner, J. 385, 388
stereo crossed feedback delay plug-in 244
stereo digital delay plug-in 239–44
stereo-linking a dynamics processor 472–4
stereo phaser with quad-phase LFOs 446–51
stereo quadrature flanger plug-in 342–5
stereo reverberation 376–7
subsequent reverberations 358
subtraction operation 17
synchronous operation 5–6

T

through-zero flanging (TZF) 330
time delay as a mathematical operator 102
transfer functions 101, 104
    evaluation of 106–12
transverse delay line 257
tremolo/panning plug-in design 489–94
trivial zeros 125

U

unitary feedback delay networks (UFDNs) 386, 390

V

vector joystick control 519–24
vibrato *see* flanger/vibrato/chorus plug-in; flanger/vibrato effect
virtual address space 22–3

Index

W

wave shaper plug-in design 497–500
wave table oscillators 301–8
wave tables: addition of 308–12
 band-limited 312–19
waveguide reverberators 389
wet/dry mix 214–15
white noise 360
'willpirkle.com' website 519
Windows operating system 501–3
wrapping of plug-in objects 503–5, 512–14

Z

z substitution 114
z transform 114–21, 133, 141–2
 of difference equations 117–18
 of impulse responses 118–19
 of signals 116–17
 zero frequencies 119–20, 145
Zöler, U. 429, 455